CW01430568

WHY TRUST A THEORY?
Epistemology of Fundamental Physics

Do we need to reconsider scientific methodology in light of modern physics? Has the traditional scientific method become outdated, does it need to be defended against dangerous incursions, or has it always been different from what the canonical view suggests? To what extent should we accept non-empirical strategies for scientific theory assessment? Many core aspects of contemporary fundamental physics are far from empirically well-confirmed. There is controversy with regard to the epistemic status of the corresponding theories, in particular cosmic inflation, the multiverse, and string theory. This collection of essays is based on the high profile workshop 'Why Trust a Theory?' (Munich, 2015) and provides interdisciplinary perspectives on empirical testing in fundamental physics from leading physicists, philosophers and historians of science. Integrating different contemporary and historical positions, it will be of interest to philosophers of science and physicists, as well as anyone interested in the foundations of contemporary science.

RADIN DARDASHTI is Junior Professor in Philosophy of Physics at the University of Wuppertal. His research focuses on the various methods used in theory development and assessment in modern physics.

RICHARD DAWID is Professor of Philosophy of Science at Stockholm University. His research focuses on epistemic issues in contemporary, fundamental physics. He is the author of *String Theory and the Scientific Method* (Cambridge University Press, 2013).

KARIM THÉBAULT is Senior Lecturer in Philosophy of Science at the University of Bristol. His research interests are principally within the Philosophy of Physics, with an emphasis on classical and quantum theories of gravity.

WHY TRUST A THEORY?

Epistemology of Fundamental Physics

Edited by

RADIN DARDASHTI
University of Wuppertal

RICHARD DAWID
Stockholm University

KARIM THÉBAULT
University of Bristol

CAMBRIDGE
UNIVERSITY PRESS

CAMBRIDGE
UNIVERSITY PRESS

University Printing House, Cambridge CB2 8BS, United Kingdom

One Liberty Plaza, 20th Floor, New York, NY 10006, USA

477 Williamstown Road, Port Melbourne, VIC 3207, Australia

314–321, 3rd Floor, Plot 3, Splendor Forum, Jasola District Centre, New Delhi – 110025, India

79 Anson Road, #06-04/06, Singapore 079906

Cambridge University Press is part of the University of Cambridge.

It furthers the University's mission by disseminating knowledge in the pursuit of education, learning, and research at the highest international levels of excellence.

www.cambridge.org
Information on this title: www.cambridge.org/9781108470957
DOI: 10.1017/9781108671224

© Cambridge University Press 2019

This publication is in copyright. Subject to statutory exception and to the provisions of relevant collective licensing agreements, no reproduction of any part may take place without the written permission of Cambridge University Press.

First published 2019

Printed and bound in Great Britain by Clays Ltd, Elcograf S.p.A.

A catalogue record for this publication is available from the British Library.

Library of Congress Cataloging-in-Publication Data
Names: Dardashti, Radin, 1983– editor. | Dawid, Richard, 1966– editor. |
Thébault, Karim, editor.
Title: Why trust a theory? : epistemology of fundamental physics / edited by
Radin Dardashti (University of Wuppertal), Richard Dawid (Stockholms Universitet),
Karim Thébault (University of Bristol).
Description: Cambridge ; New York, NY : Cambridge University Press, 2018.
Identifiers: LCCN 2018042048 | ISBN 9781108470957 (hardback : alk. paper)
Subjects: LCSH: Physics–Methodology. | Science–Methodology.
Classification: LCC QC6 .W57186 2018 | DDC 530.01–dc23
LC record available at https://lccn.loc.gov/2018042048

ISBN 978-1-108-47095-7 Hardback

Cambridge University Press has no responsibility for the persistence or accuracy of URLs for external or third-party internet websites referred to in this publication and does not guarantee that any content on such websites is, or will remain, accurate or appropriate.

In December 2015, a few days before the start of the conference "Why Trust a Theory?" that provided the basis for the present volume, Joseph Polchinski became ill, which forced him to cancel his participation. The manuscript of his talk, *String Theory to the Rescue*, was read at the conference and is part of this volume. During the weeks before his surgery in early 2016, he wrote "Why Trust a Theory?: Some Further Remarks," which is also included in this volume. Joseph Polchinski died on February 2, 2018. This book is dedicated to him.

Contents

List of Contributors   *page* ix
Preface   xi

1   Introduction   1
    RADIN DARDASHTI, RICHARD DAWID, AND KARIM THÉBAULT

Part I   Historical and Conceptual Background

2   Fundamental Theories and Epistemic Shifts: Can the History of Science
    Serve as a Guide?   13
    HELGE KRAGH

3   Scientific Speculation: A Pragmatic Approach   29
    PETER ACHINSTEIN

4   Assessing Scientific Theories: The Bayesian Approach   67
    RADIN DARDASHTI AND STEPHAN HARTMANN

5   Philosophy of Science and the String Wars: A View from the Outside   84
    MASSIMO PIGLIUCCI

Part II   Theory Assessment beyond Empirical Confirmation

6   The Significance of Non-Empirical Confirmation in Fundamental
    Physics   99
    RICHARD DAWID

7   The Dangers of Non-Empirical Confirmation   120
    CARLO ROVELLI

8   No Alternative to Proliferation   125
    DANIELE ORITI

9   Physics without Experiments?   154
     RADIN DARDASHTI

10   Scientific Methodology: A View from Early String Theory   173
     ELENA CASTELLANI

11   What Can We Learn from Analogue Experiments?   184
     KARIM THÉBAULT

12   Are Black Holes about Information?   202
     CHRISTIAN WÜTHRICH

     Part III   Cosmology and Testability

13   The Limits of Cosmology   227
     JOSEPH SILK

14   The Role of Cosmology in Modern Physics   253
     BJÖRN MALTE SCHÄFER

15   Theory Confirmation and Multiverses   275
     GEORGE ELLIS

16   Beyond Falsifiability: Normal Science in a Multiverse   300
     SEAN M. CARROLL

17   Gaining Access to the Early Universe   315
     CHRIS SMEENK

     Part IV   Prospects for Confirmation in String Theory

18   String Theory to the Rescue   339
     JOSEPH POLCHINSKI

19   Why Trust a Theory? Some Further Remarks   354
     JOSEPH POLCHINSKI

20   The Dangerous Irrelevance of String Theory   365
     EVA SILVERSTEIN

21   String/M-Theories about Our World Are Testable in the Traditional
     Physics Way   377
     GORDON L. KANE

22   Is String Phenomenology an Oxymoron?   400
     FERNANDO QUEVEDO

     Index   434

List of Contributors

Peter Achinstein Johns Hopkins University
Sean M. Carroll California Institute of Technology
Elena Castellani University of Florence
Radin Dardashti University of Wuppertal
Richard Dawid Stockholm University
George Ellis University of Cape Town
Stephan Hartmann Ludwig Maximilian University of Munich
Gordon L. Kane University of Michigan
Helge Kragh University of Copenhagen
Daniele Oriti Max Planck Institute for Gravitational Physics
Massimo Pigliucci City University of New York
Joseph Polchinski University of California, Santa Barbara
Fernando Quevedo University of Cambridge and International Centre for
Theoretical Physics, Trieste
Carlo Rovelli Marseille University
Björn Malte Schäfer Heidelberg University
Joseph Silk Johns Hopkins University
Eva Silverstein Stanford University
Chris Smeenk University of Western Ontario
Karim Thébault University of Bristol
Christian Wüthrich University of Geneva

Preface

In a 2014 letter to *Nature* entitled "Scientific method: Defend the integrity of physics," the noted cosmologists George Ellis and Jospeh Silk expressed criticism of what they described as a dangerous tendency to soften principles of scientific reasoning in contemporary fundamental physics. This letter spurred the organization of a workshop in Munich in December 2015 entitled "Why Trust a Theory? Reconsidering Scientific Methodology in Light of Modern Physics." The workshop brought together leading physicists, historians, and philosophers of science to discuss and debate a range of pressing epistemological issues that confront contemporary fundamental physics. The majority of the contributions to this book are based on talks delivered at this meeting. As such, what the reader will find is the fruits of a sustained and constructive critical engagement between the various contributors that has taken place both in print and in person. Additional contributions were solicited by the editors with the aim of ensuring as full and balanced presentation as possible of the various positions in the debate.

We are extremely grateful to the organizations that supported the original conference, without which this volume would not have existed. In particular, we are appreciative of the support of Deutsche Forschungsgemeinschaft (German Research Foundation), Foundational Questions Institute, Munich Center for Mathematical Philosophy, and Arnold Sommerfeld Center for Theoretical Physics. Particular personal thanks are due to the other members of the organizational committee, George Ellis, Dieter Lüst, and Joseph Silk; to Stephan Hartmann and Sabine Beutlhauser; and to the audience members at the conference. We are appreciative to Cambridge University Press, in particular Sarah Lambert and Nicholas Gibbons, for outstanding support throughout the editorial process, as well as the volume's copyeditor, Jill Hobbs. We are also very grateful for comments on drafts of the introductory chapter provided by Michael Talibard and Fernando Quevedo. We would like to thank Michael Kreisel for preparing the index for this volume. Finally, we would like to express our deep gratitude to the contributors for their diverse and fascinating contributions.

1

Introduction

RADIN DARDASHTI, RICHARD DAWID, AND KARIM THÉBAULT

1.1 What Is This Book?

Fundamental physics today faces increasing difficulty in finding conclusive empirical confirmation of its theories. Some empirically unconfirmed or inconclusively confirmed theories in the field have nevertheless attained a high degree of trust among their exponents and are de facto treated as well-established theories. String theory, cosmic inflation, and to a certain extent the multiverse are particularly noteworthy examples. This situation raises a number of questions that are of substantial importance for the future development of fundamental physics. The essays in this volume bring together the views of leading physicists, philosophers, and historians of science on the interconnected web of issues that are united by the theme of whether we need to reconsider in light of the particular situation that modern physics finds itself in.

One theme of this book follows on the highly visible debate over the status of string theory that has been ongoing for more than a decade and that has (rather hyperbolically) been baptised 'The String Wars'.[1] A second and related issue is the status of inflationary cosmology and the multiverse.[2] The goal of this volume is to find a constructive level of analysis for a debate that so far has played out in a fairly antagonistic and often emotional way. In particular, we believe that the best path to progress in methodological debates over modern physics will emerge from constructive engagement regarding the scientific considerations rather than the sociological points at issue. The contributions to this volume stake out and respond to the main positions in the debate.

---

[1] Two well-known very critical public books are *The Trouble with Physics* Smolin (2006) and *Not Even Wrong* Woit (2007). Notable responses are from Polchinski (2007) and Duff (2013).

[2] This controversy recently reached an apogee with Ijjas et al. (2017), a highly critical assessment of the status of inflationary cosmology, and a forceful letter in response signed by a large number of leading cosmologists.

This book's interdisciplinary approach accounts for the philosophical nature of the meta-level issues that arise in the given context. Of course, exchanges between scientific and philosophical ideas have long been of relevance for the scientific process. From the scientific revolution up until the modern day, there have been disputes, involving philosophers and scientists alike, regarding both what science is and how science should be practiced. A crucial issue running through these many centuries of debate concerns the circumstances in which we should place *trust* in our scientific theories. In particular, a persistent matter of controversy has been the specification of legitimate methodologies through which scientific theories can be established as adequate and reliable guides to the empirical world. The theme of this book is wedded in this long-standing controversy.

We hope that this book will be of interest, and also of benefit, not only to professional physicists and philosophers of physics but to scientists and philosophers of science more generally. That said, the issues at question here are too important not to also be of concern to a general-interest audience. While the focus in this volume is on the various scientific methods used in fundamental physics, the relevance of this discussion goes beyond fundamental physics. The argumentative strategies followed within the various methods discussed may be, and probably are, similarly applied in other sciences where there is need to assess theories based on scarce empirical data. The wider societal relevance of the questions about what science is and in what circumstances we should trust it are difficult to overstate. An open and rigorous interrogation of such issues is a legitimate requirement if science is to maintain its privileged status within human systems of knowledge.

1.2 Fundamental Physics in the Twenty-First Century

To set the stage, it may be helpful to briefly sum up the current state of fundamental physics. Recent decades have seen important developments and significant progress on a universal conceptual understanding of physical phenomena that goes beyond that provided by the standard model of particle physics and standard Big Bang cosmology.

At the most ambitious conceptual level, string theory has provided a scenario that brings the description of all fundamental interactions into a coherent whole. String theory started out as a perturbative theory of one-dimensional quantum objects. Particle spectra and quantum numbers of low energy theories could, on that basis, be explained in terms of the topologies and oscillation modes of the quantized string. Consistency arguments led to the identification of a number of unusual characteristics of the theory, including compactified extra dimensions and higher-dimensional extended objects, the so-called branes. The latter discovery was related to an improved understanding of the crucial role of duality relations in string theory, which provided a basis to conjecture the equivalence of seemingly very different realizations of string theory. One striking example of a duality relation, known as the AdS/CFT correspondence, asserts the empirical equivalence of string theories proper and specific kinds of gauge field theories. In recent years, it has led to a much wider but also more confusing view of what string theory amounts to at a fundamental level.

Focusing on gauge interactions, substantial conceptual work has been invested in developing models beyond the standard model of particle physics, in particular in the context of supersymmetry, grand unification, and large extra dimensions. Focusing on gravity, various approaches to quantum gravity have been developed that solve a number of formal problems which previously barred the direct quantization of general relativity.[3] An important testing bed for these ideas is consistency with the more firmly established field of black hole thermodynamics, which in turn has received significant indirect support from developments in the field of analogue gravity. Despite its diversity, contemporary work on quantum gravity is founded upon a number of widely trusted results that are expected to persist within future theories. In particular, recovery of the black hole entropy law is used as a benchmark for the description of black holes in any viable theory of quantum gravity.

In a cosmological context, attempts to explain core characteristics of the observed universe, such as flatness and the isotropy of space, have led to the development of cosmic inflation. Inflation posits a phase of exponential expansion at the earliest stages of our universe that is followed by the 'normal expansion' phase we find ourselves in today. An improved understanding of the way in which inflation and the transition from an inflationary phase to the universe we witness today can play out has led to the conclusion that models of inflation generically lead to eternal inflation. An inflationary background gives rise to an infinite series of expanding hot 'bubbles' of slower expansion. Eternal inflation is joined with the string theoretical concept of the landscape, which implies a huge number of ground states of the theory, each corresponding to a different set of characteristic numbers for the corresponding low energy effective theory. In conjunction, the landscape and eternal inflation lead to the concept of the multiverse, which conjectures a vast number of universes with a wide spectrum of parameter values. The multiverse, in turn, is claimed to provide a framework for explaining fine-tuning of parameter values in our world based on anthropic reasoning.

---

[3] Non-stringy approaches to quantum gravity have proliferated in recent years. A non-exhaustive list includes loop quantum gravity (Rovelli, 2004, Thiemann, 2007); causal set theory (Bombelli et al. 1987, Dowker 2005, Henson 2006); causal dynamical triangulation (Ambjørn et al. 2001, Loll 2001); spin foams (Baez 1998, Perez 2013); functional RG approaches (Lauscher and Reuter 2001); and group field theory (Oriti 2009).

1.3 The Empirical Problem

The remarkable conceptual innovations that have occurred in fundamental physics during recent decades share one common problem: Several decades after they were first proposed, they still have difficulties in connecting to empirical data. Two developments seem to pull physics in different directions. On the one hand, it is getting increasingly difficult to obtain conclusive empirical evidence. While collider experiments like the Large Hadron Collider (LHC) have shifted upwards the energy scales of empirical testing and have led to the discovery of the Higgs particle, the last missing building block of the standard model, the energy scales accessible by experiments will remain many orders of magnitude below the expected characteristic scale of quantum gravity for the foreseeable future. The characteristic predictions of advanced theories in high energy physics, quantum gravity, and cosmology thus lie either entirely or in part beyond the grasp of currently conceivable experiments. Some predictions can be tested by cosmological observations, but the interpretation of those observations is more complex and less univocal than in designed experiments. In some cases, it is a matter of debate to what extent the given theories are predictive at all.

On the other hand, some theories in high energy physics and cosmology are strongly believed to be true or viable by many of their exponents despite the inconclusive nature (or entire absence) of empirical confirmation. String theory and cosmic inflation, in particular, exert a strong influence on adjacent fields and in many respects are treated like well-established scientific theories. Reasons for trusting in those theories are based on observations of the respective theory's properties and the research process that led up to its current state. In recent years, sharply conflicting positions have emerged on the status of string theory and cosmic inflation as well as the legitimacy of the reasoning in support of those theories' viability. Exponents and supporters of both theories emphasise that the credit they give to the theories in question is based on carrying out further methods of theory assessment that have been part of physics all along. Critics, to the contrary, observe an exaggerated reliance on subjective criteria of non-empirical theory assessment and a deviation from the core principles of empirical science.

1.4 This Has Happened Before

While at its core the current debate over theory assessment in fundamental physics is led by scientists assessing the status of individual theories, it hinges on profoundly philosophical questions. What can count as empirical evidence? How should we define theory confirmation? How can we understand the scientific process in a wider context? The debate thus illustrates the importance of philosophical considerations in defining the norms of scientific discourse.

In recent decades, scientific reasoning in the field of physics has progressed largely without explicit reference to such meta-level issues. Physicists have felt safely embedded within a stable framework of principles that guided scientific reasoning in their field. They have seen no need to address questions of 'scientificality' to proceed with their work. This has not always been the case in the history of physics. When the paradigm of scientific reasoning was developed and established by early modern thinkers like Francis Bacon, René Descartes, Galileo Galilei, and Isaac Newton, their achievements were embedded in philosophical reasoning about human epistemic access to the external world. Significant changes of scientific methodology at later stages were at times accompanied by recurring meta-level debates on the nature of scientific reasoning.

One striking example of a period when such meta-level debates did play an important role in physics and chemistry is the debates on atomism in the late nineteenth and early twentieth centuries. At the time, opponents of atomism perceived the atomic hypothesis as an unacceptable intrusion of philosophically charged ontological reasoning into physics that was incompatible with the empiricist principles of modern science. Atomists defended their commitment to atomism by stressing the position's predictive and explanatory power. The scientific debates at the time provoked leading physicists like James C. Maxwell, Ernst Mach, Ludwig Boltzmann, and Henri Poincaré to write genuinely philosophical and highly influential texts about the scientific process and the epistemic status of scientific results.

A slightly more recent example is the debates on the epistemic and ontological implications of quantum mechanics. Early participants to the debate, from Niels Bohr and Albert Einstein to Werner Heisenberg, agreed on the point that quantum mechanics had implications for basic principles of scientific reasoning, and therefore revealed the interconnectedness of physical and philosophical analysis.

1.5 Core Discussion of This Book

The contemporary controversies regarding the epistemic status and future perspectives of theories in fundamental physics can be understood as following from this broad tradition of philosophical debate regarding the methodological foundations of science. Two overlapping main aspects may be distinguished in those controversies and will be conspicuous in the contributions to this volume.

The first aspect relates to the question of how serious the crisis of empirical testing in contemporary fundamental physics actually is. Many theoretical physicists emphasise the continuity between previous stages of physics and the current situation. They argue that physics can adhere to the very same strategies of hypothesis testing that were developed and deployed throughout the twentieth century and before. While they concede that the process may be more extensive and complex today than at earlier stages of physics, they believe that no fundamental shift of strategy is required. Exponents of this position point to the wide range of empirical data that are relevant to the appraisal of fundamental physical theories today and emphasise perspectives for strengthening the connection between those theories and empirical data in the future. Others claim that physics does face a new category of problems – one substantial enough to raise doubts about the sustainability of traditional strategies of empirical testing. In important cases, they point out, empirical testing, though possible in principle, is so far beyond the reach of current experimental methods that it is questionable whether it can ever be achieved. Notably, in the context of the multiverse scenario, theoretical claims may even be untestable in principle. If so, the question arises whether one can call such claims scientific at all. A number of authors in this text deal with the described issues from various perspectives.

This leads us to the second aspect of the debate: the status of non-empirical strategies for theory assessment. As long as empirical testing of a theory remains unavailable or insufficient, scientists may resort to other, theory-based considerations in an effort to assess a theory's prospects. Participants in the debate generally agree that such considerations are part of science, and have played a crucial role in establishing research programs, such as string physics, as preeminent within the community. However, the value of non-empirical strategies for theory assessment is hotly contested. A first question that arises in the given context is how to define non-empirical theory assessment. The contributions to this volume vary in their choices in this respect. Some authors address the entire range of arguments a physicist may cite in the absence of empirical support for a given theory. Others, in particular those using the term 'non-empirical confirmation', emphasise the importance of specifying a subset of arguments that, due to their specific structural characteristics, may carry particular epistemic significance.

Based on a given understanding of what non-empirical theory assessment entails, the authors in this volume raise a number of questions. Is the degree to which strategies of non-empirical theory assessment have generated a preference for individual theories among scientists justified? Are those strategies of purely pragmatic value or are they epistemically significant? Does the extent to which those strategies are deployed in contemporary fundamental physics amount to a qualitative shift of the scientific method or can it be understood fully within known categories of scientific theory assessment? What are the risks when such strategies become very influential in a scientific world that is bound to rely on them to an increasing degree due to the scarcity of empirical data?

1.6 Book Summary

This volume is divided into four parts. The first part discusses the relevant background for the discussion from various perspectives. **Helge Kragh** argues that there are historical precedents for the situation the field finds itself in today. By studying these analogous historical episodes, in particular the nineteenth-century vortex theory of matter, Kragh argues that we will not find guidance regarding how to proceed methodologically today, but rather a way to adequately identify the epistemic situation we are actually in. **Peter Achinstein** also relies on historical case studies to provide a philosophical evaluation of the status of empirically unconfirmed theories. He offers conceptual reasons why non-empirical arguments in their favour, even if significant, do not amount to evidence for the given theory. Empirically unconfirmed theories thus can play an important role in the scientific process and should be taken seriously under certain conditions but still must be treated as scientific speculations. Scientists use a variety of methods to assess theories: from evidence obtained through direct and indirect experiments, to analogue experiments and non-empirical methods. **Radin Dardashti** and **Stephan Hartmann** argue that Bayesian confirmation theory provides an adequate tool for both a descriptive and a normative treatment of these various methodologies. By applying Bayesian confirmation theory to various scientific methodologies, they provide examples of the fruitfulness of this approach. In the final contribution in this part of the volume, **Massimo Pigliucci** argues that, while Popper still plays a significant role as friend or foe in the current debates on the status of physical theories, his positions are often misrepresented in those debates. Drawing parallels to biology, Pigliucci points out that an overly vitriolic public appearance of debates in a scientific field carries the risk of impeding the public's ability to distinguish between science and pseudoscience.

In the second part of this volume, several papers introduce and discuss the viability of methodologies that do not rely on direct empirical evidence. In the first of these chapters, **Richard Dawid** discusses the philosophical foundations of non-empirical theory assessment and addresses several objections that have been put forward against his account. He argues that the potential significance of specific forms of non-empirical confirmation hinges on their similarity to empirical confirmation in a number of respects. Contrary to Dawid, **Carlo Rovelli** argues that the reliance on non-empirical confirmation poses a threat to one of the crucial features of science – namely, its reliability. He illustrates the danger of non-empirical theory confirmation with the example of its application to string theory. Similarly, **Daniele Oriti** considers critically the dangers of non-empirical theory confirmation, while admitting that in situations where empirical data are scarce, especially in the context of theories of quantum gravity, non-empirical theory assessment is even necessary and possibly the only methodological response to a degenerative research programme. **Radin Dardashti** analyzes the normative implications of non-empirical theory assessment for scientific practice. He identifies three problems scientists confronts when applying non-empirical theory assessment and provides possible ways to address these problems. These strategies, he argues, amount to the necessity to change scientific practice in circumstances where non-empirical theory assessment is to be applied. **Elena Castellani** considers the early development of string theory as a historical analysis of the interplay between empirical and non-empirical theory assessment. She argues based on this historical analysis that the relevant question is not one of which methodology to rely on, but rather finding and justifying the right balance between empirical and non-empirical theory assessment.

A recent alternative proposal for theory assessment relies on the use of analogue table-top experiments to test features of inaccessible target systems elsewhere. **Karim Thébault** discusses recent experiments by Steinhauer on analogue black holes and identifies the conditions under which these can provide evidence for phenomena such as black hole radiation in the inaccessible target system. The final paper of Part II discusses the role of Bekenstein entropy in assessing theories of quantum gravity. While there is no empirical confirmation of the Bekenstein formula, its recovery in theories of quantum gravity is widely considered as providing evidence in support of the theory. **Christian Wüthrich** argues that the information theoretic justification of the formula given by Bekenstein himself does not license the use of this formula in theory assessment.

In the third part of this volume, the possibilities and limitations of *empirical* confirmation in cosmology are discussed. Understanding such possibilities is crucial for identifying where, and to what extent, non-empirical and other alternative methods of theory assessment will have to supplement the available empirical assessment. **Joseph Silk** describes how the standard model of cosmology emerged and how its more phenomenal aspects have found empirical confirmation by precision data. He discusses ways in which future precision measurements could shed light on deeper conceptual claims in cosmology. Silk emphasises that, in the end, establishing hypotheses like inflation must be based on confirming precision data. **Björn Malte Schäfer** analyses of the relation between cosmology on the one hand and the theory of general relativity and the statistics of structure formation on the other hand. This provides the necessary background to assess the possible empirical viability of the various assumptions involved in the standard model of cosmology. Based on this analysis he discusses various argumentative strategies to assess the elements of the ΛCDM model. One specific controversial approach within cosmology is the multiverse idea. **George Ellis** discusses the variety of multiverse proposals and their testability. He argues that, except for some specific cases, the multiverse idea is not empirically testable. Its endorsement therefore requires a weakening of principles of scientificality. Even if non-empirical theory assessment were applicable to the multiverse idea, to Ellis, it would not provide sufficient support. Contrary to Ellis, **Sean Carroll** argues that reasoning in support of the multiverse can be fully understood in terms of conventional empirical hypothesis testing and therefore does not require a weakening of scientificality principles. According to Carroll, the evaluation of multiverse models relies on abduction, empirical success, and Bayesian reasoning – just like the evaluation of any other theory in physics. In the final chapter of this part, **Chris Smeenk** discusses the problem of underdetermination in the context of cosmology. He suggests that a theory that is consistently empirically successful indexempirical success in a given domain in effect rules out the possibility of alternative theories in that domain. He then argues that in the case of eternal inflation, empirical confirmation systematically lacks the strength to eliminate rival theories, which seriously limits the chances for conclusively establishing the theory.

In the final part of this volume, several string theorists discuss different possibilities to assess string theory. **Joseph Polchinski** starts by providing an argument for why string theorists rely on the theory despite the lack of empirical data. He considers five remarkable features of string theory which, according to his analysis, justify a considerable degree of trust in string theory's viability. He then offers a rough Bayesian assessment of the multiverse hypothesis to illustrate the claim that it is reasonable to attribute a high probability to this controversial hypothesis. While Polchinski's work reasons at a general conceptual level, the three following chapters discuss concrete models relating string theory to empirically accessible systems. **Eva Silverstein** argues that string theory provides a plethora of fruitful ideas that can be of use for empirically accessible systems. Concretely, she considers the role that string theory played in developing and motivating mechanisms for dark energy and inflation. Thus, even if it may be too early to assess the empirical viability of string theory itself, it has already produced a fruitful interpretational framework for existing empirical data. **Gordon Kane** argues that there are substantial data-based reasons to take a positive attitude towards string theory. Some concrete compactifications of string theory recover many desirable features at lower energies. Kane argues that considering predictions of those specific models as testable predictions of string theory is in line with standard principles of scientific reasoning. In the final chapter, **Fernando Quevedo** presents the current status and prospects of string phenomenology. After introducing several generic predictions of string theory, he considers ways in which observational data can constrain string theoretical model building and addresses several objections against string phenomenology. Quevedo points to a number of specific implications of string theoretical models that could be testable at energies far below the Planck scale.

References

Ambjørn, J., J. Jurkiewicz, and R. Loll. (2001). Dynamically triangulating lorentzian quantum gravity. *Nuclear Physics B 610*(1), 347–82.

Baez, J. C. (1998). Spin foam models. *Classical and Quantum Gravity 15*(7), 1827.

Bombelli, L., J. Lee, D. Meyer, and R. D. Sorkin. (1987). Space–time as a causal set. *Physical Review Letters 59*(5), 521–4.

Dowker, F. (2005). Causal sets and the deep structure of spacetime. In A. Ashtekar (Ed.), *100 Years of Relativity, Space–Time Structure: Einstein and Beyond*, pp. 445–64. Singapore, World Scientific Publishing.

Duff, M. (2013). String and m-theory: answering the critics. *Foundations of Physics 43*(1), 182–200.

Henson, J. (2006). The causal set approach to quantum gravity. In Oriti, D. (Ed.), *Approaches to Quantum Gravity*, pp. 393–413. Cambridge: Cambridge University Press.

Ijjas, A., P. J. Steinhardt, and A. Loeb. (2017, February). All strung out. *Scientific American*, 72–75.

Lauscher, O., and M. Reuter. (2001). Ultraviolet fixed point and generalized flow equation of quantum gravity. *Physical Review D 65*(2), 025013.

Loll, R. (2001). Discrete Lorentzian quantum gravity. *Nuclear Physics B: Proceedings Supplements 94*(1), 96–107.

Oriti, D. (2009). The group field theory approach to quantum gravity. In D. Oriti (Ed.), *Approaches to Quantum Gravity*, pp. 310–31. Cambridge: Cambridge University Press.

Perez, A. (2013). The spin foam approach to quantum gravity. *Living Reviews in Relativity 16*(3), 1–218.

Polchinski, J. (2007). Pop goes the universe. *American Scientist Online 95*, 32–39.

Rovelli, C. (2004). *Quantum Gravity*. Cambridge: Cambridge University Press.

Smolin, L. (2006). *The Trouble with Physics*. London: Allen Lane.

Thiemann, T. (2007). *Modern Canonical Quantum General Relativity*. Cambridge: Cambridge University Press.

Woit, P. (2007). *Not Even Wrong*. New York: Basic Books.

Part I

Historical and Conceptual Background

2

Fundamental Theories and Epistemic Shifts: Can the History of Science Serve as a Guide?

HELGE KRAGH

The recent debate about fundamental physical theories with no or little connection to experiment and observation concerns primarily the relationship between theoretical physics and philosophy. There are reasons to believe that a more enlightened perspective on the debate can be obtained by also taking into regard the history of physics and the history of science generally. Possibly unknown to many physicists, there are several historical precedents, cases that are somewhat analogous to the present one and from which much can be learned. Apart from outlining what I consider to be the essence of the current debate, this chapter briefly discusses the general role that the history of science can play in science and the philosophy of science. It refers to some noteworthy lessons from past physics, of which one particular case, the nineteenth-century vortex theory of matter, is singled out as a possible analogy to the methodological situation in string physics. While I do not suggest that these earlier cases are substantially similar to the ones concerning string theory and the multiverse, I do suggest that there are sufficient similarities on the level of methodology and rhetoric to make them relevant for modern physicists and philosophers.

2.1 The Very Meaning of Science

In May 2008, there appeared in *New Scientist* an article with the provocative question: "Do we need to change the definition of science?" (Matthews, 2008). Six years later, *Nature* published the article "Scientific method: Defend the integrity of physics" (Ellis & Silk, 2014). Both articles discussed essentially the same question, namely whether certain recent developments in theoretical physics belong to science proper. For more than a decade there has been an ongoing and often heated dispute in the physics community, and also in some corners of the philosophical community, concerning the scientific status of theories such as superstring physics and multiverse hypotheses. These theories are cultivated by a fairly large number of professional physicists and, by ordinary sociological standards, are undoubtedly to be counted as scientific. But are they also scientific from an epistemic point of view, or does their status as branches of physics require an extension or revision of the traditional meaning of science?

The classical demarcation problem between science and non-science (which may or may not include pseudoscience) has taken a new turn with the appearance of fundamental and highly mathematical theories that may not be experimentally testable in the ordinary sense. So why believe in them? According to the philosopher Dudley Shapere (2000, pp. 159–61), "physics is in fact approaching, or perhaps has reached, the stage where we can proceed without the need to subject our further theories to empirical test." He asks, "Could empirical enquiry, which has guided up to a certain point science in its history, lead at that point to a new stage wherein empiricism itself is transcended, outgrown, at least in a particular domain?" More than a few physicists would presently respond affirmatively to Shapere's question. It should be noted that the demarcation problem and the traditional criteria of falsifiability and empirical testability are discussed not only by physicists but also in some other branches of science. For example, biologists have questioned these criteria and suggested, in striking analogy to the debate concerning multiverse physics, that methodological norms of what constitutes good science are not only irrelevant but actually detrimental to the progress of their science (Yang, 2008).

What it is all about can be summarized in the notion of "epistemic shifts," meaning claims that the basic methodological and epistemological rules of science are in need of revision (Kragh, 2011). These rules may be appropriate for most science and have been appropriate for all science until recently, but in some areas of modern physics they are no longer adequate and should therefore be replaced by other norms for the evaluation of theories. A proposed shift in epistemic standards may be of such a drastic nature that it challenges the very meaning of science as traditionally understood. In this case it effectively implies a new demarcation line separating what counts as science and what does not. This is what Steven Weinberg (2007) alluded to when he, referring to the string-based multiverse, said that "we may be at a new turning point, a radical change in what we accept as a legitimate foundation for a physical theory."

Another way of illustrating the notion of an epistemic shift is to compare it to Thomas Kuhn's idea of revolutions separated by different paradigms. Richard Dawid (2013, p. 124) speaks of the debate in the physics community as "a paradigm shift regarding the understanding of scientific theory assessment." According to the original version of Kuhn's philosophy of science, paradigm shifts include different criteria for what counts as acceptable science and also for evaluating theories. Rival paradigms carry with them rival conceptions of science; for this reason alone they are incommensurable. In principle, no rational argument can decide whether one paradigm is superior to a competing paradigm. The rhetoric of epistemic shifts has become part of modern physics. "We are in the middle of a remarkable paradigm shift in particle physics," asserts one physicist, referring to the anthropic string landscape (Schellekens, 2008, p. 1). According to another physicist, the multiverse promises "a deep change of paradigm that revolutionizes our understanding of nature" (Barrau, 2007, p. 16).

The purpose of this chapter is not to reexamine the recent debate concerning string theory and multiverse cosmology, but rather to look at it through the sharp lenses of the history of science. Although knowledge of the history of the physical sciences is of no direct relevance to the ongoing debate, it is of some indirect relevance. It may serve the purpose of correcting various mistakes and placing the subject in a broader historical perspective. Physicists may think that super-strings and the multiverse have ushered in a uniquely new situation in the history of science, but they are mistaken. There have been several cases in the past of a somewhat similar nature, if not of quite the same scale. I modestly suggest that modern fundamental physics can in some sense learn from its past. Before turning to this past I shall briefly review what is generally and for good reasons considered the most important of the traditional standards of theory evaluation – namely that a theory must be testable.

2.2 Testability

To speak of the "definition" of science is problematic. There simply is no trustworthy methodological formulation that encapsulates in a few sentences the essence of science and is valid across all periods and all disciplines. Nonetheless, some criteria of science and theory choice are relatively stable, enjoy general acceptance, and have been agreed upon since the early days of the scientific revolution (Kuhn, 1977). Almost all scientists subscribe to the belief that testability is more than just a desideratum that scientists have happened to agree upon and that suited science at a certain stage of development. They consider it a *sine qua non* for a theory being scientific that it must be possible to derive from it certain consequences that can be proved right or false by means of observation or experiment. If there are no such consequences, the theory does not belong to the domain of science. In other words, although empirical testability is not a sufficient criterion for a theory being scientific, it is a necessary one. Einstein was a great believer in rationalism and mathematical simplicity and yet he was convinced that "Experience alone can decide on truth" (Einstein, 1950, p. 17). He is followed in this belief by the large majority of modern physicists, who often go to great lengths to argue that their theories, however speculative and mathematical they may appear to be, do connect with empirical reality. Lee Smolin (2004, p. 70) echoed Einstein when he concluded about the opposing views of string theory and loop quantum gravity that, "Because this is science, in the end experiment will decide."

Physicists working with string theory, multiverse cosmology, or related areas of fundamental physics are routinely accused of disregarding empirical testability and of replacing this criterion with mathematical arguments. These accusations are not quite fair (Johannson & Matsubaru, 2009; Dawid, 2013, p. 154). On the one hand, most physicists in these fields readily accept the importance of testability, admitting that empirical means of assessment have a higher epistemic status than non-empirical means. On the other hand, they stress the value of the latter methods, which sometimes may be the only ones available. At the same time they maintain that their theories have – or in the near future will have – consequences that at least indirectly can be tested experimentally. They have not really abandoned the commonly accepted view of experiment as the final arbiter of physical theory. "The acid test of a theory comes when it is confronted with experiments," two string theorists say (Burgess & Quevedo, 2007, p. 33). Unfortunately, the necessary experiments are in most cases unrealistic for the time being, but what matters to them is that predictions from the theories are not beyond empirical testability in principle.

Although one can identify a consensus view concerning testability, it is to some extent rhetorical and of limited practical consequence. It is one thing to agree that theories of physics must be testable, but another thing to determine the meaning of the concept of testability where there is no corresponding consensus. Everyone agrees that actual and present testability, involving present instrument technologies or those of a foreseeable future, is preferable, but that is where the agreement ends. Some of the questions that physicists and philosophers have discussed are the following:

(1) Should it be required that a theory is actually testable, or will testability in principle, such as in the form of a thought experiment, suffice?
(2) Should a theory result in precise and directly testable predictions, or will indirect testability do?
(3) If a theory does not result in precise predictions, but only in probability distributions, is it then testable in the proper sense?
(4) Will a real test have to be empirical, by comparing consequences of the theory with experiments or observations, or do mathematical consistency checks also count as tests?
(5) Another kind of non-empirical testing is by way of thought experiments or arguments of the *reductio ad absurdum* type. A theoretical model may lead to

consequences that are either contradictory or so bizarre that they are judged unacceptable. How should such arguments enter the overall evaluation picture?

(6) At what time in the development of a theory or research program can one reasonably demand testability? Even if a theory is not testable at present, in a future version it may result in testable predictions.

(7) How should testability be weighed in relation to other epistemic desiderata? For example, is an easily testable theory with a poor explanatory record to be preferred over a non-testable theory with great explanatory power? What if the testable theory is overly complicated and the non-testable one is mathematically unique and a paragon of simplicity?

(8) Should predictions of novel phenomena be counted as more important than predictions or retrodictions of already known phenomena?

I shall not comment further on these issues except to point out that some of them are particularly relevant with regard to theories, such as string theory or multiverse scenarios, that are not actually testable in the ordinary way. Indirect testability (#2) may mean that if a fundamental and well-established background theory with great empirical success results in a prediction that cannot be tested directly, then the success of the background theory functions as an indirect test. The existence of multiple universes can in this sense be said to be tested by quantum mechanics as a background theory. "Accepting quantum mechanics to be universally true means that you should also believe in parallel universes," says Max Tegmark (2007, p. 23). As to #4, there are indeed physicists who have appealed to mathematical consistency as a kind of test of string theories; if admitted as a test, it is part of what Dawid (2013) calls non-empirical theory assessment. By comparison, issues #6, 7, and 8 are of a general nature and relevant to all scientific theories, whether modern or old. They can be easily illustrated by means of concrete cases from the history of science, but this is beyond the scope of the present chapter.

2.3 Physics, Philosophy, and History

The fundamental question of the demarcation between science and non-science is of a philosophical nature, rather than a scientific nature. It rests on certain standards and criteria that nature herself does not provide us with and cannot be determined by purely scientific means. The standards and criteria do not need to be part of a philosophical system or even to be explicitly formulated, but they nevertheless belong to the realm of philosophical discourse. The borderline between physics and philosophy has shifted over time, typically with physics appropriating areas that were traditionally considered subjects of philosophical speculation. At least, this is how some physicists like to see the historical development. They question whether there is any need for external norms of science of a philosophical nature, suggesting that such norms are unnecessary and may even be harmful to the progress of science. Only the scientists themselves can decide what the boundary is between science and non-science. If this is the view of most physicists (which I believe it is), it is to some extent counteracted by a growing group of philosophers of physics who ask questions that have traditionally been seen as within the remit of physicists.

According to Barrau (2007, p. 13), "If scientists need to change the borders of their own field of research it would be hard to justify a philosophical prescription preventing them from doing so." Susskind (2006, p. 194) agrees, adding that "Good scientific methodology is not an abstract set of rules dictated by philosophers." However, the question of what constitutes a legitimate scientific theory cannot be left entirely to the scientists. The seductive claim that science is what scientists do is circular; moreover, it presupposes that all scientists have the same ideas of what constitutes science. But if there were such a methodological consensus in the physics community there would be no controversy concerning the scientific legitimacy of research areas such as superstrings and the multiverse. The history of science strongly suggests that certain methodological prescriptions, such as testability of theories, are almost universally accepted; but it also shows that in some cases the consensus breaks down. The current discussion about string theory and the multiverse is evidently such a case.

Insofar that philosophical ideas about science are intended to relate to real science (and not to be purely normative), they must agree with or at least confront the large pool of data provided by the history of science. This is particularly important for philosophical views concerning the dynamics of science or the development of science in its temporal dimension. Consider a philosophical rule that has the consequence that the change from the Bohr–Sommerfeld quantum theory to the new quantum mechanics in the mid-1920s was not progressive. Such a rule is simply not credible. It may be tempting to consider the history of science as providing an empirical database for the testing of philosophical theories about science, in somewhat the same way that experiments and observations function as tests of scientific theories. However, this is problematic for several reasons (Schickore, 2011). Foremost among these reasons is that historical case studies exhibit a great deal of variation over time, discipline, and culture; they deal with a particular case at a particular time, with the result that the methodological lessons one can draw from two important cases in the history of science are rarely the same. Philosophers, by contrast, aim at saying something general about science and its development.

There have been several attempts among philosophers of science to extract rules of scientific development from the history of science or to formulate rules strongly guided by history. None of them have been very successful. Larry Laudan (1977, p. 160) suggested that "our intuitions about historical cases can function as decisive touchstones for appraising and evaluating different models of rationality, since we may say that it is a necessary condition of rationality that it squares with (at least some of) our intuitions." Laudan's standard intuitions included that after 1840 it was irrational to believe that light consisted of particles and similarly, after 1920, that the chemical atoms had no parts. Such examples are rather trivial and of no real use. The more interesting cases in the history and philosophy of science are precisely those about which there are no standard intuitions of rationality. Thus, with Augustin Fresnel's wave theory of light from 1821, light came to be conceived as transverse waves rather than material corpuscles, but the transition did not occur instantly. Was it rational to believe in the late 1820s in the corpuscular theory of light? Likewise, while the atom was traditionally seen as an indivisible and simple body, the discoveries of radioactivity and the electron in the 1890s indicated that it had an inner structure. Was it rational to believe in 1900 in the non-composite atom?

A more recent and more ambitious research project on the testing of theory dynamics included sixteen historical case studies ranging from the Copernican revolution in astronomy to the electroweak unification in particle physics (Donovan, Laudan, & Laudan, 1988). These cases were compared with philosophical rules such as "the appraisal of a theory depends on its ability to solve problems it was not invented to solve" and "the standards for assessing theories do not generally vary from one paradigm to another." Although the project resulted in interesting analyses, it failed in establishing a non-trivial philosophical methodology on the basis of the history of science. Among its obvious weaknesses was that the project rested on a rather arbitrary selection of case studies; had other cases been chosen the result would have been different.

The failure of this approach to philosophy of science does not imply that philosophers can or should avoid engaging in historical studies. Indeed, if philosophers want to retain contact with real science and how it changes in time, they must take the history of science seriously; they must investigate real science in either its present or past forms. By far most of our knowledge of how science has developed comes from the past and is solidly documented in the historical sources. The history of science is an indispensable but, by its very nature, incomplete guide to understanding contemporary science. By investigating cases from the past and relating them to cases from the present, philosophers as well as active scientists may obtain a broader and more enlightened view of current science.

Historical arguments and analogies have a legitimate albeit limited function in the evaluation of current science. Even a physicist with no interest in or knowledge of the history of his or her field of research cannot avoid being guided by the past. In science, as in other areas of human activity, it would be silly to disregard the historical record when thinking about the present and the future. At the same time, such guidance should be based on historical insight and not, as is often the case, on arbitrary selections from a folk version of history. Generally speaking, the history of science is so diverse and so complex that it is very difficult to draw from it lessons of operational value for modern science.

In a paper published in 1956, the brilliant and controversial astrophysicist Thomas Gold, one of the fathers of the steady state theory of the universe, argued that cosmology was a branch of physics, and hence a science. But he dismissed the idea of a methodology particular to physics or to the sciences generally. "In no subject is there a rule, compliance with which will lead to new knowledge or better understanding," Gold (1956, p. 1722) wrote. "Skilful observations, ingenious ideas, cunning tricks, daring suggestions, laborious calculations, all these may be required to advance a subject. Occasionally the conventional approach in a subject has to be studiously followed; on other occasions it has to be ruthlessly disregarded." Undoubtedly with the ongoing controversy between the steady state theory and relativistic evolution cosmology in mind, Gold further reflected on the lessons of the history of science with regard to the methods of science. But he considered history to be an unreliable and even treacherous guide:

Analogies drawn from the history of science are frequently claimed to be a guide [to progress] in science; but as with forecasting the next game of roulette, the existence of the best analogy to the present is no guide whatever to the future. The most valuable lesson to be learned from the history of scientific progress is how misleading and strangling such analogies have been, and how success has come to those who ignored them. (p. 1722)

Gold's cynical and anarchistic view is not without merit, but it seriously underestimates the power of history and the value of insight based on the history of science. Although the development of the history of science is not governed by law and does not exhibit much regularity, neither is it an accidental series of events comparable to a game of roulette.

2.4 Some Lessons from Past Physics

As mentioned, empirical testability is an almost universally accepted criterion of science. But even with respect to this most sacred of the defining features of science, we find in the history of science a few exceptions. It is, after all, not a necessary ingredient of science. Dawid (2013, p. 97) argues that the role played by non-empirical theory assessment in modern fundamental physics is a continuation of earlier tendencies to be found in post–World War II particle physics. This is undoubtedly correct – think of the development of S-matrix or "bootstrap" theory in the 1950s and 1960s – but in my view there is no reason to restrict the historical perspective to the era of quantum and relativity physics. There are also inspiration and instruction to be found in other and earlier examples from the history of physics.

During the early decades of the nineteenth century Romantic natural philosophy (known as *Naturphilosophie*) made a great impact on physics and the other sciences in Northern Europe (Cunningham & Jardine, 1990; Kragh, 2011, pp. 26–34). In this grand attempt to revolutionize science and base it on an entirely new foundation, speculations and aesthetic sentiments were not just considered legitimate parts of science, they were necessary parts and even more fundamental than empirical investigations. The philosopher Friedrich Schelling, the spiritual leader of the *Naturphilosophie* movement, even founded a *Journal of Speculative Physics* as a means of promoting and communicating the new science. At the time the term "speculation" did not have the pejorative meaning it typically has today but rather was largely synonymous with "intuition." It was a fundamental assumption of the new speculative physics that mind and nature coexisted as a unity, such that one was unable to exist without the other.

Schelling and those who followed his thinking were not necessarily against experiments, but they thought that measuring the properties of objects and phenomena was of no great importance since it provided no understanding of the inner working of nature. In some cases natural philosophers went so far as to completely deny that observation and experiment could lead to any real insights into nature's secrets. The sort of nature that could be empirically investigated was regarded as a dull wrapping that contained and obscured the real, non-objective nature. The only way to recognize the latter was by taking the route of speculative physics – that is, to be guided by the intuitive mind of the genius. The laws of nature were thought to coincide with the laws of reason; they were true a priori and for this reason it made no sense to test them by means of experiment.

Before dismissing Romantic natural philosophy as nothing but pseudoscientific and metaphysical nonsense, it should be recalled that some of the greatest physicists of the time were much influenced by this movement. Examples include H. C. Ørsted and Michael Faraday, the two celebrated pioneers of electromagnetism; another example is J. Ritter, the discoverer of ultraviolet radiation. Nonetheless, one cannot conclude from the case that good physics can flourish in the absence of experimental testing of theories. Neither Ørsted, nor Faraday, nor Ritter subscribed to Schelling's more extreme ideas and especially not to his disrespect of experiment. Ørsted's belief in a unity of electric and magnetic forces was rooted in the Romantic philosophy, but it was only when he verified it experimentally in 1820 that he turned it into a scientific discovery.

More than a century later we meet a very different version of rationalistic physics in the context of "cosmophysics," an ambitious attempt to formulate a complete and final theory of the universe and all what is in it. The leading cosmophysicists of the 1930s were two of Britain's most reputed scientists, Arthur Eddington and E. Arthur Milne. Although their world systems were quite different, both aimed at reconstructing the very foundation of physics; they did so by basing physics on a priori principles from which the laws of nature could be uniquely deduced by pure reason. Experimental tests played but an insignificant role, being subordinated logical and mathematical arguments. Milne seriously believed that when his system of world physics (as he called it) was completed there would be no contingent elements at all in the laws of nature; it would then turn out that the laws were no more arbitrary than the theorems of geometry. A mathematician knows whether a theorem is true or not without consulting nature. Likewise, Milne (1948, p. 10) wrote that "it is sufficient that the structure [of world physics] is self-consistent and free from contradiction."

Eddington's idiosyncratic fundamental theory promised a way to deduce unambiguously all the laws and constants of nature from epistemic and mathematical considerations. In his bold attempt to unify cosmology and the quantum world, mathematics played a role no less elevated than in Milne's theory (Eddington, 1936, p. 3; Durham, 2006):

> It should be possible to judge whether the mathematical treatment and solutions are correct, without turning up the answer in the book of nature. My task is to show that our theoretical resources are sufficient and our methods powerful enough to calculate [of nature] the constants exactly – so that the observational test will be the same kind of perfunctory verification that we apply to theorems in geometry.

Of course, neither Milne nor Eddington could afford the luxury of disregarding experiments altogether. But they argued that experiments did not reveal the true laws of nature and consequently could not be used to test the laws. Eddington famously calculated the precise values of many of the constants of nature such as the fine-structure constant, the proton-to-electron mass ratio, and the cosmological constant. When experiments failed to agree with the predicted values he arrogantly maintained that the theory was correct; any discrepancy between theory and experiment must lie with the latter.

The theories of Milne, Eddington, and their few followers shared the same fate as the revolutionary Romantic natural philosophers: they were unproductive mistakes and are today relegated to the long list of grand failures in the history of science. All the same they are of some relevance in that aspects of the same aspirations and rationalist methods can still be found in modern physics. The most extreme version is probably the Platonic "mathematical universe hypothesis" proposed by Max Tegmark (2014). Likewise, the history of string theory includes examples that show at least some similarity to the earlier ideas of cosmophysics. Referring to the theory of superstrings, John Schwarz (1998, p. 2) wrote, "I believe that we have found the unique mathematical structure that consistently combines quantum mechanics and general relativity. So it must almost certainly be correct."

Unfortunately the prediction of supersymmetric particles remained unverified, but this did not worry Schwarz too much: "For this reason, even though I do expect supersymmetry to be found, I would not abandon this theory if supersymmetry turns out to be absent" (p. 2).

Thus one can conclude from the history of physics that fundamental theories, to be successful from a physical (and not merely mathematical) point of view, must have some connection to empirical reality. The historical record of such theories suggests that empirical testability is a necessary condition for progress. But this is as far as the historical argument can go. Because one can observe some regularity in the past – say, that all physically progressive theories have been actually testable – there is no guarantee that the regularity will continue in the future.

Many of the arguments in string theory and multiverse physics rely implicitly on two philosophical principles that can be traced back to Leibniz in the late seventeenth century. One is the doctrine of a pre-established harmony between the mathematical and physical sciences, making pure mathematics the royal road to progress and unification in physics (Kragh, 2015). The other is the principle of plenitude, which essentially states that whatever is conceived as possible must also have physical reality. The plenitude principle is a metaphysical claim that translates potential existence into real existence. In its more modern formulation it is often taken to mean that theoretical entities exist in nature insofar that they are consistent with the fundamental laws of physics. Since numerous universes other than ours are consistent with the equations of string theory they must presumably exist (Susskind, 2006, p. 268).

The ontological plenitude principle has played a most important role in the history of science and ideas, including modern theoretical physics from Dirac's positron to Higgs's boson. Although in many cases it has been dramatically fruitful, it cannot be justified by reference to its historical record. For every example of success, there is one of failure. If the former are better known than the latter, it is because history is written by the victors. In this case, as in many others, the history of science is ambiguous. It does not speak unequivocally in favor of either the principle of plenitude or a pre-established relationship between mathematics and physics; nor does it speak unequivocally against the doctrines.

2.5 A Victorian Analogy to String Theory?

Yet another case from the past deserves mention: the vortex theory of atoms that attracted much scientific attention during the latter part of the nineteenth century. In particular from a methodological point of view, but not of course substantially, there is more than a little similarity between this long-forgotten theory and the string theory of contemporary physics. Based on a hydrodynamic theory of vortices proposed by Hermann von Helmholtz, in 1867 William Thomson (the later Lord Kelvin) argued that atoms and all atomic phenomena might be understood in terms of permanent vortex rings and filaments moving in a continuous, all-pervading medium or "fluid." The medium was generally identified with the commonly assumed world ether, meaning that the discrete atoms were reduced to particular states of motion in the continuous ether. Dualism was replaced by monism. For details about the history of the vortex theory, see Kragh (2002), which includes references to the sources of the quotations in this section.

For about two decades the research program initiated by Thomson was vigorously cultivated by a minor army of mostly British physicists and mathematicians. Although the mathematically complex theory did not easily lead to results that could be compared to experiments, the vortex physicists stressed that it was empirically useful and, at least in principle, testable in the ordinary sense. Indeed, it was applied to a wide range of physical and chemical phenomena, including spectroscopy, the behavior of gases, chemical bonds, the periodic system of the elements, and even gravitation. Considered a truly fundamental theory, ultimately it was expected to provide a physical explanation of the riddle of gravity. Thomson's collaborator Peter Tait went even further. "The theory of vortex-atoms must be rejected at once if it can be shown to be incapable of explaining this grand law of nature," he wrote in 1876. In spite of many attempts to derive gravity from the properties of vortex atoms, no explanation came forward – but neither was it conclusively shown that the theory could not account for Newton's law of gravitation. After all, lack of verification is not falsification. Proponents of the theory suggested optimistically that, when the theory was developed into still more advanced versions, it would eventually solve the problem.

Despite the vortex theory's connection to a variety of physical and chemical phenomena, its empirical record was far from impressive. And yet, although quantitative verification was missing, there were just enough suggestive qualitative agreements to keep the theory alive as more than just a mathematical research program. A characteristically vague defense of the vortex theory was offered by the American physicist Silas Holman: "The theory has not yet, it is true, been found capable of satisfactorily accounting for several important classes of phenomena, ... but this constitutes no disproof. The theory must be judged by what it has accomplished, not by what we have not yet succeeded in doing with it." In any case, what justified the theory and made it so attractive was not so much its ability to explain and predict physical phenomena but rather its methodological and aesthetic virtues. Albert Michelson believed that the vortex theory "ought to be true even if it is not," and Oliver Lodge similarly described it as "a theory about which one may almost dare to say that it deserves to be true." It was also these virtues – the theory's purity and lack of arbitrary hypotheses – that greatly appealed to Maxwell. "When the vortex atom is once set in motion, all its properties are absolutely fixed and determined by the laws of motion of the primitive fluid, which are fully expressed in the fundamental equations," he wrote in 1875. "The method by which the motion of this fluid is to be traced is pure mathematical analysis. The difficulties of the method are enormous, but the glory of surmounting them would be unique."

Mathematics played a most essential role in how the vortex theory was developed and perceived. In fact, a large part of the development was driven by interest in pure mathematics and quite unrelated to physical phenomena. Papers on the subject appeared equally in journals devoted to mathematics and physics. In his 1867 paper Thomson pointed out that the calculation of the vibration frequencies of a vortex atom presented "an intensely interesting problem of pure mathematics." To his mind and to the minds of many other vortex theorists, the difficulties of deducing observable phenomena from the vortex theory were a challenge rather than an obstacle. The difficulties only added to the appeal of the theory. The hope that progress would come through mathematics was a persistent theme in the history of the theory. It was routinely argued that it was not *yet* understood sufficiently to be physically useful. According to Tait, theoretical investigations of the kind that related to real phenomena would "employ perhaps the lifetimes for the next two or three generations of the best mathematicians in Europe." Although this was a formidable difficulty, there was no reason for despair: "it is the only one which seems for the moment to attach to the development of this extremely beautiful speculation; and it is the business of mathematicians to get over difficulties of this kind."

The reliance on mathematics was a mantra among the advocates of the vortex theory of matter and ether. Here is yet another expression of the mantra, this time from the British physicist Donald Mcalister:

The work of deduction is so difficult and intricate that it will be long before the resources of the theory are exhausted. The mathematician in working it out acquires the feeling that, although there are still some facts like gravitation and inertia to be explained by it, the still unexamined consequences may well include these facts and others still unknown. As Maxwell used to say, it already explains more than any other theory, and that is enough to recommend it. The vortex-theory is still in its infancy. We must give it a little time.

Although the term "theory of everything" did not exist at the time of the vortex theory (it seems to date from 1985), this is what the theory aimed to be. As late as 1895 – just a few years before the arrival of quantum theory – the mathematical physicist William Hicks discussed the vortex theory as by far the best candidate for what he called an ultimate theory of pure science. The aim of such a theory, he said, was "to explain the most complicated phenomena of nature as flowing by the fewest possible laws from the simplest fundamental data." Science, he went on, "will have reached its highest goal when it shall have reduced ultimate laws to one or two, the necessity of which lies outside the sphere of our cognition." When the laws had been found, "all physical phenomena will be a branch of pure mathematics." Hicks believed that the vortex theory was at least a preliminary version of the ultimate theory, a complete version of which would require even more mathematical investigation. "It is at present a subject in which the mathematicians must lead the attack."

The mathematical richness of the vortex theory might be considered a blessing, since it effectively protected the theory from being falsified, but it was also a curse. It made G. F. FitzGerald exclaim that "it seems almost impossible but that an explanation of the properties of the universe will be found in this conception." The final vortex theory that he and others dreamed of could in principle explain everything, including the properties of the one and only universe. But could it also explain why the numerous other conceivable states of the universe, all of them consistent with the theory's framework, did not exist? Could it explain why the speed of light has the value it has rather than some other value? The theory could in principle explain the spectral lines and the atomic weight of chlorine, but had chlorine had any other spectrum and other atomic weight the theory could account for that as well. In short, the ratio between the theory's explanatory and predictive force was embarrassingly large.

Let me end with one more quotation: "I feel that we are so close with vortex theory that – in my moments of greatest optimism – I imagine that any day, the final form of the theory might drop out of the sky and land in someone's lap. But more realistically, I feel that we are now in the process of constructing a much deeper theory of anything we have had before and that ... when I am too old to have any useful thoughts on the subject, younger physicists will have to decide whether we have in fact found the final theory!" The quotation is not from a Victorian physicist in the late nineteenth century but from an interview Edward Witten gave in the late twentieth century. The only change I have made is to substitute "vortex" for "string" in the first sentence. It is probably unnecessary to elaborate on the methodological and rhetorical similarities of the vortex theory of the past and the string theory of the present.

I find the case of the vortex theory to be instructive because it exhibits on a meta-level some of the features that are met in much later fundamental theories of physics. Among these are the seductive power of mathematics and the no less seductive dream of a final theory. It illustrates that some of the general problems of contemporary physics are not specifically the result of the attempts to unify quantum mechanics and general relativity. But the dissimilarities are no less distinct than the similarities. For example, string theory is cultivated on a much larger scale than the vortex theory, which largely remained a British specialty. Another difference is the ideological and religious use of the vortex theory in Victorian Britain, where it was customary to see the initial vortical motion in the ether as a result of God's hand. As Thomson pointed out in his 1867 address, vortex atoms could have come into existence only through "an act of creative power." As far as I know, the religious dimension is wholly absent from string theory (but not always from multiverse cosmology).

2.6 Conclusion

The primary function of the history of science is to describe and understand past science irrespective of whether it connects with the modern development. It is not to assist or guide contemporary science, a function for which historians of science are generally unfit. There is no reason to believe that modern string theorists would perform better if they were well acquainted with earlier episodes in the history of physics. While this is undoubtedly true in a technical sense, historical reflection has something to offer in a broader sense, when it comes to the general understanding of the present debate about fundamental physics. In my view, what can be learned from the past relates mostly to the philosophical aspects of the debate, to the novelty of the epistemic situation, and to the rhetoric used in the presentation of modern fundamental physics to a general public.

By adopting a historical perspective it becomes clear that many of the claims concerning the uniqueness of the present situation are exaggerated. It is not the first time in history that scientists have seriously questioned the traditional norms of science and proposed revisions of what should pass as theory assessment. Apart from the precedents mentioned here, there are several other cases of intended epistemic shifts (Kragh, 2011). Physics may presently be at "a turning point" or "a remarkable paradigm shift," but if so it is not for the first time. Knowledge of the history of science, and the history of fundamental physics in particular, cannot but result in an improved and more balanced picture of what is currently taking place in areas such as string theory and multiverse cosmology.

References

Barrau, A. (2007). Physics in the multiverse. *Cern Courier* (20 October).

Burgess, C., & Quevedo, F. (2007). The great cosmic roller-coaster ride. *Scientific American* **297** (10), 29–35.

Cunningham, A., & Jardine, N., eds. (1990). *Romanticism and the Sciences*. Cambridge: Cambridge University Press.

Dawid, R. (2013). *String Theory and the Scientific Method*. Cambridge: Cambridge University Press.

Donovan, A., Laudan L., & Laudan, R., eds. (1988). *Scrutinizing Science: Empirical Studies of Scientific Change*. Dordrecht: Kluwer.

Durham, I. (2006). Sir Arthur Eddington and the foundation of modern physics. Arxiv:quant-ph/0603146.

Eddington, A. (1936). *Relativity Theory of Protons and Electrons*. Cambridge: Cambridge University Press.

Einstein, A. (1950). On the generalized theory of relativity. *Scientific American* **182** (4), 13–17.

Ellis, G., & Silk, J. (2014). Defend the integrity of physics. *Nature* **516**, 321–3.

Gold, T. (1956). Cosmology. *Vistas in Astronomy* **2**, 1721–26.

Johannson, L. G., & Matsubaru, K. (2009). String theory and general methodology: A mutual evaluation. *Studies in History and Philosophy of Modern Physics* **42**, 199–210.

Kragh, H. (2002). The vortex atom: A Victorian theory of everything. *Centaurus* **44**, 32–114.

Kragh, H. (2011). *Higher Speculations: Grand Theories and Failed Revolutions in Physics and Cosmology*. Oxford: Oxford University Press.

Kragh, H. (2015). Mathematics and physics: The idea of a pre-established harmony. *Science and Education* **24**, 515–27.

Kuhn, T. (1977). *The Essential Tension: Selected Studies in Scientific Tradition and Change*. Chicago: University of Chicago Press.

Laudan, L. (1977). *Progress and Its Problems: Towards a Theory of Scientific Growth*. London: Routledge & Kegan Paul.

Matthews, R. (2008). Some swans are grey. *New Scientist* **198**, 44–7.

Milne, E. A. (1948). *Kinematic Relativity*. Oxford: Clarendon Press.

Schellekens, A. N. (2008). The emperor's last clothes? Overlooking the string theory landscape. *Reports on Progress in Physics* **71**, 072201.

Schickore, J. (2011). More thoughts on HPS: Another twenty years later. *Perspectives on Science* **19**, 453–81.

Schwarz, J. H. (1998). Beyond gauge theories. Arxiv:hep-th/9807195.

Shapere, D. (2000). Testability and empiricism. In E. Agazzi & M. Pauri, eds., *The Reality of the Unobservable* (pp. 153–64). Dordrecht: Kluwer Academic.

Smolin, L. (2004). Atoms of space and time. *Scientific American* **290** (1), 66–75.

Susskind, L. (2006). *The Cosmic Landscape: String Theory and the Illusion of Intelligent Design*. New York: Little, Brown and Company.

Tegmark, M. (2007). Many lives in many worlds. *Nature* **448**, 23–4.

 (2014). *Our Mathematical Universe*. New York: Alfred E. Knopf.

Weinberg, S. (2007). Living in the multiverse. In B. Carr, ed., *Universe or Multiverse?* (pp. 29–42). Cambridge: Cambridge University Press.

Yang, A. (2008). Matters of demarcation: Philosophy, biology, and the evolving fraternity between disciplines. *International Studies in the Philosophy of Science* **22**, 211–25.

3

Scientific Speculation
A Pragmatic Approach

PETER ACHINSTEIN

"Hypotheses, whether metaphysical or physical, or based on occult qualities, or mechanical, have no place in experimental philosophy."

– Isaac Newton

"I think that only daring speculation can lead us further, and not accumulation of facts."

– Albert Einstein

3.1 Introduction

During the history of science, controversies have emerged regarding the legitimacy of speculating in science. At the outset, I will understand speculating as introducing assumptions without knowing that there is evidence for those assumptions. If there is evidence, the speculator does not know that. If there is no such evidence, the speculator may or may not know that. The speculator may even be introducing such assumptions implicitly without realizing that fact. In any of these cases (under certain conditions to be specified later), the individual is speculating. I will use the term "speculation" to refer both to the activity of speculating and to the product of that activity – the assumptions themselves. Which is meant should be clear from the context. In this chapter, I propose to do three things: (1) to clarify and expand the initial characterization of speculation; (2) to ask whether and under what conditions speculating in science is a legitimate activity; and (3) assuming that speculating is or can be legitimate, to consider how, if at all, speculations are to be evaluated. Although philosophers and scientists have expressed strong and conflicting opinions on the subject of the second task, little has been written about the other two, particularly the first.

In Section 3.2, I offer three examples of speculations from the history of physics. In Section 3.3, I introduce three influential contrasting views about whether and when speculating is legitimate in science. In Sections 3.4–3.10, I focus on the basic definitional question, attempting to show exactly how the concept of speculation can be defined using various concepts of evidence, my own and Bayesian ones. In Section 3.11, I discuss and reject the three contrasting views about speculation introduced in Section 3.3. In Sections 3.12 and 3.13, I defend a different view, a pragmatic one suggested by James Clerk Maxwell, one of the great speculators in physics, who had very interesting philosophical ideas about speculation.

3.2 Three Speculations from Physics

Let me begin with three examples from the history of physics, together with claims of their detractors who reject or at least criticize them not because they are false or refuted, but because they are speculations.

3.2.1 Thomas Young's Wave Theory of Light

In 1802, Thomas Young published "On the Theory of Light and Colours" [1], in which he resuscitated the wave theory of light by introducing four basic assumptions: (1) A rare and highly elastic luminiferous ether pervades the universe; (2) a luminous body excites undulations in this ether; (3) the different colors depend on the frequency of the vibrations; and (4) bodies attract this medium so that the medium accumulates within them and around them for short distances. With these and other assumptions, Young shows how to explain various observed properties of light.

In 1803, Henry Brougham, a defender of the particle theory of light, wrote a scathing review of Young's paper, in which he says:

> As this paper contains nothing which deserves the names either of experiment or discovery, ... it is in fact destitute of every species of merit.... A discovery in mathematics, or a successful induction of facts, when once completed, cannot be too soon given to the world. But ... an hypothesis is a work of fancy, useless in science, and fit only for the amusement of a vacant hour.... [2]

Brougham defends the Newtonian particle theory of light on the grounds that it was inductively supported by experiments, and he rejects Young's wave theory on the grounds that it is mere speculation.

3.2.2 William Thomson: Baltimore Lectures on Molecules and the Wave Theory of Light

In 1884, Sir William Thomson, Lord Kelvin, delivered a series of lectures at Johns Hopkins University on "molecular dynamics and the wave theory of light." His aim was to provide a molecular interpretation for the luminiferous ether postulated by the wave theory. He assumes that there is an ether and that its properties can be described mechanically. He writes:

It seems probable that the molecular theory of matter may be so far advanced sometime or other that we can understand an excessively fine-grained structure and understand the luminiferous ether as differing from glass and water and metals in being very much more finely grained in its structure. [3]

He proceeds by offering various mechanical models of the ether to explain known optical phenomena, including rectilinear propagation, reflection, refraction, and dispersion.

In his 1906 classic *The Aim and Structure of Physical Theory* [4], Pierre Duhem excoriates Thomson for presenting a disorderly series of contradictory models (as, he claims British minds, incapable of Continental [meaning French] orderliness, are wont to do), for invoking occult causes, and for not producing a "system of principles, which aim to represent as simply, as completely, and as exactly as possible a set of experimental laws" (p. 19). Duhem writes:

The multiplicity and variety of the models proposed by Thomson to represent the constitution of matter does not astonish the French reader very long, for he very quickly recognizes that the great physicist has not claimed to be furnishing an explanation acceptable to reason, and that he has only wished to produce a work of imagination. [5]

Again the complaint is that we have a theory, or set of them, that are pure speculations, and ones of the worst kind, since they lack order and simplicity.

3.2.3 String Theory

Characterized by some of its proponents as a "Theory of Everything," string theory attempts to unify general relativity and quantum mechanics into a single framework by postulating that all the particles and forces of nature arise from strings that vibrate in 10-dimensional spacetime and are subject to a set of simple laws specified in the theory. The strings, which can be open with endpoints or closed loops, vibrate in different patterns, giving rise to particles such as electrons and quarks.

The major problem, or at least one of them, is that there are no experiments that show that strings and 10-dimensional spacetime exist. The theory is generally regarded, especially by its critics, as being entirely speculative. Steven Weinberg, once an enthusiastic supporter of string theory (as the "final theory"), in 2015 writes:

String theory is ... very beautiful. It appears to be just barely consistent mathematically, so that its structure is not arbitrary, but largely fixed by the requirement of mathematical consistency. Thus it has the beauty of a rigid art form – a sonnet or a sonata. Unfortunately, string theory has not yet led to any predictions that can be tested experimentally, and as a result theorists (at least most of us) are keeping an open mind as to whether the theory actually applies to the real world. It is this insistence on verification that we mostly miss in all the poetic students of nature, from Thales to Plato. [6]

As these examples illustrate, speculations are assumptions normally introduced in the course of activities such as explaining, unifying, predicting, or calculating. Young sought to explain, or at least to see whether it is possible to explain, known phenomena of light by a theory other than the particle theory. Kelvin was attempting to provide a molecular account of the ether, and in terms of this, to explain the known optical phenomena. String theorists want to explain and unify the four known fundamental forces and calculate the fundamental constants of nature. In the course of doing so, they introduce speculative assumptions.

There are two sorts of speculations I want to distinguish. The first, and most common, are made by speculators who, without knowing that there is evidence (if there is), introduce assumptions under these conditions: (a) They believe that the assumptions are either true, or close to the truth, or possible candidates for truth that are worth considering; and (b) they introduce such assumptions when explaining, predicting, unifying, calculating, and the like, even if the assumptions in question turn out to be incorrect.[1] I will call (a) and (b) "theorizing" conditions. Assumptions introduced, without knowing that there is evidence for them, but in a way satisfying these conditions, I will call *truth-relevant* speculations. They are represented by the three examples cited previously.

In other cases, assumptions, without evidence, are introduced in the course of explaining, predicting, unifying, and so on, but their introducers do not believe that they are true, or close to the truth, or even possible candidates for truth. Indeed, it is often believed that they are false and cannot be true. A good example, which I will discuss in Section 3.12, is Maxwell's imaginary fluid hypothesis, introduced in his 1855 paper "On Faraday's Lines of Force." Here to represent the electromagnetic field Maxwell describes an incompressible fluid flowing through tubes of varying section. The fluid is not being proposed as something that exists or might exist. It is, as Maxwell says, purely imaginary. Its purpose is to provide a fluid analogue of the electromagnetic field that will help others understand known electrical and magnetic laws by employing an analogy between these laws and ones governing an imaginary fluid.

Another prominent example of this second type of speculation is atomic theory as viewed by some nineteenth-century positivists. They employed the assumptions

---

[1] Anti-realists can substitute "empirically adequate" for "true" and "correct." I don't want to provide an account of speculation just for realists.

of atomic theory not as ones they believed to be true, or close to it, or as possible candidates for truth, but as fictions useful for explaining, predicting, and unifying certain observable phenomena.[2] For them, as for Maxwell in the imaginary fluid case, no evidence is given for the truth of the assumptions introduced. Indeed, evidence is irrelevant, since truth is. I will call these *truth-irrelevant* speculations.

Both truth-relevant and truth-irrelevant speculations contain assumptions about objects and their behavior for which there is no known evidence. Both are introduced for purposes of explaining, predicting, and organizing phenomena.[3] That is why I call them both speculations. The difference between them stems from "theorizing" condition (a). Truth-relevant speculations satisfy it; truth-irrelevant speculations do not. However, both kinds are anathema to writers such as Brougham, who demands inductive proof based on experiment, and Duhem, who rejects atomic theory construed either realistically or as a useful fiction. According to their detractors, one must refrain from using both truth-relevant and truth-irrelevant speculations. In the former case one is to do so until one determines that sufficient evidence exists to believe the assumptions, in which case they are no longer speculations. In the latter case one is to find assumptions supported by such evidence. By contrast, according to proponents such as Maxwell, speculations of both kinds are legitimate in science and can be evaluated. They do not need to be avoided or rejected simply on the grounds that they are speculations.

My discussion of truth-irrelevant speculations and their evaluation appears in Section 3.12. The main focus of this chapter is on truth-relevant speculations. Until Section 3.12, when I speak of speculations I will mean just these. (Some readers might prefer to restrict the term "speculation" to these, using a different term, such as "imaginary construction," for truth-irrelevant ones. Because of similarities noted

---

[2] As with Maxwell's incompressible fluid, the explanations are not meant to be causal. In Maxwell's case, we explain not what causes the phenomena, but what they are, as well as unify them, by invoking an analogy between these phenomena and others, real or imagined. (See Section 3.12.) In the atomic case, according to some positivists, we explain not what causes, for example, Brownian motion, but how the observed Brownian particles are moving. We explain that they are moving as if they are being randomly bombarded by molecules (without committing ourselves to the claim that they are being so bombarded). For my own account of explanation in general, and noncausal explanation in particular, see *The Nature of Explanation* (New York: Oxford University Press, 1983). For a much more recent account of noncausal explanations, see Marc Lange. *Because without Cause: Non-Causal Explanations in Science and Mathematics* (New York: Oxford University Press, 2016).

[3] There are non-speculative cases when assumptions are introduced without knowing that there is evidence for them – for example, introducing an assumption known to be false in the course of giving a *reductio* argument, or in the course of giving an historical account of a discarded theory, or just to see whether it is consistent with what we know. But here the assumption does not satisfy "theorizing" condition (b) required for both truth-relevant and truth-irrelevant speculations. It is not introduced with the purpose of explaining, predicting, and so on, but, in the first case, just with the purpose of showing that it is false; in the second case, just doing some history of science; and in the third case, just determining consistency. Nor, in such cases, does the assumption satisfy the "theorizing" condition (a) required for truth-relevant speculations. It is not introduced with the idea that it is true, or close to the truth, or a candidate for truth worth considering. To be sure, in such cases the epistemic situation of the introducer may change, and the assumption may come to be treated by the introducer in a way satisfying (a) and/or (b). But that is a different situation.

earlier, I will continue to classify them both as speculations, while recognizing an important difference between them.) Truth-relevant speculations have sparked the most controversy among scientists and philosophers. I will turn to three contrasting views about them next.

3.3 Speculation Controversies

The conflicting views I have in mind are as follows:

3.3.1 *Very Conservative*

The idea can be simply expressed: "Don't speculate." I will take this to mean: Don't introduce an assumption into a scientific investigation, with the idea that it is or might be true or close to it, if you don't know that there is evidence for it. Earlier we saw such a view expressed in Brougham's response to Thomas Young's speculations about light. Two other scientists who express this idea as part of their general scientific methodology are Descartes and Newton – both of whom demand certainty when assumptions are introduced in scientific investigations.

In his Rule 3 of "Rules for the Direction of the Mind," Descartes writes, "we ought to investigate what we can clearly and evidently intuit or deduce with certainty, and not what other people have thought or what we ourselves conjecture. For knowledge can be obtained in no other way" [7]. He continues: those who, "on the basis of probable conjectures venture also to make assertions on obscure matters about which nothing is known, ... gradually come to have complete faith in these assertions, indiscriminately mixing them up with others that are true and evident." And in Rule 12, he writes: "If in the series of things to be examined we come across something which our intellect is unable to intuit sufficiently well, we must stop at that point, and refrain from the superfluous task of examining the remaining items." Indeed, Descartes's view is considerably stronger than "don't speculate" (in the sense of speculation I briefly characterized earlier). His view of "evidence" requires proof with mathematical certainty. And it requires more than knowing that there is such a proof: It demands knowing what the proof is.

Newton, at the end of the *Principia*, claims to have proved the law of gravity (not with the certainty of mathematical proof, but in his sense of an empirically established truth: "deduced from the phenomena"). He admits, however, that he has "not yet assigned a cause to gravity," a reason why the law of gravity holds and has the consequences it does. He says he will not "feign" a hypothesis about this cause, "for whatever is not deduced from the phenomena must be called a hypothesis, and hypotheses, whether metaphysical or physical, or based on occult qualities, or mechanical, have no place in experimental philosophy. In this experimental philosophy, propositions are deduced from the phenomena and are made general by induction."

Finally, when defenders of this "very conservative" view say, "Don't speculate," I will take them to mean at least that scientists should not make public their speculations. Perhaps they would allow scientists to indulge in speculation in private. (Descartes seems to disallow even that.) But, at a minimum, scientists should avoid publishing their speculations or communicating them in other ways to the scientific community, an injunction violated by Thomas Young, Lord Kelvin, and string theorists.

3.3.2 Moderate

The slogan of this view is a modification of one expressed by former President Ronald Reagan: "Speculate, but verify." In the mid-nineteenth century, William Whewell formulated the idea succinctly in this passage: "[A]dvances in knowledge are not commonly made without the previous exercise of some boldness and license in guessing" [8]. In the twentieth century, it was Karl Popper's turn: "According to the view that will be put forward here, the method of critically testing theories and selecting them according to the results of test, always proceeds on the following lines. From a new idea, put up tentatively, and not yet justified in any way – an anticipation, a hypothesis, a theoretical system, or what you will – conclusions are drawn by means of logical deduction" [9].

The general idea expressed by these and other so-called hypothetico-deductivists is that the correct scientific procedure is to start with a speculation, which is then tested by deriving consequences from it, at least some of which can be established, or disproved, experimentally. Even if you do not know that there is evidence for h, you can introduce h into a scientific investigation as a (truth-relevant) speculation, provided you then proceed to test h to determine whether evidence supports h. The only constraint Popper imposes on the speculation is that it be bold (e.g., speculating that Newton's law of gravity holds for the entire universe, and not just for the solar system). Bolder speculations are easier to test and falsify. According to Whewell, scientists are usually capable of putting forth different speculations to explain a set of phenomena, and this is a good thing: "A facility in devising [different] hypotheses, therefore, is so far from being a fault in the intellectual character of a discoverer, that it is, in truth, a faculty indispensable to his task" [10]. For these writers there are, then, no constraints on the character of the speculation, other than (for Popper) boldness and (for Whewell) multiplicity.

The constraints emerge in the testing stage, concerning which Whewell and Popper have significantly different views. Whewell believes that speculative theories can be verified to be true by showing that they exhibit "consilience" (they can explain and predict a range of different phenomena in addition to the ones that prompted the theories in the first place) and "coherence" (they contain assumptions that fit together, that are not ad hoc or otherwise disjointed, especially as new assumptions are added when new phenomena are discovered). Popper believes that speculative theories cannot be verified, only falsified by deriving consequences from them that can be tested experimentally and shown to be false. If the speculative theory withstands such attempts to falsify it, all we can claim is that it is well tested, not that it is verified or true. But, for our purposes, the important claim for both theorists is that speculation is not enough for science; empirical testing must be conducted. In practical terms, we might put it like this: If you want to do science, it is fine, even necessary, to speculate. But that must be followed by testing. Don't publish your speculations without at least some progress in testing, even if that amounts only to saying how experiments should be designed.

3.3.3 Very Liberal

The slogan here is "Speculate like mad, even if you cannot verify." The most famous proponent of this idea is Paul Feyerabend, who proposes adopting a "principle of proliferation: invent and elaborate theories which are inconsistent with the accepted point of view, even if the latter should happen to be highly confirmed and generally accepted ..., such a principle would seem to be an essential part of any critical empiricism" [11]. Feyerabend believes that introducing speculations, particularly ones that are incompatible with accepted theories, is the best way to "test" those theories critically by finding alternative explanations that might be better than those offered by the accepted theories. He places no restrictions on such speculations, other than that they be worked out and taken seriously. On his view, you may, and indeed are encouraged to, publish your speculations, even when you have no idea how to test them empirically. Feyerabend would award high marks to Thomas Young, Lord Kelvin, and string theorists for inventing and elaborating speculations about light waves, a mechanical ether, and strings, even if they produced no testable results, or, indeed, even if they produced results incompatible with what are regarded as empirically established facts.[4]

---

[4] I began the paper with two quotes, one from Newton and one from Einstein. It is clear that Newton's "official" view about speculation puts him in the very conservative camp. (His practice, as I note in Section 3.11, was somewhat different.) Where to put Einstein is less clear. Perhaps he should be placed somewhere between the "moderate" and "very liberal" camps, more toward the latter. Like Whewell, he believed that fundamental theories in physics "cannot be extracted from experience but must be freely invented." But, unlike Whewell,

3.4 What Counts as Evidence?

How the conflicting claims in the previous section are to be evaluated depends crucially on what is to count as evidence, since the characterization of speculation I have given so far does, too. We need to get clear about which concept or concepts of evidence we should use in understanding what it is to be a speculation, and what implications this will have for views, including the three given previously, about whether speculating is legitimate and how, if at all, speculations are to be evaluated.

A standard Bayesian idea is that something is evidence for a hypothesis if and only if it increases the probability of the hypothesis:

(B) e is evidence that h if and only if $p(h/e) > p(h)$.[5]

I will call this B- (for Bayesian) evidence in what follows. Depending on what sort of Bayesian you are, probability here can be construed either subjectively or objectively. In the former case, we obtain a subjective concept of evidence; in the latter, an objective one.

Elsewhere, using numerous counterexamples, I have argued that definition (B), whether understood subjectively or objectively, provides neither a necessary nor a sufficient condition for evidence – as the latter concept is employed in the sciences.[6] I replace this definition with several, the most basic of which I call "potential evidence." It defines evidence using a concept of (objective epistemic) probability and a concept of explanation.[7] The idea is that for e to be potential evidence that h, there must be a high probability (at least greater than 1/2) that, given e, there is an explanatory connection between h and e. There is an explanatory connection between h and e, which I shall write as $E(h, e)$, if and only if either h correctly explains why e is true, or e correctly explains why h is true, or some hypothesis correctly explains why both h and e are true. In what follows I will call this and the related concepts I will introduce A- (Achinsteinian) evidence:

---

he thought that theories cannot be empirically verified by showing that they explain and predict a range of phenomena. Theories are "underdetermined" by the evidence. And as my initial Einstein quote suggests, and as Einstein in his actual practice confirmed, speculation is crucial even in the absence of empirical test or knowledge of how to test.

[5] Many of those who write about evidence use the symbol h for "hypothesis." I will follow this practice, but will also use the terms "hypothesis" and "assumption" interchangeably, and h for both.

[6] Briefly, here is an example that questions the sufficiency of (B): The fact that I bought one ticket out of 1 million sold in a fair lottery is not evidence that I won, although it increases the probability. Here is an example against the necessity of (B): A patient takes medicine M to relieve symptoms S, where M works 95% of the time. Ten minutes later he takes medicine M', which is 90% effective but has fewer side effects and destroys the efficacy of the first medicine. In this case, I claim that the patient taking M' as he did is evidence that his symptoms will be relieved, even though the probability of relief has decreased. For a detailed discussion of these and other counterexamples, possible Bayesian replies to them, and my responses, see *The Book of Evidence* (New York: Oxford University Press, 2001), chapter 4.

[7] Both concepts are explicated in the books cited in footnotes 2 and 6.

(A) Potential evidence: Some fact e is potential evidence that h if and only if $p(E(h,e)/e)$ > 1/2; e is true; and e does not entail h.

The concept of probability here (objective epistemic probability) measures the degree of reasonableness of believing a proposition. I claim that evidence, and hence reasonableness of belief, are "threshold" concepts with respect to probability. When the threshold has been passed, e provides a good reason to believe h – where the degree of reasonableness increases with the degree of probability $p(E(h,e)/e)$.

I will briefly mention three other concepts of evidence that are defined in terms of "potential evidence."

(A′) Veridical evidence: Some fact e is veridical evidence that h if and only if e is potential evidence that h, h and e are both true, and in fact there is an explanatory connection between h and e.

Using (A′) we can define a concept of evidence that is relativized to the epistemic situation ES of some actual or potential agent:

(A″) ES-evidence: Some fact e is ES-evidence that h (relative to an epistemic situation ES) if e is true and anyone in ES is justified in believing that e is veridical evidence that h.

An epistemic situation is a type of abstract situation in which one knows or believes that certain propositions are true, and one is not in a position to know that other propositions are true, even if such a situation does not in fact hold for any person.[8]

Finally, as a counterpart to the Bayesian concept that is subjective, I offer this:

(A‴) Subjective evidence: Some fact e is person P's subjective evidence that h if and only if P believes that e is veridical evidence that h, and P's reason for believing that h is true is that e is true.

An example of all four types of A-evidence is given in note 9 below.[9] The first three are objective concepts, in the sense that whether e is evidence that h does not depend on whether anyone in fact believes that e is evidence that h. Only the last, A‴, is subjective.

---

[8] A person in a given epistemic situation ES may not know that some propositions $P1, \ldots, Pn$ believed in that situation are true. But if, on the basis of $P1, \ldots, Pn$ such a person is to be justified in believing that e is veridical evidence that h, then the person must be justified in believing $P1, \ldots, Pn$. For more on epistemic situations and ES-evidence, see *The Book of Evidence*, chapter 1.

[9] In 1883, Heinrich Hertz performed experiments on cathode rays in which he attempted to deflect them electrically. He was unable to do so, and concluded that they are not charged. Fourteen years later, J. J. Thomson claimed that Hertz's experiments were flawed because the air in the cathode tube used was not sufficiently evacuated, thus blocking any electrical effects. In Thomson's experiments, when greater evacuation was achieved, electrical effects were demonstrated. Hertz's experimental results constituted his subjective evidence that cathode rays are not charged. They were also ES-evidence for this hypothesis, since, given Hertz' epistemic situation in 1883, he was justified in believing that the results constituted veridical evidence. But because the experimental setup was flawed, unbeknownst to him, they constituted neither potential nor veridical evidence. By contrast, Thomson's experimental results constituted evidence in all four of these senses for the hypothesis that cathode rays are electrically charged.

I much prefer the definitions supplied by the (A's) to that supplied by (B). But since there are so many adherents to the Bayesian definition, I will make reference to that as well as to the A's in my discussion of scientific speculation. Let's see how far we can get with either type of definition. Later I will show why, in order to offer a more complete definition of speculation, the basic Bayesian definition (B) needs to be upgraded in a way that makes it much closer to (A).

3.5 Truth-Relevant Speculations

With these concepts of evidence in mind, I return to the task of clarifying what it is to be a truth-relevant speculation. In the case of such a speculation, a speculator P, without knowing that there is evidence for assumption h, introduces h under "theorizing" conditions (a) and (b) of Section 3.2, p. 32. In saying that P is speculating when introducing h under condition (a), I don't mean that P believes true (or close to the truth, or a candidate for truth worth considering) just a conditional statement of the form "if h, then. . . ." I mean that P believes this with respect to h itself. Of course, the speculative assumption introduced may itself be a conditional whose antecedent is h. But then it is the entire conditional that is the speculation, not h by itself.[10]

If P introduces h in a way that satisfies the "theorizing" conditions (a) and (b), we might say that

(Spec) h is a (truth-relevant) speculation for P if and only if P does not know that there is evidence that h.

One way to know that there is evidence that h is to know what that evidence is and to know that this is evidence that h, i.e., to know of some fact e that it is evidence that h. But it is also possible to know that there is evidence that h without knowing what that evidence is. If authoritative textbooks all tell me that that there is evidence for the existence of the top quark without telling me what that evidence is, and I introduce the assumption that the top quark exists and do so in a way that satisfies the "theorizing" conditions, then my assumption is not a speculation for me, since I know there is evidence for its existence.

---

[10] What about "thought experiments," understood as describable by conditionals whose antecedents will not be, or could not be, satisfied? Some are not speculations at all, since the thought experimenter knows that there is evidence for the conditional expressing the thought experiment. An example is Newton's thought experiment involving many moons revolving around the earth, the lowest of which barely grazes the highest mountains on the earth. Newton provides evidence that if such a moon did exist and lost its inertial motion, it would fall at the same rate of acceleration as bodies do near the earth. If Newton did not know that there is evidence for such a conditionally expressed thought experiment, but if he then introduced it while "theorizing," it would have been a truth-relevant speculation.

If you introduce a hypothesis *h* under the "theorizing" conditions, and you don't know that there is evidence that *h*, then if *h* is true you don't know that it is. Speculating in this way entails lack of knowledge of the truth of the speculation. However, the converse is not true. If you introduce *h* under the "theorizing" conditions without knowing that *h* is true, it doesn't follow that you are speculating, at least on the concept of speculation I am proposing, since you may know that there is evidence that *h* is true. On this concept, lack of knowledge regarding *h* is a necessary but not a sufficient condition for speculating that *h*. By extension, if you introduce *h* under the "theorizing" conditions, and unbeknownst to you *h* is false, the fact that *h* is false doesn't make *h* a speculation for you, since you may know facts that you know to be evidence that *h* is true.[11]

A good deal more needs to be said about various ideas associated with (Spec). I will do so in the remainder of this section and the next two.

In (Spec), *h* stands for an individual assumption. Now, as is the case with the three examples of speculations from physics introduced at the beginning of this chapter, theories usually contain *sets* of assumptions, some of which may be speculations, some of which are not. It is not my claim that each assumption in a speculative theory is necessarily a speculation.

I have spoken of speculators *introducing* assumptions. As noted earlier, usually this is done in the course of "theorizing" activities such as explaining, predicting, calculating, and so on. Occasionally, though, it is done with little, if any, "theorizing" by writing these ideas down and calling them "assumptions," "speculations," or "hypotheses," that the speculator believes to be true or possible and worth further investigation at some point. Newton does this in his Queries in the *Opticks* when he speculates about the particle nature of light, though even there he theorizes a bit by offering a few arguments against the rival wave theory. I will call both cases speculating while "theorizing," even if the theorizing is intended for a future occasion.

Among the speculations introduced in a theory consisting of many assumptions, some may be intended as literally true, some may be thought of as approximations with different degrees of closeness to the truth, and still others may be regarded as just possibilities worth considering. (In the book *Speculation*, [see note 15 below], I argue that, despite Newton's claims to the contrary, his law of gravity was a speculation, but one he believed to be literally true. His assumption that the only gravitational force acting to produce the orbit of a given planet was the gravitational force of the sun acting on the planet was also a speculation, but an assumption he

---

[11] An example: Let *e*, which you know to be true, be that you own 95% of the tickets in a fair lottery. Let *h* be that you will win the lottery. Assume that *e* is (potential) evidence that *h*, and that you know this. Suppose that, unbeknownst to you, *h* is false, and you introduce *h* in the course of "theorizing," doing so believing that *h* is true. Since you know that there is evidence that *h* – indeed, very strong evidence – you are not speculating, even though *h* is false.

regarded as only approximately true. And the speculation that "the center of the system of the world is at rest" – which Newton explicitly classified as a speculation (or "hypothesis") – he perhaps regarded as a possibility.

Whether speculations are introduced boldly with the idea that they are true, or more cautiously with the idea that if not true they are close to it, or even more cautiously with the idea that they are possibilities worth considering, does not affect their speculative status if the speculator does not know that there is evidence for them. What it does affect is the question "Evidence for what?" Evidence that h is a possibility, that h is close to the truth, or that h is true, will usually be different. If I introduce the assumption that h is a possibility – meaning that it is not precluded by known laws or facts – without knowing that there is evidence for the claim that it is not so precluded, then I am speculating when I claim that h is a possibility.

Accordingly, we could modify (Spec) by introducing distinctions between types of speculations: speculating that h is true, speculating that h is close to the truth, speculating that h is a possibility, and perhaps others. But I will not do so. When some assumption h is classified as a speculation, the focus of the scientific community is usually on the truth of h, even if the speculator is introducing h only as a possibility. When Brougham criticized Young for producing a speculation ("a work of fancy"), it is not very convincing for Young to reply: "No, I am not; I am only saying that the wave theory is a possibility, that it is consistent with Newtonian mechanics, and for that claim I can provide evidence." Even if this reply correctly represents Young's intentions, the main interest of scientists, including Brougham, is on the question of whether the theory is true. Scientists want to know if there is evidence that light is a wave motion in the ether, not simply whether this is a possibility. So, in such cases I will retain (Spec), and say that even if the speculative assumption h was introduced by P with the idea that it is a possibility, h is a speculation for P *with respect to the truth of h*, since P does not know that there is evidence that h is true.

What about "closeness to truth"? Newton, in defending Phenomenon 1 pertaining to the Keplerian motions of the moons of Jupiter, introduces the assumption that the orbits of these moons are circular. The assumption itself is literally false, which Newton realized, but it is a good approximation. Since Newton had evidence that this is a good approximation, the latter claim was not a speculation for him. In view of such cases we could keep (Spec) as is but add another type of speculation for cases in which no evidence is known for "h is close to the truth." Or we could simply retain (Spec) without adding other types, and understand "true" in a broad way to include "true or close to it." I prefer the latter.

There is a view about the introduction of speculations that I want to reject. It is based on a distinction that philosophers once regarded as important (perhaps some still do) between the "context of discovery" (when one first gets the idea of the hypothesis) and the "context of justification" (when one is attempting to test or defend the hypothesis by providing evidence). Those wedded to this distinction claim that speculations appear in the first context, but not the second. On the view I am defending, whether something is a speculation does not depend on whether it is introduced when one first gets the idea or afterward when one is (or is not) attempting to defend it. For speculations, what matters is only that they are introduced in the course of "theorizing" activities with the idea that they are truths or close to it, or at least possibilities worth considering.

Finally, in accordance with (Spec), a "theorizing" assumption h introduced by P can be a speculation even if P does not believe that h is the sort of assumption for which evidence is possible. Assumption h might be regarded by P as "metaphysical," "theological," or of some other sort for which there can be no evidence. In such a case, it follows trivially that P does not know that there is evidence that P, so that if P introduces h under the "theorizing" conditions, then h is a speculation for P. At the other extreme, what if P regards h as self-evident, having and needing no evidence? Whether self-evident propositions having and needing no evidence truly exist is controversial. But if there are, and if P knows that h is one of them and "theorizes" using h, then, since P knows that h is "self-evident," and therefore true, I will say that P is not speculating.

3.6 Examples to Clarify the Scope of (Spec)

The three examples of speculations noted in Section 3.2 – Young's 1802 assumptions about the wave nature of light, Kelvin's assumptions about the molecular nature of the ether, and string theory's assumptions about strings and 10-dimensional spacetime – all satisfy (Spec). Let me mention two other types of cases that will help clarify the scope of (Spec).

Case 1

Suppose that I have read in a usually but not completely reliable newspaper that there is evidence that 10-dimensional spacetime exists. And suppose the newspaper is right in saying this. Because the newspaper is not completely reliable, I do not know that there really is such evidence, though I have a good reason to believe there is. In accordance with (Spec), if, while "theorizing," I introduce the assumption that such a spacetime exists, I am speculating.

Case 2

Suppose I read in *Science* magazine, a very reliable source, that evidence for 10-dimensional spacetime has been discovered, and I use the assumption that 10-dimensional spacetime exists in theorizing in a way that satisfies "theorizing" conditions (a) and (b). In fact, unlike in Case 1, what *Science* reports is false: What it took to be evidence is not so, and indeed no evidence exists. According to (Spec), when having read the article I introduce the 10-dimensional space assumption in my theorizing, I am speculating. I am justified in believing that there is evidence that 10-dimensional spacetime exists, but in fact there is no such evidence, so I don't know that there is. I am justified in my belief that I am not speculating, but it turns out that I am speculating.

In view of cases such as (1) and (2), the concept of speculation I have introduced is a strong one, since it rules something as a speculation in a situation in which the speculator has a good reason to believe there is evidence for the speculation but not enough to know that there is. This doesn't imply that any assumption that I put forth while "theorizing" and whose truth I don't know is a speculation for me. That would be much too strong. It means only that any such assumption is a speculation if I don't know that there is evidence for the assumption.[12] So, if I assume that h is true, but don't know that it is, *h* will not be a speculation for me if I know there is evidence that *h*. We might call this the "no knowledge of the existence of evidence" concept of speculation.

A weaker concept might require only that, given my epistemic situation, I am not justified in believing that there is evidence. If I am so justified, I would not be speculating. But suppose I am justified in believing that there is evidence that *h*, even though there is no such evidence. Then, on this weaker concept of speculation, if I "theorize" using *h* I am not speculating. Under these circumstances, I am inclined to say that since there is no evidence that *h*, I am speculating, even though I think there is such evidence and even though I think that I am not speculating.[13]

I will continue to use the stronger "no knowledge" concept. If, like some scientists, you reject speculation, you are not rejecting the "theorizing" use of all assumptions whose truth you don't know, but only the use of assumptions for which you don't know that evidence exists. This permits speculations to cover a broad group of cases, ranging from simple hunches, where you have no idea whether evidence exists or even what would count as evidence, to cases in which you have good reason to think that something is (or that there is) evidence but don't know that this is so.

3.7 Relativizations

In accordance with (Spec), an assumption *h* may be a speculation for one person but not another, depending on what the person knows or doesn't know. But even if we focus on one theorizer and one assumption being introduced, we get different

---

[12] This includes a range of possibilities including cases in which *e* is true and I know that it is, and *e* is evidence that *h*, but I don't know that *e*, or anything else I know to be true, is evidence that *h*.

[13] This leaves open the question of which concept of evidence to employ, a question I take up in the next section.

answers to the question "Is *h* a speculation for *P*?" depending on which concept of evidence is used. For the Bayesian concept, understood subjectively or objectively, if *P* knows that there is some *e* that increases the probability of *h*, then *h* is not a speculation for *P*. This is pretty weak fare, and I will suggest strengthening it in a moment. My A-concepts of evidence yield different results depending on which concept is chosen. Veridical evidence A′, which requires the truth of *h*, classifies *h* as a speculation for "theorizer" *P* whenever *h* is false, even if *P* knows facts *e* that constitute evidence that *h* in the other (A)-senses, and *P* is justified in believing that *e* also constitutes veridical evidence that *h*. This is perhaps too strong a sense of speculation. So, if we use one of my (A)-concepts, I suggest either potential, or ES-, or subjective evidence. All of these can be used in connection with (Spec), though they give somewhat different classifications for speculations.

For example, Isaac Newton believed that the fact that (*e*) the planets lie on the same plane and rotate around the sun in the same direction is evidence that (*h*) God exists and designed it that way [12]. (He claimed this because, he said, this cannot be due to "mechanical causes, since comets go freely into very eccentric orbits and into all parts of the heavens.") Newton would have denied vehemently that he was speculating in this case, since he regarded *h* as "deduced from the phenomena." Many would reject Newton's claim that *e* is evidence that *h*, and say that when Newton introduces *h* to explain *e* he is indeed speculating. When Newton claims that *e* is evidence that *h*, he should be understood to be using the concept of subjective evidence. And, his denial that he is speculating should be understood as using the same concept of evidence in (Spec). In the subjective sense (reflecting what he believed he was doing or intended to be doing), Newton was not speculating. Newton's critics who claim that he is speculating should be understood as using either the concept of ES-evidence or potential evidence in (Spec).[14] They are saying that in one or the other of these senses of evidence, Newton was indeed speculating (whether or not he believed he was). If speculation is defined in terms of evidence, or lack thereof, and if, as I suppose, different concepts of evidence are in use, then different classifications should be expected. For the sake of argument, I will suppose that those supporting one of the conflicting views regarding speculation noted in Section 3.3 – either for it or against it - would be for or against speculation defined in at least one, and perhaps all, of these ways.

Now two questions need to be addressed. (1) Using the Bayesian definition (B), how much increase in probability must *e* give to *h* so that *P*'s knowing that *e* is B-evidence that *h* prevents *h* from being classified as a speculation for *P*?

---

[14] With ES-evidence, the critics might say, their claim can be understood as relativized to their own epistemic situation, or indeed even to Newton's.

Alternatively, using my definitions supplied in the (A's), how probable must it be that an explanatory connection exists between h and e, given e, so that knowing that e is A-evidence that h prevents h from being classified as a speculation for P? (2) What sorts of considerations will increase the probability of an assumption h? Alternatively, what sorts of considerations will make it such that there is a high probability that there is an explanatory connection between h and e, given e?

I will not attempt to give question (1) a detailed answer. We could talk about degrees of speculation, depending on how high the probabilities are. But let's simplify the discussion, make things more precise, and set the bar reasonably high by requiring that the probability involved be greater than $\frac{1}{2}$. If we can suppose, as I do, that evidence must provide a good reason for believing a hypothesis, and that this requires that the evidence makes the hypothesis at least more probable than not, this will set a minimal standard for what is and is not a speculation. Assumption h is a speculation for you if you introduce h, under "theorizing" conditions, without knowing that there is evidence that provides a good reason for believing h. Using the Bayesian idea, we upgrade (Spec) and say this: h is a speculation for person P, who introduces h (under "theorizing" conditions), if and only if P does not know that there is some fact that increases h's probability so that the latter is greater than 1/2 (i.e., more likely than not). Using my definition of evidence, this upgrading idea is unnecessary, since my concepts of evidence require a probability greater than 1/2.

I regard question (2) as the central one. In the following section I will look at various answers proposed to this question. These answers are based on the use of an objective, rather than a subjective, concept of evidence, and hence of speculation. They propose ways of obtaining evidence that are supposed to hold no matter what particular individuals believe about what is evidence for what. This is not at all to reject the idea that there are subjective concepts of evidence (e.g., my subjective A-concept, or the Bayesian one understood subjectively). Nor is it to reject the idea that there are subjective concepts of speculation (e.g., ones that accord with [Spec] in which evidence is understood in terms of my subjective A-concept or in terms of a subjective Bayesian one). My claim is that there are objective concepts as well, and that these are the most important and interesting ones, especially in understanding the controversies noted in Section 3.3 about the legitimacy of speculating and ways to evaluate speculations. Later, on the basis of the answers proposed for question (2), I will argue that if an objective B-concept is employed in understanding what it is to be a speculation, it will need to be augmented by incorporating an explanatory idea present in the A-concepts. To simplify the discussion, the objective A-concept I will focus on is potential evidence, since the other A-concepts are defined using it. The question now is how to get objective evidence in one of these senses.

3.8 How to Get Evidence

The answers proposed by scientists and philosophers over the ages have varied considerably, and have sparked lively debates. I will begin with three historically influential, contrasting accounts: Newtonian inductivism and two versions of hypothetico-deductivism – due to William Whewell and Peter Lipton – that introduce fairly strong conditions for obtaining evidence. Then I will turn to some weaker, but more controversial, accounts. It is not my claim that these exhaust the views of how to get evidence, in either an objective Bayesian sense or mine – only that they should suffice to give us a sufficiently broad basis for discussing the idea of speculation. I will treat them as proposals for sufficient conditions for obtaining evidence, even though their defenders (in the first three cases) also regard them as necessary.

1. **Newtonian inductivism.** Briefly, following Newton's four rules of reasoning in the *Principia*, to get evidence for a causal law (such as Newton's universal law of gravity), you attempt to establish that a cause satisfying certain conditions exists (in Newton's case, that there is a gravitational force whose magnitude varies as the product of the masses of the bodies and inversely as the square of the distance between them), and that all objects of certain sorts (e.g., all bodies) satisfy a general law invoking that cause. For Newton, it suffices to establish that the cause exists by inferring that it does from the same observed effects it produces on various objects. And from these observed effects one infers, by induction, that all other objects of these sorts also satisfy the general law. This part of the argument Newton calls "analysis." The second part, "synthesis," consists in taking the law and showing how it can explain and predict phenomena other than the ones initially used to provide a causal-inductive argument to the law. Newton's "Phenomena," both the initial ones and those later explained, constitute evidence for the law.[15]

2. **Whewellian hypothetico-deductivism.** William Whewell offers a sophisticated version. To show that *e* is evidence for a system of hypotheses *H* you show that *H* explains *e*; that *H* predicts and explains new phenomena of a different sort from those in *e* - phenomena that are later established to be the case by observation and experiment ("consilience"); and that the system of hypotheses is "coherent" and remains so over time as new phenomena are discovered. On this view, *e* together with the successfully predicted and explained new phenomena constitute evidence for *H*.[16]

---

[15] For a discussion of Newton's rules and his inductivism, see my *Evidence and Method* (New York: Oxford University Press, 2013), chapter 2. A critical account of how he uses his rules to generate his law of gravity is given in my *Speculation: Within and About Science* (New York: Oxford University Press, 2018).

[16] *The Philosophy of the Inductive Sciences, Founded upon Their History* (London: Routledge, reprinted 1996); parts of chapter 5 are reprinted in Achinstein, *Science Rules*.

According to both of these influential accounts, if you want to obtain evidence for a hypothesis or system of hypotheses in the way each proposes, you must show more than simply that the hypothesis or system explains or entails some set of observed phenomena. For Whewell, you must show that it predicts and explains phenomena of different types, not just the sort you started with. For Newton, you need to do this as well, but also provide a justified causal argument that the cause inferred exists and operates in a certain way, and a justified inductive argument that this claim can be extended to all bodies of certain sorts. Proponents of these accounts would say that if you have satisfied the conditions they require, then the probability of the hypothesis or system will be increased (indeed, to the point of certainty), thus satisfying the Bayesian definition of evidence (B). Also, according to these accounts, if you establish the hypothesis in the way proposed you will have shown that there is a high probability of an explanatory connection between the hypothesis and the evidence, thus satisfying my concept of potential evidence, given by definition (A).

The next account is similar to Whewell's in certain respects, but weaker.

3. **Lipton's "inference to the best explanation."**[17] According to this account, to show that e is evidence for some hypothesis or system H, you must show that H offers what Peter Lipton calls the "loveliest" explanation for e. For Lipton, the "loveliest" explanation is one that "would, if correct, be the most explanatory or provide the most understanding" of those phenomena under consideration [13]. Such an explanation would be simple, unifying, and deep (however these terms are to be understood). This account does not require Whewellian "consilience" or Newtonian causal-inductive reasoning.

The last three accounts I will mention are weaker than those above. They are meant to provide only sufficient, but not necessary, conditions for obtaining evidence. And they are meant to provide sufficient conditions for obtaining at least some evidence for a hypothesis, not necessarily strong or conclusive evidence.

4. **Meta-inductive evidence.** The idea here is that you can get evidence for a hypothesis if you can show that your hypothesis is of a certain general type and that other hypotheses of that general type have been successful in the past. For example, in defense of his molecular–kinetic theory of gases, Maxwell writes that it is a completely mechanical theory, and that (he claims) mechanical theories have worked well in astronomy and electricity. These facts would constitute evidence for the theory.

---

[17] See Peter Lipton, *Inference to the Best Explanation*, 2nd ed. (London: Routledge, 2004). The father of the doctrine and of the expression "inference to the best explanation" is Gilbert Harman, in his "The Inference to the Best Explanation," *Philosophical Review* 64 (1965), pp. 519–33.

5. **"Only-game-in-town" evidence.** You can get evidence for a hypothesis *h* if you can show that *h* is the "only game in town" for explaining some set of phenomena *e*. It is the only hypothesis scientists have been able to come up with to explain *e*, after thinking about the matter and rejecting known alternatives for various good reasons. Then the fact that *h* is the only game in town is some evidence in favor of it.

6. **Evidence from authority.** You can get evidence for a hypothesis *h* by showing that the authorities or experts in the field relevant for *h* believe that *h* is true. The fact that the authorities believe *h* is evidence that *h* is true.[18]

Defenders of the first three accounts presented in this section – the Newtonian, Whewellian, and Liptonian ones – maintain that if you satisfy the conditions they advocate, then, given *e*, you will have shown that it is highly probable that there is an explanatory connection between *h* and *e*, in which case you will have shown that e constitutes A-evidence that *h*. For the sake of argument, let us also suppose that, according to all six of the accounts, if you satisfy the conditions they specify, you will show that *e* increases the probability of *h* so that the latter becomes greater than 1/2 and *e* constitutes (upgraded) B-evidence for *h*.

3.9 Speculation and Evidence

I characterized (truth-relevant) speculating as introducing assumptions without knowing that there is evidence for those assumptions, and doing so under "theorizing" conditions I specified. In Section 3.4, I supplied definitions for A- and B-evidence and, using these, offered a more precise account of what it means for an assumption to be a speculation for someone with respect to the truth of the hypothesis. These definitions generate somewhat different concepts of speculation. In Section 3.8, I noted various views about how to obtain evidence that satisfies the definitions in Section 3.4. In what follows, I examine these views in light of the concepts of evidence I have introduced.

Let's begin with the last three views: "meta-inductive evidence," "only-game-in-town evidence," and "evidence from authority." For the meta-inductive case, let *h* be that gases are composed of spherical molecules that obey Newton's laws

---

[18] Richard Dawid defends versions of views 4 and 5, but not 6. On his versions, evidence is not evidence for the truth, or even for the empirical adequacy, of a hypothesis, but for what he calls its "viability." A hypothesis is viable relative to a given field, and to certain types of experiments that can be performed within that field, if it predicts the results of those experiments. Dawid claims that with his restricted sense of evidence, meta-inductive evidence and only-game-in-town evidence can provide evidence for "viability." Among the meta-inductive cases he has in mind are ones where the theory is the "only game in town." If such theories have tended to be "viable," then this fact counts as evidence for the viability of the particular "only-game-in-town" theory. (Dawid, "The Significance of Non-Empirical Confirmation in Fundamental Physics," (this volume, Chapter 6))

(Maxwell's first kinetic theory hypothesis – to be discussed in Section 3.12). Let *e* be the fact that *h* is a mechanical hypothesis, and that (as Maxwell noted) mechanical hypotheses have been successful in astronomy and (he thought) electricity. But, using my A-concept of "potential evidence," it is not probable that, given *e*, there is an explanatory connection between *h* and *e*. (It is not probable that the reason that gases are composed of spherical molecules is that mechanical hypotheses have been successful, or that the reason that mechanical hypotheses have been successful is that gases are composed of spherical molecules, or that some hypotheses correctly explain both why gases are composed of spherical molecules and why mechanical hypotheses have been successful.[19]) Analogous claims are justified in the case of "only-game-in-town" and "authoritative" evidence.

What follows about speculation? If you employ the A-definition of potential evidence, and if you theorize using *h* and know that there are only "meta-inductive facts" in favor of *h*, then *h* is a speculation for you. Matters get trickier in the case of "authoritative" and "only-game-in-town" facts in favor of *h*. Let's start with the former. Suppose you know that the following is the case:

(e) The recognized authorities announce that they have discovered (potential) evidence that *h* is true, and that for this reason they believe that *h* is true, even though they do not announce what this evidence is.

Suppose that the evidence they have discovered really is potential evidence that *h*. Finally, suppose that these authorities are so good that you can rightly claim to know that such potential evidence for h exists, even though you don't know what it is. If you use *h* in theorizing, then you are not speculating, since you know (from authority) that there is potential evidence that *h*. This is the case even though *e* is not potential evidence that *h* (it fails to satisfy the explanatory connection condition). Thus *e* is not "authoritative" *evidence* for h. Instead, in this case at least, if you know that the authoritative fact *e* is true, then you know there is (non-authoritative) potential evidence for *h*.

But suppose that the evidence the authorities have in mind is not actually potential evidence that *h*, though they believe it is, and perhaps, given their epistemic situation, they are even justified in doing so. Then, even though the authorities believe they have discovered potential evidence that *h* is true, they haven't. And they, as well as you, are speculating (in a sense of [Spec] employing potential evidence) when they and you use *h* in theorizing. So, in general, knowing the truth of an "authoritative" fact such as *e* is not sufficient to prevent *h* from being a speculation. Nor, for analogous reasons, is knowing that *h* is the "only game in town."

---

[19] This holds as well for meta-inductive evidence of the sort Dawid has in mind. See note 18.

What happens if we use an objective Bayesian B-concept of evidence (one that is upgraded to require that *e* is true and that *e* increase *h*'s objective probability so that the latter is greater than 1/2, but not to require satisfaction of the explanatory connection condition)? Such a Bayesian needs to argue first that meta-inductive, only-game-in-town, and authoritative facts do, or at least can, increase the probability of a hypothesis, so that they really do or can constitute (objective Bayesian) evidence.[20] Next the Bayesian has to argue that these kinds of facts can increase the probability of a hypothesis sufficiently to make the probability greater than 1/2, so that the hypothesis is not a speculation (in an objective sense). Elsewhere I have given arguments to show that at least "only-game-in-town" facts fail to raise the probability of a hypothesis, so that they don't even satisfy the Bayesian definition of evidence [14]. Furthermore, even if the Bayesian can show that meta-inductive facts can increase the probability that some theory of the type in question is true or viable, it doesn't follow from this that the probability of the particular theory in question has increased. Maxwell argued from the success of mechanical theories in astronomy and electricity to the probable success (or the increase in probability) of *some type* of mechanical theory of gases. He did not claim that his particular theory of spherical molecules acting only by impact was probable, or that its probability was increased.[21] "Authoritative" facts – ones stating just that the authorities believe that h is true – are subject to Laudan's "pessimistic" induction: "Authorities" (those deemed to be such) from Ptolemy to Aristotle to Newton, to the present time have gotten it wrong so often that the only inference possible just from their success rate is that they will get it wrong again – not that what they believe now is made more probable by the fact that they believe it [15].

For the sake of argument, suppose now (contrary to what I claimed earlier) that the fact that *h* is the "only game in town," or that *h* is a theory of a type that has been successful, or that a vast majority of the authorities believe *h*, could raise the probability of *h* beyond 1/2. This would satisfy the upgraded objective Bayesian concept of evidence. Moreover, since the probability of *h* on *e* would be greater than 1/2, a necessary condition for potential evidence would be satisfied as well. However, the major claim I want to make is this: So raising the probability of *h*, and thus making it reasonable to believe *h*, is not enough for what scientists want or should want of evidence. Even if *h* is the only game in town, even if *h* is a

---

[20] If the Bayesian insists on using subjective concepts of probability and evidence, then for any given individual, any of these facts can increase his degree of belief in a hypothesis, depending on the individual's system of degrees of belief. The matter is entirely subjective, so long as his system of degrees of belief is probabilistically consistent. This will yield only a subjective concept of speculation, not an objective one. Again, my discussion is based on the claim that there are objective concepts of evidence and speculation as well, and that these are central in understandigng the controversies over speculating.

[21] It could be that the probability of Maxwell's particular mechanical theory is zero, given what is known, even though the probability that some mechanical theory is true is greater than zero.

hypothesis of a general type that has been successful in other areas, even if the authorities believe *h* to be true, and even if these facts provide a good reason to believe the hypothesis, or a better reason than without them, when scientists seek evidence for *h* they want something more. I turn to this demand next.

3.10 Scientific Speculation

Part of Newton's evidence for his law of gravity (given in his "Phenomena," Book 3) is that the planets in their orbits about the sun sweep out equal areas in equal times. For Newton this is one among a set of facts that provide a good reason for believing that the law of gravity is true. Now consider a Newtonian who says that the fact that the "incomparable Mr. Newton" believes the law of gravity to be true is a good reason to believe this law is, in fact, true. What scientists seek when they want evidence for a hypothesis is something along the lines offered by Newton, not by my imagined Newtonian. *They want evidence that provides a reason why e or h is true, not just a reason to believe that one or both are true.* Newton's evidence is the sort scientists subscribing to the "mechanical philosophy" wanted because, they believed, it is based on a probable explanatory connection between *e* and *h*: Probably the reason why the planets sweep out equal areas in equal times is that they are subject to a force governed by the law of gravity. My imagined Newtonian's "evidence," even if it gives a reason to believe Newton's law, does not satisfy the condition that there is a probable explanatory connection between *e* and *h*. (It is not probable that the reason why the planets sweep out equal areas in equal times is that Newton believes this is true, or conversely.) My Newtonian's reason is an authoritative one of the sort noted in Section 3.8. Even if only-game-in-town and meta-inductive reasons, like authoritative ones, are or can be reasons to believe that a hypothesis is true, they are not reasons based on a probable explanatory connection between *e* and *h*.

Now suppose that, given Newton's "Phenomena," the probability of an explanatory connection between these phenomena and his law of gravity is *not* very high, and that Newton did not know that there are other phenomena that would make it high.[22] In accordance with my A-concept of potential evidence, if Newton used the law of gravity in theorizing (which indeed he did, especially in showing how to explain the tides and other phenomena), he would then be speculating, since the "Phenomena" he cited would not be evidence for the law, and he didn't know about other phenomena that would be. Suppose, further, that a follower of Newton uses the law in theorizing, believing that Newton has evidence that provides a good reason to believe the law, although this follower has no idea what this evidence is.

---

[22] In *Speculation* (see note 15), chapter 4, I argue that this is indeed so.

His reason for believing the law to be true is authoritarian: Mr. Newton believes it. Now we can see how an important difference between my objective concept of A-evidence and the objective (and upgraded) Bayesian one makes an important difference over what counts as a speculation. According to the Bayesian, the follower of Newton would have evidence that the law of gravity is true ("Mr. Newton believes it," which let us suppose raises the probability of the law to a number greater than $\frac{1}{2}$). So, on the upgraded Bayesian concept of evidence, the follower of Newton would not be speculating.[23]

By contrast, according to my A-concept of evidence, the fact that Newton believes the law of gravity is not evidence that the law is true (because of the lack of a probable explanatory connection between the law's being true and Newton believing it). So, under the present scenario, since Newton did not have A-evidence that the law is true, our imagined follower of Newton would not know that there is evidence that the law is true, since his belief that it is true is based solely on the false assumption that Newton has evidence for the law. For him, as well as for Newton, the law would be a speculation, despite their protestations to the contrary.

What remains from the list of "ways to obtain evidence" in Section 3.8 are Newtonian inductivism, the Whewellian version of hypothetico–deductivism, and Lipton's "inference to the best explanation." Unlike the other three, all of these accounts involve the idea of some explanatory connection between h and e, for example, that the reason that e is true is that h is. Of the "explanatory connection" accounts, I find Lipton's version the most problematic. Suppose you can show that a system of hypotheses H, if true, offers the "loveliest" explanation of phenomena reported in e. Have you shown that e constitutes either A- or (upgraded) B-evidence for H? And if so, have you shown that the evidence supplied is sufficiently strong so that H cannot be regarded as a speculation? Have you shown, for example, that, given e, the probability of H, or the probability of an explanatory connection between H and e, is greater than 1/2? To do so, you have to show, at a minimum, that beauty tracks probable truth – that the more beautiful a theory is, the more probable it is, and that if its beauty surpasses some threshold, its probability is sufficient for belief. Lipton simply assumes that this is so, without giving any argument.[24] This does not mean that beauty is irrelevant in the assessment of a theory, but rather that,

---

[23] With an upgraded "subjective" Bayesian concept of evidence – one that employs a subjective concept of evidence and requires that the subjective probability of h be increased by e so that it is greater than 1/2, but does not impose the explanation requirement – we also get the result that the follower of Newton would not be speculating even if Newton was.

[24] In *Evidence and Method*, pp. 104–10, I examine possible arguments and reject them. In Chapters 2 and 3 of *Speculation* I consider various arguments in favor of the claim that simplicity – one of Lipton's favorite types of beauty – is an epistemic virtue. None of them succeeds. For a critical discussion of arguments for the more general claim that beauty tracks probable truth, see the article by Gregory Morgan and my reply in G. Morgan, ed., *Philosophy of Science Matters: The Philosophy of Peter Achinstein* (New York: Oxford University Press, 2011).

if it is a virtue, it is a non-epistemic one: It does not, and should not, affect the reasonableness of believing a theory.

Of the remaining two "explanatory connection" accounts, I prefer the Newtonian to the Whewellian one. However, I will not pursue this issue here.[25] For the sake of argument, I will suppose that both Newtonian and Whewellian evidence e can make it highly probable that an explanatory connection exists between h and e (thus satisfying my A-concept of potential evidence). If you have A-evidence for h, then h is not a speculation for you. What is missing in only-game-in-town, meta-inductive, and authoritative "evidence" e (even if such "evidence" could make a hypothesis [more] reasonable to believe), and what is present in Newtonian and Whewellian evidence, is the idea that e and h are (probably) explanatorily related. With this, we have what I call "evidence," or perhaps, more accurately, "explanatory evidence." Without it, we don't. Explanatory evidence is what Brougham was demanding of Young's wave theory of light, Duhem of Kelvin's molecular theory of the ether, and Weinberg of string theory. These critics claim that the theories in question are speculations because their proponents lack such evidence.

This explanatory idea is built into my concept of potential evidence, but it is not built into the Bayesian B-concept, even if the latter is upgraded to require that e raise the probability of h to a point greater than 1/2. So, we can incorporate this idea into a Bayesian concept by saying that when one seeks explanatory B-evidence for a hypothesis h, one seeks an e that increases the probability that there is an explanatory connection between h and e so that this probability is greater than 1/2. That is,

e is explanatory B-evidence for h only if $p(E(h,e)/e) > p(E(h,e))$,

$$\text{and } p(E(h,e)/e) > \frac{1}{2}.{}^{26}$$

If probability here is construed objectively, we obtain an objective concept of evidence. Now we have closed the gap very considerably between my concept of ("potential") A-evidence and the Bayesian concept. The only important difference between my concept and explanatory B-evidence (construed objectively) is that the latter requires that evidence increase the probability of an explanatory connection, while the former does not.[27]

Finally, suppose that, in the course of a scientific investigation, P introduces an assumption h under previously noted "theorizing" conditions. Then, we can say that

---

[25] For a critical discussion of Whewell, see my *Evidence and Method*.
[26] Since $E(h,e)$ entails h, the usual Bayesian idea $p(h/e) > p(h)$ follows.
[27] For arguments against requiring that evidence increase this probability, see my *The Book of Evidence*, chapter 4.

(Scientific Spec): *h* is a (truth-relevant) scientific speculation for *P* (with respect to the truth of *h*) if and only if *P* does not know that there is explanatory evidence that *h* (A-evidence, or explanatory B-evidence).[28]

Even if meta-inductive, only-game-in-town, and authoritative facts could raise the probability of a hypothesis and make it highly probable, these "facts" would not constitute explanatory evidence for the hypothesis. If these "facts" are the only ones known that make the hypothesis highly probable, and if it is not known that any others do so and constitute A-evidence or explanatory B-evidence, then the hypothesis, if introduced in a scientific investigation and believed to be true, is a scientific speculation.

It is this concept of speculation – (Scientific Spec) – that I will utilize in what follows. I will take it to be the concept in question when disputes occur about when, if at all, to introduce truth-relevant speculations. I employ this concept in Section 3.12 when I address the issue of how to evaluate such speculations.

3.11 Speculation Controversies Revisted

All the views in Section 3.3 can be understood in terms of the concept of (Scientific Spec). Both Newton (espousing the "very conservative" view) and Whewell (the "moderate" one) are talking about evidence and speculation in an objective sense, not a subjective one. Newton's causal-inductive "Rules for the Study of Natural Philosophy" and Whewell's "consilience" and "coherence" are supposed to yield evidence that is independent of any particular person's beliefs about what counts as evidence.[29] Moreover, both Newton and Whewell require "explanatory" evidence, since the proposition inferred from the evidence will explain that evidence and other phenomena as well. They are not talking about non-explanatory "evidence" – whether meta-inductive, only-game-in-town, or authoritarian. In addition, both have in mind evidence that provides a good reason for believing a hypothesis, rather than simply increasing its probability. Using A- or explanatory B-evidence in defining speculation will yield the kind of activity that Newton opposes and Whewell supports. Feyerabend, a champion of the "very liberal" view that encourages unbridled speculation ("proliferation") in the absence of evidence, presents no particular view of evidence or of how to get evidence.

---

[28] This, of course, leaves open the question of what makes an investigation a "scientific" one – a topic long debated. What I am saying is that however this is to be understood, whether an assumption introduced in such an investigation is a speculation depends on whether the speculator knows that explanatory evidence for it exists.

[29] To be sure, a subjectivist about evidence could adopt Newtonian or Whewellian ideas by saying that *e* is evidence for him, but not necessarily for others, if and only if *e* and *h* satisfy Newtonian inductivism or Whewellian consilience and coherence. But this is not the position of Newton or Whewell. Their concepts of evidence are not tied to individuals.

I will construe his self-described "anarchist" view as encouraging speculation, where this at least includes speculation in the sense of (Scientific Spec).

Understanding these conflicting claims using (Scientific Spec), how shall we respond to them? As for the "very conservative" view – the view that Newton puts by saying "hypotheses have no place in experimental philosophy"– although both Newton and Descartes promulgate this idea in their official principles, they violate it in their practice. In the *Principia*, Newton introduces two propositions he himself calls "hypotheses." (Hypothesis 1 is "the center of the system of the world is at rest" [16].) And at the end of his book *Opticks* there are speculations about the particle nature of light that are clearly hypotheses in his sense. Here, after offering empirical proofs of various laws of geometrical optics, Newton raises the question of what light is and offers a speculative hypothesis: "Are not the rays of light very small bodies emitted from shining substances? For such bodies will pass through uniform mediums in right lines without bending into the shadow, which is the nature of rays of light" [17]. These are Newtonian speculations in the sense given by (Scientific Spec).

Descartes, as well, speculates in his *Principles of Philosophy*. After proving his general laws of motion (to his satisfaction) – laws regarding inertial motion and motion after impact – he introduces assumptions about the vortex motion of celestial bodies in the universe, saying "I will put forward everything that I am going to write [in what follows] just as a hypothesis" [18]. Descartes then proceeds to "suppose that all the matter constituting the visible world was originally divided by God into unsurpassably equal particles of medium size ... that each turned round its own centre, so that they formed a fluid body, such as we take the heavens to be; and that many revolved together around various other points ... and thus constituted as many different vortices as there now are stars in the world." These speculations also satisfy (Scientific Spec).

As these examples show, proof of a proposition or set of propositions about the behavior of some objects often leads to more questions about the nature of those objects and their behavior. It is natural for scientists to raise these questions and, when they do so, to think of possible answers. Refusing to do so is tantamount to stifling curiosity. How else could progress in science occur? Propositions don't usually come to the mind fully proved, and even when and if they do, additional questions will come to mind, some of which, at least, are accompanied with answers that are not proved. This happened with both Descartes and Newton, and is part of both everyday and scientific thinking. Indeed, humans could not survive without it. Despite pronouncements in their "official" methodologies to eschew speculations altogether, perhaps a more charitable interpretation of the positions of Newton and Descartes is this: Speculations are allowed, provided that they are clearly labeled as such, and provided they are not inferred to be true or probable on the grounds that, if true, they could explain, or help to explain, various known phenomena. If these conditions are met, why object to publicly communicating speculations, as was done by both Newton and Descartes?

What about the "moderate" position ("Speculate, but verify") typically taken by hypothetico-deductivists? There are several questions here. First, to raise a practical question, how long a period of time should be allowed between the speculation and the verification? Taking string theory as an example, and construing "verification" as providing (Newtonian or Whewellian) scientific evidence sufficient for belief, how long on the "moderate" view should string theorists be given to provide such verification of their theory: 10 years, 20 years, or 100 years? Is the view simply that until verification occurs, the speculation is not to be believed? Or is it that until at least some reasonable prospect or plan for verification emerges, the theory should not be seriously pursued (developed, worked out, taught in the classroom) by scientists or even regarded as "scientific"? Or, as a practical question, is the answer a subjective one, to be decided by the interests, finances, and patience of individual scientists?

More importantly, until it is tested, how, if at all, is a speculation to be evaluated? That is the main question I want to raise in the case of both the "moderate" view and Feyerabend's "very liberal" one ("Speculate like mad"). Hypothetico-deductivists, as well as Feyerabendians, provide little information here. The "context of discovery," where, according to hypothetico-deductivists, speculation is supposed to take place has few, if any, constraints placed upon it. Popper wants the speculations to be "bold." Hypothetico-deductivists want the speculations at least to be potential explanations of the data or potential solutions to the problems prompting the speculation. Feyerabend is more liberal: He asserts that a speculation can be inconsistent with the data and can even reject the problems prompting the speculation. A speculation, as such, is not subject to any standards of evaluation.

I reject all three views, not only because scientists do and must speculate, but because, as I will argue next, speculations can be evaluated in various ways other than by testing, and doing so is entirely appropriate. Giving "free rein to the imagination" – a favorite slogan of hypothetico-deductivists and Feyerabendians – sounds good, but isn't always the best policy.

3.12 How Should Speculations Be Evaluated? Maxwell's Speculative Strategies

The speculations I have in mind – ones satisfying (Scientific Spec) – are not subject to universal standards that determine in general what is to count as a good speculation. Instead, they are subject to pragmatic standards that depend on the aims and epistemic situation of the speculator and the evaluator, which can vary from one scientist and context to another. To see how this is supposed to work in practice, I will invoke the case of James Clerk Maxwell, a speculator par excellence, who had compelling philosophical ideas about speculating. My aim in what follows is not to present evaluations of Maxwell's speculations, but to show what Maxwell regarded, and reasonably so, as criteria that can be used in such evaluations.

Beginning in 1855 and for more than 20 years Maxwell made numerous speculations about electricity and molecules. Three are of particular concern in what follows, because they represent three different ways to speculate, and can be judged using different standards of evaluation. Maxwell himself had labels for these methods. The first he called an "exercise in mechanics," the second a "method of physical speculation," and the third a "method of physical analogy." In the three examples that I will mention Maxwell begins with known phenomena that no established theory has explained, following which he introduces speculations, either truth-relevant ones for which scientific evidence is unknown or insufficient or else truth-irrelevant ones. Although the methods he employs are different, he expresses three very general aims that are common to all. One aim is to present a physical, rather than purely mathematical, way of understanding the known phenomena. A second aim is to proceed in a very precise way, using mathematics to present the physical ideas. A third is to work out the speculation in considerable detail, drawing various consequences. Maxwell's speculative methods represent different ways to accomplish these common aims, and the results are to be evaluated differently. In previous sections I have concentrated on truth-relevant speculations, so I will begin with these.

3.12.1 Exercise in Mechanics

One type of speculation that Maxwell employs is illustrated by what he calls "an exercise in mechanics." In 1860, in a ground-breaking work on the kinetic–molecular theory of gases entitled "Illustrations of the Dynamical Theory of Gases" [19], Maxwell introduces a series of speculative assumptions, including (1) that gases are composed of spherical molecules that move with uniform velocity in straight lines, except when they strike each other or the sides of the container; (2) that they obey Newtonian laws of dynamics; (3) that they exert forces only at impact and not at a distance; and (4) that they make perfectly elastic collisions. He then works out these assumptions mathematically so as to explain various known gaseous phenomena, and to derive new theoretical results, including his important distribution law for molecular velocities.

Just before publishing the paper Maxwell writes to Stokes in 1859:

I do not know how far such speculations may be found to agree with facts ..., and at any rate as I found myself able and willing to deduce the laws of motion of systems of particles acting on each other only by impact, I have done so as an exercise in mechanics. [20]

Maxwell does not claim that the assumptions he makes are true or close to it. His aim in this paper is to see whether a dynamical explanation of observed gaseous phenomena is even *possible, that could be true, given what is known* – an explanation that invokes bodies in motion exerting forces subject to Newton's laws. For this purpose he employs a set of simplified assumptions about molecules constituting gases. His basic question in the paper is whether these assumptions about constituents of gases could be used to explain known gas laws and could be developed mathematically to yield some interesting theoretical claims about molecular motion. However, when Maxwell explicitly classifies his assumptions as speculations (as he does in the quotation given earlier), he is thinking of them as being speculations with respect to truth, not possibility. They are (truth-relevant) speculations that Maxwell introduces for explanatory purposes without knowing that there is evidence for the truth, or the closeness to truth, of the assumptions he introduces.

How should such speculations be evaluated? This can be done from different perspectives, not just from the perspective of truth or evidence. One is purely historical, recognizing the significance of Maxwell's speculations in the development of kinetic–molecular theory. A different perspective, his own, is obtained by focusing on what Maxwell was trying to do when he introduced the speculative assumptions – namely, to see whether a molecular theory of gases is possible. To do so, he showed how such a theory could offer mechanical explanations of pressure, volume, and temperature of gases and of known laws relating these and other properties. In addition, he showed how such a theory could be extended by deriving consequences, such as his distribution law for molecular velocities, his most important new result. From this perspective, speculative assumptions are evaluated by considering whether and how well, if true, they would correctly explain various properties of, and laws governing, gases. For Maxwell, how well they would do so depends on whether they are worked out in considerable detail using mathematics, and whether they provide a physical and not merely mathematical way to think about gases. Such a perspective was important during Maxwell's time: Although substantially developed mechanical theories had been introduced in other areas of physics and astronomy, this had not yet been accomplished for gases, at least not with the depth and precision that Maxwell demanded.

To be sure, another perspective of evaluation is that of Newton: Did Maxwell have any evidence for these assumptions? If not, he should have discarded them. Maxwell is urging a much more pragmatic approach. Speculations, like everything else, can be evaluated in different ways. From the Newtonian perspective of "proof" or "deduction from the phenomena," all speculations are bad. That perspective is supposed to be paramount and dwarf or even disallow others. If Maxwell had no evidence for his assumptions, he should not have speculated, at least not publicly. If he had evidence, it was not a speculation. But why should the Newtonian perspective be the only one from which to judge Maxwell's speculations, especially since Newton himself violated the requirements of that perspective? Why not just label the assumptions as speculations, and say that the fact that they can explain known gas laws, and yield interesting new results, is not evidence or proof that they are correct? This, of course, is exactly what Maxwell did.

3.12.2 Physical Speculation

In 1875, 15 years after the publication of his first kinetic theory paper, Maxwell published "On the Dynamical Evidence of the Molecular Constitution of Bodies" [21]. The paper contains various truth-relevant speculations about molecules, but Maxwell regards his methodology as different from that in his "exercise in mechanics." Calling it a "method of physical speculation," he writes:

When examples of this method of physical speculation have been properly set forth and explained, we shall hear fewer complaints of the looseness of the reasoning of men of science, and the method of inductive philosophy will no longer be derided as mere guess-work. [22]

Maxwell does not spell out the method, but illustrates its use in the paper itself. The method contains some elements present in his "exercise in mechanics," but it adds a crucial component, which I call "independent warrant." This consists of giving reasons for making the speculative assumptions in question, beyond simply that if one makes them then certain phenomena can be explained mechanically. Some of the reasons for some assumptions include appeals to experimental results and may rise to the level of what I have called scientific evidence for those assumptions. Some reasons do not rise to that level, including ones that appeal to the success of similar assumptions used in other domains and to the fundamentality and simplicity of some of the assumptions.

Maxwell supplies reasons of all three sorts in his paper. For example, he claims that there is "experimental proof that bodies may be divided into parts so small that we cannot perceive them." In this paper he does not say what experiments he has in mind. But in his 1871 book *Theory of Heat* he cites experiments showing that heat is not a substance (caloric) but rather a form of energy. (In all probability he has in mind experiments by Joule in the 1840s.) He proceeds to give empirical reasons why the energy must be kinetic rather than potential energy. And he concludes:

The motion we call heat must therefore must be a motion of parts too small to be observed separately .... We have now arrived at the conception of a body as consisting of a great many parts, each of which is in motion. We shall call any one of these parts a molecule of the substance. [23]

Maxwell seems to believe that he has explanatory evidence at least for the claim that molecules in motion exist in bodies. Since he uses the expression "experimental proof," he appears to regard this evidence as sufficient to justify believing that they do. He also provides an empirical reason for thinking that the molecules satisfy a general virial equation derived by Clausius from classical mechanics as applied to a system of particles constrained to move in a limited region of space, and whose velocities can fluctuate within certain limits; the reason is that this equation works for macroscopic bodies in an enclosure. This may not rise to the level of explanatory evidence, let alone proof, but Maxwell thinks it provides at least some empirical reason for supposing the equation works for unobservable molecules in an enclosure as well.

Before introducing the Clausius equation, Maxwell makes the important general assumption that molecules constitute mechanical systems subject to Newtonian laws. Part of his reason for this assumption is methodological, having to do with the fundamentality of mechanical explanations. He writes:

When a physical phenomenon can be completely described as a change in the configuration and motion of a material system, the dynamical explanation of that phenomenon is said to be complete. We cannot conceive any further explanation to be either necessary, desirable, or possible, for as soon as we know what is meant by the words configuration, motion, mass and force, we see that the ideas which they represent are so elementary that they cannot be explained by means of anything else. [24]

Another part of the reason that Maxwell offers for assuming that molecules are subject to mechanical laws is that such laws have been successful in astronomy and (he thinks) electrical science. The claims of fundamentality and historical success for mechanical principles are nowhere near proof of, or scientific evidence for, the truth of the assumptions Maxwell makes about molecules. Nevertheless, they are among the reasons he offers for making the kind of assumptions about molecules that he does. Maxwell clearly regards the theory he develops as a speculation (he uses the term "physical speculation"). Moreover, it is one in accordance with the objective ideas in (Scientific Spec), since even though he thinks he has "experimental proof" for some assumptions, for others he does not. In the latter case, he has reasons for introducing such assumptions, working them out, and trying to develop experiments that could confirm them. But he does not have, or know that there is, explanatory evidence for them.

Again we can evaluate his speculations from the perspective of his epistemic situation and the questions he was raising. We can ask whether, given his epistemic situation, Maxwell was justified in concluding that he had explanatory evidence for certain assumptions, and whether the reasons he offered for other assumptions, even if they did not rise to the level of explanatory evidence, were good ones, and how good they were. And, as in the case of the earlier 1860 paper, we can also provide non-epistemic evaluations – for example, ones pertaining to the historical importance of his results.

3.12.3 Physical Analogy

Finally, I turn to a truth-irrelevant speculation. In 1855, Maxwell published a paper, "On Faraday's Lines of Force" [25], in which he imagines the existence of an incompressible fluid flowing through tubes of varying section. His aim in developing this idea is to construct a physical analogue of electric and magnetic fields. In the electrical case, the velocity of the imaginary fluid at a given point represents the electrical force at that point, and the direction of flow in the tube represents the direction of the force. Particles of electricity are represented in the analogue as sources and sinks of fluid, and the electrical potential as the pressure of the fluid. Coulomb's law, according to which the electrical force due to a charged particle a distance r from the particle varies as $1/r^2$, is represented by the law that the velocity of the fluid at a distance r from the source of fluid varies as $1/r^2$. The bulk of Maxwell's 75-page paper is spent working out this analogy mathematically by showing how to derive equations governing the imaginary fluid that are analogues of ones governing electrical and magnetic fields.

Why does Maxwell proceed in this way? He wants to understand and unify a range of known electrical and magnetic phenomena. He notes that one way to do so is to explain why they occur by introducing a hypothesis that provides a physical cause for these phenomena. But he has no such physical hypothesis to offer. In its place he wants to introduce a different way to understand and unify the phenomena – a way that explains not *why they occur* but *what they are*, without providing a cause. One way to do the latter is to describe the known phenomena using concepts that are to be applied more or less literally to the phenomena – for example, describing known electrical phenomena using concepts such as charged particle, force, and motion. Another way is to draw an analogy between these phenomena and some others that are known or can be described. In Maxwell's time, fluids were much better known than electrical phenomena. So, thinking of electricity as being like a fluid, and charged particles as being like sources and sinks of fluid, and of the electric force at a point as being like the velocity of that fluid at a point, might help one to understand what electrical phenomena are, without understanding why they occur. This is so, even though the fluid in question doesn't exist.

Some analogies involve truth-relevant speculations in the sense that I have defined. For example, in an article on molecules in *Encyclopedia Britannica,* Maxwell draws an analogy between molecules in a gas and bees in a swarm: They are similar in some of their motions. This involves a truth-relevant speculation. Maxwell is introducing an assumption about the existence and behavior of molecules with the idea that molecules exist and behave like bees in a swarm, without knowing whether sufficient evidence exists for the claim that molecules exist or for the claim that they behave like bees in a swarm. Other analogies, such as the one in his 1855 paper, involve speculations that are not truth-relevant. Here Maxwell is talking about something that does not (and could not) exist: his imaginary fluid. He introduces a range of assumptions about that fluid, without the idea that these assumptions are or might be true – indeed, with the idea that they cannot be true. Maxwell is, in effect, making up a story about an imaginary fluid, one that is not to be taken as true but one that will serve, he hopes, as a useful analogy for understanding what electrical phenomena are. This analogue story is a truth-irrelevant speculation. Viewed in this way, Maxwell's story is like a Biblical story (under some pragmatic interpretations of the Bible). No claim is made that the "characters" in the story (the imaginary fluid; Cain and Abel) exist. Even so, in both cases, the behavior of certain things that are real and that we know about can be understood by analogy with the behavior of the "characters" in the story.

How is such a speculation to be evaluated? Here major differences emerge between those such as Maxwell, who take a pragmatic view, and those such as Descartes, Newton, Brougham, and Duhem, who, at least in their official methodologies, take a nonpragmatic view of any explanatory assumptions introduced in a scientific investigation. From the latter perspective, you evaluate solely on the basis of what you regard as an ideal and how close the theorizing in question has come to satisfying that ideal. The Newtonian ideal is to introduce an assumption that is true and to establish its truth by causal-inductive reasoning from observed phenomena. From this perspective, Maxwell's physical analogy is a nonstarter. Using words that Brougham, a devout Newtonian, employs against Young's speculation about the wave nature of light, it is "a work of fancy, useless in science .... fit only for the amusement of a vacant hour."

From Maxwell's point of view, the question is whether the imaginary fluid analogy provides a noncausal physical way of understanding and unifying certain electrical and magnetic phenomena for which no causal explanation has yet been discovered; a way that is worked out mathematically; and, most importantly, a way that may help others, including physicists, in understanding and unifying the phenomena. This is a pragmatic perspective because it says that if you can't get what you and others might regard as ideal (e.g., a true theory experimentally verified that causally explains and unifies electrical and magnetic phenomena), do what you can that will be useful (in this case, provide a physical analogy that will help explain and unify the phenomena in a noncausal way). From this perspective, you evaluate the speculation, in part at least, by seeing whether it is or was useful in producing the kind of understanding sought. Here, perhaps unsurprisingly, I adopt Maxwell's pragmatism:

> To conduct the operations of science in a perfectly legitimate manner, by means of methodized experiments and strict demonstration, requires a strategic skill which we must not look for, even among those to whom science is most indebted for original observations and fertile suggestions. It does not detract from the merit of the pioneers of science that their advances, being made on unknown ground, are often cut off, for a time, from that system of communications with an established base of operations, which is the only security for any permanent extension of science. [26]

When you are so "cut off," Maxwell proposes that you proceed to speculate in various ways, "on unknown ground," including by the use of physical analogies such as his imaginary fluid one.[30]

3.13 Be Pragmatic

All three strong positions on speculation discussed in Sections 3.3 and 3.11 make a bold claim about standards to be used in evaluating a speculation. All three hold that for ideas considered as a speculation, and not as something for which there is explanatory evidence, there are no such standards. According to the ("official") Newtonian view, scientists should not speculate at all. The only standards of evaluation are those used in determining whether there is explanatory evidence for the truth of a hypothesis. Under both the "moderate" and "very liberal" views, few, if any, constraints are imposed on speculations. (Proof is a different matter.)

What I am claiming is that there are legitimate ways to evaluate a speculation – as a speculation – independently of standards requiring explanatory evidence. Moreover, different perspectives are possible for such an evaluation: ones pertaining to the speculator, with his knowledge and interests; and ones pertaining to an evaluator, with different knowledge and interests. It is one thing to evaluate Maxwell's imaginary fluid analogue from his perspective in 1855, when no confirmed electromagnetic theory was known that would unify and explain observable electrical and magnetic phenomena in physical terms. It is another to evaluate his account from our twenty-first-century perspective. If we were to engage in constructing an imaginary physical analogue for electrical and magnetic phenomena, even if it were somewhat different from the one Maxwell proposed, perhaps we would deserve Brougham's epithet "a work of fancy, useless in science."

---

[30] For more details on Maxwell's speculative strategies, see my *Evidence and Method*, chapter 4.

My view is pragmatic because it allows speculations to be introduced with different purposes and different epistemic and non-epistemic situations in mind, and it allows an evaluation of those speculations based on those purposes and situations. In Section 3.2, I noted that Duhem criticized Kelvin's speculations about the molecular nature of the ether as being a "work of imagination," not "acceptable to reason." Duhem also criticized both Kelvin and Maxwell for theorizing in physical, rather than in purely abstract mathematical, terms. He called this type of mentality "British," and he clearly thought it was inferior to the abstract "Continental" (or "French") mind. Theorizing, whether speculative or not, should always be done in the "Continental" style, he thought. (Whether this perspective is reasonable is another matter!) Maxwell, too, recognized that there are different sorts of scientific minds:

[S]ome minds ... can go on contemplating with satisfaction pure quantities presented to the eye by symbols, and to the mind in a form which none but mathematicians can conceive.

Others require that these quantities be represented physically. He concludes:

For the sake of persons of these different types, scientific truth should be presented in different forms, and should be regarded as equally scientific, whether it appears in the robust form and the vivid colouring of a physical illustration or in the tenuity and paleness of a symbolic expression. [27]

Maxwell himself strongly preferred physical explanations and sought to provide mechanical ones in his theorizing because he regarded them as fundamental. But he recognized that there are other ways to theorize that can be just as good or better for certain minds. There are other legitimate perspectives from which to theorize and evaluate the result. As far as speculation itself is concerned, yes, it would be great to have "methodized experiments and strict demonstration." But the "pioneers of science" (including Maxwell) are often "cut off, for a time" from methodized experiments and strict demonstration. Their "advances, being made, on unknown ground," are speculative, but advances nonetheless.

Finally, if you are a pragmatist, and you want to provide "methodized experiments and strict demonstration," how long do you wait? For a pragmatist, that is mostly a practical question, the answer to which depends on your interests, your temperament, your time, and your money. It also depends on how likely it is that such experiments can be performed, and when, given what is known. Suppose, however, that for some empirical reasons – say, energies required for proper detection are not and may never be achievable (perhaps string theory), or signals from objects postulated can never reach us (multiverse theory) – methodized experiments and strict demonstration will never be possible. Suppose, that is, that there will be no known evidence sufficient to justify belief in the speculation. Should you just give up? Should you stop speculating? Newtonians following the official party line will say: Of course. Speculations, especially those that will always remain so, "have no place in experimental philosophy." Pragmatists, however, in the spirit of Maxwell's first paper on molecules, can say:

Not so fast. Here is a way the world might be, even if neither I nor anyone else can present evidence sufficiently strong to believe in the existence of molecules (or strings, or multiverses, or whatever speculative entities are postulated), indeed perhaps even if there never will be such evidence. If, for that reason, you choose to stop working on, or even considering, the theory introduced, that is your pragmatic choice. If you reply that this is not science, my response will again be Maxwellian: Not so fast. Science encompasses many activities, including speculation. Ideally, the latter will lead to testing, but it may not. From a pragmatic viewpoint one can evaluate a speculation without necessarily testing it. You don't have to know how to test the speculation for it to be a good one, or even good science. And, perhaps in the most extreme cases, if you have reasons to suppose that it is not testable and will not become so, you may still be able to give reasons, whether epistemic or non-epistemic, for making the speculation, even if these do not amount to evidence sufficient to believe it. Such reasons can be evaluated from a pragmatic scientific perspective.[31]

References

[1] *Philosophical Transactions of the Royal Society* 92 (1802), pp. 12–48

[2] *Edinburgh Review*, 1 (1803), pp. 450, 455

[3] Reprinted in Robert Kargon and Peter Achinstein, eds., *Kelvin's Baltimore Lectures and Modern Theoretical Physics* (Cambridge, MA: MIT Press, 1987), p. 14

[4] Duhem, trans. Philip P. Wiener (Princeton, NJ: Princeton University Press, 1954, 1982).

[5] Duhem, Chapter 4, sec. 5.

[6] Steven Weinberg, *To Explain the World* (New York: Harper, 2015), p. 14.

[7] Peter Achinstein, ed., *Science Rules* (Baltimore: Johns Hopkins Press, 2004), p. 19.

[8] Whewell, in Achinstein, *Science Rules*, p. 155.

[9] Karl Popper, *The Logic of Scientific Discovery* (New York: Basic Books, 1959), p. 32.

[10] Achinstein, *Science Rules*, p. 154.

[11] "Against Method: Outline of an Anarchistic Theory of Knowledge," reprinted in part in Achinstein, *Science Rules,* p. 377.

[12] *The Principia: Mathematical Principles of Natural Philosophy*, trans. I. Bernard Cohen and Anne Whitman (Berkeley: University of California Press, 1999), Book 3, "General Scholium."

[13] See Lipton, p. 58; Achinstein, *Evidence and Method*, p. 95.

[14] Peter Achinstein, *Particles and Waves* (New York: Oxford University Press, 1991), Chapter 9.

---

[31] This chapter, with slight modifications, is from the book *Speculation* cited in note 15. For very helpful criticisms and suggestions, I am grateful to Justin Bledin, Richard Dawid, Steven Gimbel, Fred Kronz, and Richard Richards.

[15] Larry Laudan, "A Confutation of Convergent Realism," *Philosophy of Science* 48 (March 1981), pp. 19–49.

[16] Achinstein, *Science Rules*, p. 74.

[17] Newton, *Opticks*, (New York, 1979), p. 370.

[18] Achinstein, *Science Rules*, p. 46.

[19] *The Scientific Papers of James Clerk Maxwell*, 2 vols., edited by W. D. Niven (New York: Dover, 1965), vol. 1, pp. 377–409.

[20] Elizabeth Garber, Stephen G. Brush, and C. W. F. Everitt, eds., *Maxwell on Molecules and Gases* (Cambridge MA: MIT Press, 1986), p. 279.

[21] Maxwell, *Scientific Papers*, vol. 2, pp. 418–38.

[22] *Scientific Papers*, vol. 2, p. 419.

[23] *Theory of Heat* (New York: Dover, 2001), pp. 304–5.

[24] *Scientific Papers*, vol. 2, p. 418.

[25] *Scientific Papers*, vol. 1, pp. 155–229.

[26] *Scientific Papers*, vol. 2, p. 420.

[27] *Scientific Papers*, vol. 2, p. 220.

4

Assessing Scientific Theories
The Bayesian Approach

RADIN DARDASHTI AND STEPHAN HARTMANN

4.1 Introduction

Scientific theories are used for a variety of purposes. For example, physical theories such as classical mechanics and electrodynamics have important applications in engineering and technology, and we trust that this results in useful machines, stable bridges, and the like. Similarly, theories such as quantum mechanics and relativity theory have many applications as well. Beyond that, these theories provide us with an understanding of the world and address fundamental questions about space, time, and matter. Here we trust that the answers scientific theories give are reliable and that we have good reason to believe that the features of the world are similar to what the theories say about them. But why do we trust scientific theories, and what counts as evidence in favor of them?

Clearly, the theories in question have to successfully relate to the outside world. But how, exactly, can they do this? The traditional answer to this question is that established scientific theories have been positively tested in experiments. In the simplest case, scientists derive a prediction from the theory under consideration, which is then found to hold in a direct observation. Actual experimental tests are, of course, much more intricate, and the evaluation and interpretation of the data is a subtle and by no means trivial matter. One only needs to take a look at the experiments at the Large Hadron Collider (LHC) at the European Organization for Nuclear Research (CERN) near Geneva, and at the huge amount of workforce and data analysis involved there, to realize how nontrivial actual experiments can be. However, there is no controversy among scientists and philosophers of science about the proposition that the conduct and analysis of experiments is the (only?) right way to assess theories. We trust theories because we trust the experimental procedures to test them. Philosophers of science have worked this idea out and formulated theories of confirmation (or corroboration) that make the corresponding intuition more precise. We discuss some of these proposals in the next two sections.

While empirical testing is certainly important in science, some theories in fundamental science cannot (yet?) be tested empirically. String theory is a case in point. This arguably most fundamental scientific theory makes, so far, no empirically testable predictions, and even if it would make any distinct predictions, referring to predictions that are not also predictions of the theories it unifies (i.e., the Standard Model and general relativity), these predictions could not be confirmed in a laboratory because the required energies are too high. The question, then, is what we should conclude from this. Are fundamental theories such as string theory not really scientific theories? Are they only mathematical theories? Many people would refrain from this conclusion and argue that string theory does, indeed, tell us something about our world. But why should we believe this? Are there other ways of assessing scientific theories, ways that go beyond empirical testing?

Several well-known physicists think so (see, for example, Polchinski, this volume) and in a recent book, the philosopher Richard Dawid (2013) defends the view that there are non-empirical ways of assessing scientific theories. In his book, he gives a number of examples, including the so-called No Alternatives Argument, which we discuss in detail in Section 4.4. Another example is analogue experiments, examined in Section 4.5. There is, however, no agreement among scientists regarding the viability of these nonstandard ways of assessing scientific theories. In a similar vein, these new methods do not fit into traditional philosophical accounts of confirmation (or corroboration) such as Hempel's hypothetico-deductive (HD) model or Popper's falsificationism. Interestingly, these deductivist accounts exclude indirect ways of assessing scientific theories from the beginning as, in those accounts, the evidence for a theory under consideration must be a deductive consequence of it and it must be observed.

There is, however, an alternative philosophical framework available: Bayesian confirmation theory. We will show that it can be fruitfully employed to analyze potential cases of indirect confirmation such as the ones mentioned previously. This will allow us to investigate under which conditions indirect confirmation works, and it will indicate which, if any, holes in a chain of reasoning have to be closed if one wants to make a confirmation claim based on indirect evidence. By doing so, Bayesian confirmation theory helps the scientist (as well as the philosopher of science) better understand how, and under which conditions, fundamental scientific theories such as string theory can be assessed and, if successful, trusted.

The remainder of this chapter is organized as follows. Section 4.2 considers the traditional accounts of assessing scientific theories mentioned earlier and shows that they are inadequate to scrutinize indirect ways of assessing scientific theories. Section 4.3 introduces Bayesian confirmation theory and illustrates the basic mechanism of indirect confirmation. The following two sections present a detailed

Bayesian account of two examples, namely the No Alternatives Argument (Section 4.4) and analogue experiments (Section 4.5). Section 4.6 provides a critical discussion of the vices and virtues of the Bayesian approach to indirect theory assessment. Finally, Section 4.7 concludes with a summary of our main results.

4.2 Trusting Theories

Why do we trust a scientific theory? One reason for trusting a scientific theory is certainly that it accounts for all data available at the time in its domain of applicability. But this is not enough: We also expect a scientific theory to account for all future data in its domain of applicability. What, if anything, grounds the corresponding belief? What grounds the inference from the success of a theory for a finite set of data to the success of the theory for a larger set of data? After all, the future data are, by definition, not available yet, but we would nevertheless like to work with theories which we trust to be successful in the future.

Recalling Hume's problem of induction (Howson, 2000), Popper argues that there is no ground for this belief at all. Such inferences are not justified, so all we can expect from science is that it identifies those theories that do not work. We can only falsify a proposed theory if it contradicts the available data. Popper's theory is called *falsificationism* and it comes in a variety of versions. According to *naive falsificationism*, a theory T is corroborated if an empirically testable prediction E of T (i.e., a deductive consequence of T) holds. Note that this is only a statement about the theory's past performance, with no implications for its future performance. If the empirically testable prediction does not hold, then the theory is falsified and should be rejected and replaced by an alternative theory. As more sophisticated versions of falsificationism have the same main problem relevant for the present discussion as naive falsificationism, we do not have to discuss them here (see Pigliucci or Carroll, this volume). All that matters for us is the observation that, according to falsificationism, a theory can be corroborated only empirically. Hence, a falsificationist cannot make sense out of indirect ways of assessing scientific theories. These possible new ways of arguing for a scientific theory have to be dismissed from the beginning because they do not fit the proposed falsificationist methodology. We think that this is not a good reason to reject the new methods. It may well turn out that we come to the conclusion that none of them work, but the reason for this conclusion should not be that the new method does not fit our favorite account of theory assessment.

Hempel, an inductivist, argued that we have grounds to believe that a well-confirmed theory can also be trusted in the future. Here is a concise summary of the main idea of the hypothetico-deductive (HD) model of confirmation that Hempel famously defended:

General hypotheses in science as well as in everyday use are intended to enable us to anticipate future events; hence, it seems reasonable to count any prediction that is borne out by subsequent observation as confirming evidence for the hypothesis on which it is based, and any prediction that fails as disconfirming evidence. (Hempel, 1945, p. 97)

Note that the HD model shares an important feature with Popper's falsificationism: Both are deductivist accounts; that is, the evidence has to be a deductive consequence of the tested theory. Thus indirect ways of confirmation also do not fit the HD model and must be dismissed for the same reason falsificationism has to dismiss them.

The HD model has a number of (other) well-known problems, which eventually led to its rejection in the philosophical literature (see, however, Schurz, 1991 and Sprenger, 2011). The first problem is the *tacking problem*: If E confirms T, then it also confirms T $\wedge$ X. Note that X can be a completely irrelevant proposition (such as "pink dragons like raspberry marmalade"). This is counterintuitive, as we do not expect E to confirm the conjunction T $\wedge$ X, but only T. The second problem has to do with the fact that the HD model cannot account for the intuition that some evidence may confirm a theory more than some other evidence. The HD model lacks an account of degrees of confirmation, but can justify only the qualitative inference that E confirms T (or not). Bayesian confirmation theory, discussed in the next section, accounts for these and other problems of more traditional accounts of confirmation. It also has the resources and the flexibility to model indirect ways of assessing scientific theories.

4.3 Bayesian Confirmation Theory

In this section, we give a brief introduction to Bayesian confirmation theory (BCT). For book-length introductions and discussions of the topic, we refer the reader to Earman (1992), Howson and Urbach (2006), and Sprenger and Hartmann (2019). For recent surveys of the field of Bayesian epistemology, see Hájek and Hartmann (2010) and Hartmann and Sprenger (2010).

Let us consider an agent (e.g., a scientist or the whole scientific community) who entertains the propositions T: "The theory under consideration is empirically adequate" and E: "The respective evidence holds" before a test of the theory is performed.[1] In this case the agent is uncertain as to whether the theory is empirically

---

[1] "Empirical adequacy" is a technical term made popular in the philosophical literature by Bas van Fraassen. In his book *The Scientific Image*, he writes that "a theory is empirically adequate exactly if what it says about the observable things and events in the world is true – exactly if it 'saves the phenomena'" (van Fraassen, 1980, p. 12). Note that empirical adequacy is logically weaker than truth. A true theory is empirically adequate, but not every empirically adequate theory is true. We could make our arguments also using the term "truth" instead of "empirical adequacy," but decided to stick to the weaker notion in the following discussion.

adequate; she also does not know that the evidence will hold. The easiest way to represent her attitude toward these two propositions is to assign a probability to them. Bayesians request that rational agents assign a prior probability distribution P over the propositional variables they consider. In our case the agent considers the binary propositional variables T (with the values T and $\neg$T) and E (with the values E and $\neg$E).[2]

Next we assume that a test is performed and that the evidence holds. As a result, the probability of E shifts from $P(E) < 1$ to $P^*(E) = 1$, where P^* denotes the new "posterior" probability distribution of the agent after learning the evidence. To make sure that P^* is coherent, meaning that it satisfies the axioms of probability theory, the agent has to adjust the other entries in the posterior probability distribution. But how can this be done in a rational way? For the situation just described, Bayesians argue that the posterior probability of T should be the old conditional probability:

$$P^*(T) = P(T|E). \tag{4.1}$$

This identification, which is sometimes called "Bayes' theorem" or "conditionalization," can be justified in various ways such as via Dutch Book arguments (Vineberg, 1997), epistemic utility theory (Pettigrew, 2016) and distance-minimization methods (Diaconis & Zabell, 1982; Eva & Hartmann, 2018), which we do not consider here. Once we accept Bayes' theorem as a diachronic norm, the right-hand side of Eq. (4.1) can be expressed differently using the definition of the conditional probability. As $P(T|E)P(E) = P(T, E) = P(E, T) = P(E|T)P(T)$, we obtain

$$P^*(T) = \frac{P(E|T)\, P(T)}{P(E)}. \tag{4.2}$$

This equation expresses the posterior probability of T, $P^*(T)$, in terms of the *prior probability* of T, $P(T)$; the *likelihood* of the evidence, $P(E|T)$; and the *expectancy* of the evidence, $P(E)$.

According to BCT, E *confirms* T if and only if $P^*(T) > P(T)$, so that the observation of E raises the probability of T. Likewise, E *disconfirms* T if and only if $P^*(T) < P(T)$, so that the observation of E lowers the probability of T. The evidence E is *irrelevant* for T if it does not change its probability, so that $P^*(T) = P(T)$.

Note that in the standard account of BCT, the prior probability distribution P (and therefore also the posterior probability distribution P^*) is a *subjective* probability distribution, such that different agents may disagree on it. It may therefore

---

[2] We follow the notation of Bovens and Hartmann (2004) and denote propositional variables in italic script and their values in roman script. Note further that we sometimes use the letter "E" (in roman script) to refer to the evidence directly and not to the proposition it expresses, and likewise for the theory T. We submit that this does not cause any confusion.

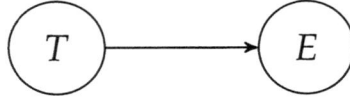

Figure 4.1 A Bayesian network representing the direct dependence between the variables *E* and *T*.

happen that one agent considers E to confirm T while another agent considers E to disconfirm T (or to be irrelevant for T). That is, in BCT, confirmation is defined only relative to a probability distribution, which implies that confirmation is not an objective notion in BCT.

Using BCT, especially Eq. (4.2), has a number of desirable features. Consider, for example, the situation where the evidence is a deductive consequence of the theory in question. In this case, $P(T|E) = 1$ and therefore $P^*(T) = P(T)/P(E)$. As $P(E)$ expresses the expectancy of the evidence before a test is performed, a rational agent will assign a value $P(E) < 1$. Hence, $P^*(T) > P(T)$ and therefore E confirms T. This is in line with our expectations, and it is in line with the hypothetic-deductive (HD) model of confirmation mentioned in the last section. Note further that the more E confirms T, the lower the expectancy of E is. Again, this is in line with our intuition: More surprising evidence confirms a theory better than less surprising (or expected) evidence. As BCT is a quantitative theory, it allows us to account for this intuition, whereas qualitative theories such as the HD model and Popper's falsificationism do not have the resources to do so. They answer only the yes–no question of whether E confirms (or corroborates) T.

Let us develop BCT a bit further. In many cases, there is a *direct dependency* between T and E. We mentioned already the case where E is a deductive conse-quence of T. In other cases, there is a direct probabilistic dependency between E and T (because of the presence of various uncontrollable disturbing factors). In these cases, $P(T|E) < 1$, but $P^*(T)$ may, of course, still be greater than $P(T)$. The direct dependence between theory and evidence is depicted in the Bayesian network in Figure 4.1. Here the nodes *E* and *T* represent the respective propositional variables and the arc that connects them indicates the direct dependency.[3]

Note, however, that BCT can also deal with cases where the evidence is indirect – that is, when the evidence is not a deductive or inductive consequence of the theory in question. In these cases, the correlation between *E* and *T* is accounted for by a third ("common cause") variable *X* as depicted in the Bayesian network in Fig-ure 4.2. Here is an illustration: We take it that having yellow fingers (E) is evidence of having heart disease (T). However, there is no direct dependence between the two

---

[3] For an introduction to the theory of Bayesian networks and their use in epistemology and philosophy of science, see Bovens and Hartmann (2004, Section 3.5).

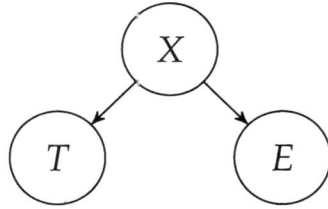

Figure 4.2 A Bayesian network representing the (indirect) dependence between the variables E and T, mediated by a common cause variable X.

variables E and T: For example, painting your fingers green will (unfortunately) not lower your probability of having heart disease.[4] Instead, the positive correlation between E and T and the fact that the observation of E confirms T result from the presence of a common cause X, the possible tobacco use (smoking) of the respective agent. Note that X and T and X and E are positively correlated, such that observing E confirms X, which in turn confirms T. Hence, for the Bayesian network depicted in Figure 4.2, E confirms T.[5]

To elaborate a bit further on this, the common cause X (of E and T) has the following property: If the value of X is not known, then E and T are dependent. But once the value of X is known (i.e., once the variable X is instantiated), E and T become independent. One also says that X *screens off* E from T.

We will see later that the common cause structure can be used to model cases of indirect confirmation. Here, similar to the previous example, the evidence probes a third variable, which in turn probes the theory in question. The powerful formal machinery of Bayesian networks (and conditional independence structures) makes sure that all of this can be made precise. It is important to note that deductivist accounts of confirmation lack something analogous to a common cause structure, which is their major drawback.

In the next two sections, we show in detail how these ideas can be applied to analyze cases of indirect confirmation.

4.4 Illustration 1: The No Alternatives Argument

Scientists have been defending and developing string theory for several decades, despite the lack of direct empirical confirmation. What is more, no one expects

---

[4] For more on this, see Woodward (2005).
[5] Note that the confirmation relation is symmetric: E confirms T if and only if T confirms E. Note further that the confirmation relation is not necessarily transitive. Consider, for example, the situation where, in addition to the arcs in the Bayesian network in Figure 4.2, there is an arc from T to E (or vice versa) that represents a sufficiently strong negative correlation between E and T. This negative correlation can then outweigh the positive correlation from the path via X.

this situation to change in the foreseeable future. This raises a question: Why do scientists put so much trust in a theory that is not (and perhaps cannot be) assessed by empirical data? What grounds this enormous trust in string theory? In his recent book *String Theory and the Scientific Method*, Dawid (2013) provides a rationale for three non-empirical ways of arguing for a scientific theory that lack direct empirical support. One of them is the No Alternatives Argument (NAA), which was subsequently analyzed in the framework of BCT by Dawid, Hartmann, and Sprenger (2015). Our presentation follows this discussion.

The NAA relies on an observation at the meta-level; it is not a prediction of the theory itself, but rather an observation about the status of the theory. It is only in this sense non-empirical. The evidence, which is supposed to do the confirmatory work, is the observation that the scientific community has, after considerable effort, not yet found an alternative to a hypothesis H that solves some scientific problem **P**. Let us denote the corresponding proposition by F_A ("the scientific community has not yet found an alternative to H"). Let us furthermore define T as the proposition that the hypothesis H is empirically adequate. The propositional variables T and F_A take as values the previously described propositions and their respective negations.

To show that the meta-level observation F_A confirms T in the Bayesian sense, one has to show that

$$P(T|F_A) > P(T). \tag{4.3}$$

Now, as F_A is neither a deductive nor an inductive consequence of T, there can be no direct probabilistic dependence between the two variables. Following the strategy suggested in the last section, we therefore look for a common cause variable that facilitates the dependence. But which variable could that be? Here Dawid, Hartmann, and Sprenger (2015) introduce the multi-valued variable Y, which has the values

Y_k: There are k distinct alternative theories that solve **P**,

where k runs from 0 to some maximal value N. Y_k is a statement about the *existing* number of theories able to solve the scientific problem in question. It is easy to see that Y screens off F_A from T: Once we know the value of Y, learning F_A does not tell us anything new about the probability of T. To assess it, all that matters is that we know how many equally suitable candidate theories there are. Y facilitates the probabilistic dependence between F_A and T, since if there *are* only a small number of alternative theories, this would provide an explanation for why the scientists have not yet found any; that is, it would explain F_A. In addition, if there *are* only a few alternative theories, that should probabilistically impact our trust in the available theory. This relies on an innocuous assumption, namely that there is at least one theory which is empirically adequate. If this is the case, the number of alternatives

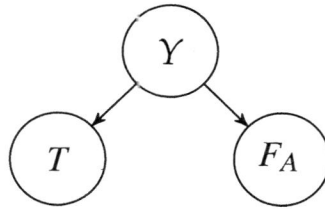

Figure 4.3 A Bayesian network depicting the No Alternatives Argument.

will probabilistically influence our trust in the theory. After the introduction of the variable Y and its inclusion in the Bayesian network depicted in Figure 4.3, one can show, given certain reasonable assumptions, that eq. (4.3) holds.[6]

This may suggest a very simple method to confirm scientific theories. However, this method relies on how well F_A functions as a probe of Y. Note that several complicating factors can arise. First, there might be another explanation for why the scientific community has not yet found an alternative. For instance, Dawid, Hartmann, and Sprenger (2015) introduce an additional node into the Bayesian network representing the difficulty of the scientific problem **P**. The observation of F_A then may provide support only for the difficulty of the problem. However, the difficulty of the problem is probabilistically independent of the empirical adequacy of the hypothesis, indicated by the variable T, and hence the observation of F_A may not confirm the theory in question. Second, our argument relies on it being possible to establish F_A in the first place. However, it is a nontrivial task to find agreement among the members of the scientific community about the existence or nonexistence of alternative solutions to some scientific problem **P**. Even if there is agreement, this is only probative of the existing number of alternatives, indicated by the value of the variable Y, provided that the scientific community has attempted to explore the space of alternative theories (see Oriti, this volume) and have considered all the problems one may encounter in doing so (see Dardashti, this volume). This may have as a normative consequence a requirement to change the way physics is practiced. Theory proliferation is not a common practice, but it may be required for a successful application of non-empirical theory assessment.

4.5 Illustration 2: Analogue Experiments

When scientists are concerned with black holes, neutron stars, or the whole universe, experiments are hard to come by. In these cases it has been suggested that

---

[6] One of these assumptions is that the agent is uncertain about the number of alternatives. If the agent were certain (i.e., if she knew the value of Y), then T and F_A would be probabilistically independent and F_A would not confirm T. For example, an anti-realist who adopts the underdetermination thesis to show that the number of alternatives is infinite (and therefore sets $P(Y_\infty) = 1$) will not find the NAA convincing.

one may be able to use so-called analogue experiments.[7] The idea is to model the experimentally inaccessible *target system* via a table-top experiment, the *source system*, that has specifically been built to model the equations that are assumed to hold in the target system. Among the choices of source systems, one finds fluid systems, Bose–Einstein condensates (BECs), and optical lattices. The underlying physical laws in these source systems are, therefore, significantly different from the laws governing the inaccessible target system. This raises the question of what one can learn about a target system from analogue experiments (see Thébault, this volume). More specifically, one may ask whether the evidence obtained from manipulating the source system is also confirmatory evidence for the inaccessible target system.

The most-discussed examples of analogue experiments concern black hole Hawking radiation. The thermal radiation of black holes was predicted in the 1970s by Hawking. It has played an important role in foundational debates ever since (see Wüthrich [this volume] for a critical discussion), but lacks any direct empirical support. This motivated a first proposal of an analogue model of black hole Hawking radiation based on a fluid system by Unruh in the early 1980s. This model and many of the subsequently proposed analogue models[8] are very difficult to implement experimentally. After several partial successes in the last decade, Jeff Steinhauer (2016) finally announced that he has observed quantum Hawking radiation in a BEC analogue model.

Steinhauer's claim, however, goes even further. He stated that his findings provide "experimental confirmation of Hawking's prediction about the thermodynamics of the black hole" (Steinhauer in Haaretz, August 2016). Thus the evidence obtained in the experiment is taken to be evidence not only for the existence of Hawking radiation in BECs, but also of black hole Hawking radiation. This is in stark contrast to theoretical physicist Daniel Harlow's attitude regarding Steinhauer's experiment, namely that it is "an amusing feat of engineering that won't teach us anything about black holes" (*Quanta Magazine*, November 8, 2016). This example illustrates that scientists disagree about whether the observation of Hawking radiation in a BEC confirms black hole Hawking radiation. We submit that a Bayesian analysis can shed some light on this question.

To do so, we follow the analysis given in Dardashti, Hartmann, Thébault, and Winsberg (2018) where the technical details can be found. As a first step, we identify the relevant propositional variables and specify the probabilistic dependencies that hold between them. Let us denote by T_{BH} the binary propositional variable that takes the following values:

---

[7] See Barceló et al. (2011) for a comprehensive review article on analogue experiments of gravity.
[8] See Barceló et al. (2011, Ch. 4) for a discussion of classical and quantum analogue systems.

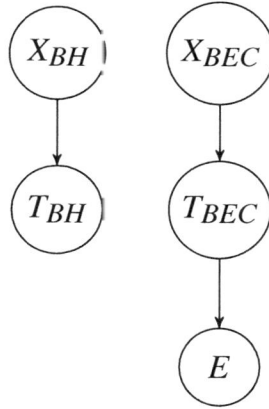

Figure 4.4 The Bayesian network of the BEC model and the black hole model (independent case).

T_{BH}: The model provides an empirically adequate description of the black hole system $\mathcal{T}$ within a certain domain of conditions.

$\neg T_{BH}$: The model does not provide an empirically adequate description of the black hole system $\mathcal{T}$ within a certain domain of conditions.

The domain of conditions encodes the conditions necessary for the respective context, namely the application of the model to derive Hawking radiation. Analogously, we define the binary propositional variable T_{BEC} that is associated with the BEC system and takes the following values:

T_{BEC}: The model provides an empirically adequate description of the BEC system $\mathcal{S}$ within a certain domain of conditions.

$\neg T_{BEC}$: The model does not provide an empirically adequate description of the BEC system $\mathcal{S}$ within a certain domain of conditions.

Furthermore, we introduce the propositional variable E, which has the values E: "The empirical evidence for the source system holds" and $\neg E$: "The empirical evidence for the source system does not hold."

To better understand how the probabilistic dependence between the two systems can occur, we need to introduce another layer of variables related to the domain of conditions mentioned previously. The domain of conditions depends on the various background assumptions involved in developing the model. The empirical adequacy of each model in the respective context therefore depends on whether the various background assumptions involved hold. These background assumptions rely on the knowledge of the modeler and involve both theoretical and experimental knowledge, both implicit and explicit. For instance, I may assume in my description of the fluid model that the fluid is inviscid and barotropic. If the experiment turns out to agree with the predictions of that model, then that outcome not only supports the proposition regarding the empirical adequacy of the model, but also suggests that the assumptions were justified. Note that this does not entail that the model itself is actually viscosity-free or that barotropicity is a realistic assumption, but only that it was a justified assumption within the respective domain of conditions. Let us denote the variables representing the set of background assumptions by X_{BH} and X_{BEC}, respectively. They take the following values:

X_{BH}/X_{BEC}: The background assumptions are satisfied in the model of system $\mathcal{S}/\mathcal{T}$.

$\neg X_{BH}/\neg X_{BEC}$: The background assumptions are not satisfied in the model of system $\mathcal{S}/\mathcal{T}$.

If the background assumptions were probabilistically independent of the other variables, the relevant Bayesian network would be represented by Figure 4.4. In that case, there would be no probabilistic dependence, which does not seem to be unreasonable given that the assumptions in the context of black hole physics, and whether they are justified, seem to be independent of the assumptions involved in the context of BEC physics. However, it has been argued by Unruh and Schützhold (2005) that Hawking radiation may, under certain conditions, be considered a universal phenomenon. Here "universal" is meant in the sense that the phenomenon does not depend on the degrees of freedom at very high energies. The possible universality of the Hawking phenomenon now relates directly to one of the elements of the background assumptions involved in the black hole system, namely the possible influence of the trans-Planckian physics on the thermality of the radiation. The semi-classical derivation of Hawking radiation assumes that the trans-Planckian physics, whatever that might be, does not have an effect on whether Hawking radiation occurs. As the phenomenon relies on the physics at very high frequencies, a domain where the semi-classical approach is not applicable, the assumption is considered to be problematic. It is referred to as the "trans-Planckian problem." Note, however, that the analogue model contains a similar assumption regarding its independence from the underlying high energy theory. Based on these considerations, we introduce an additional variable U corresponding to the universality claim (see Figure 4.5). It has the following values:

U: Universality arguments in support of common background assumptions hold.

$\neg$U: Universality arguments in support of common background assumptions do not hold.

One can now see how U can play the role of the common cause variable mentioned in Section 4.3: If the universality assumption is true, then that will directly impact

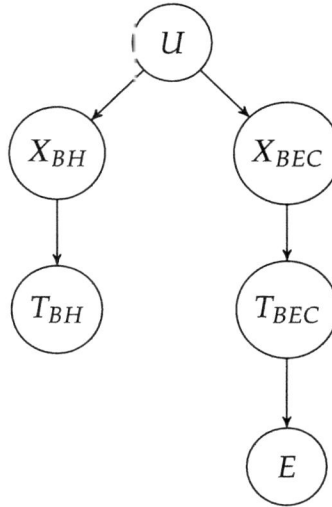

Figure 4.5 The Bayesian network of the BEC model and the black hole model (with universality).

the legitimacy of the corresponding background assumptions. Also, evidence for each analogue model of black hole Hawking radiation will provide empirical support for the universality claim itself, as each analogue model has a different high energy theory. Using the Bayesian network from Figure 4.5 one can then show that, under certain reasonable conditions, E confirms T_{BH}: $P(T_{BH}|E) > P(T_{BH})$.

Together with a number of plausible assumptions about the prior distribution, the Bayesian network depicted in Figure 4.5 provides a *possible* rationale for Steinhauer's strong claim about the empirical confirmation of black hole Hawking radiation. At the same time, it illustrates the more general problem of analogue experiments: To probe an inaccessible target system via analogue experiments, additional arguments establishing the probabilistic dependence need to be established. If these cannot be established, the analogue experiment cannot provide inductive support for the black hole system and may in this sense be just an "amusing feat of engineering." This, of course, does not rule out other (e.g., heuristic) roles it may have (see Thébault, this volume).

4.6 Discussion

In the previous two sections, we have shown how BCT can be used to evaluate ways of indirectly assessing scientific theories. Constructing Bayesian network models of the corresponding testing scenarios allowed us to explore the conditions under which these new ways of assessing scientific theories can be successful and establish trust in a theory. In this section, we discuss a number of concerns related to the Bayesian approach and suggest how a Bayesian can respond to them.

1. The Bayesian approach is too flexible.

   The advantages that come with the flexibility of the Bayesian approach also bear the danger of too much flexibility ("anything goes"). Is it not always possible to come up with a complicated Bayesian network that does what one wants it to do? If this were so, then more needs to be done to make a convincing claim in favor of the confirmatory value of some piece of indirect evidence.

   In response, the following can be said: First, a model is always only as good as its assumptions. Sometimes called the GIGO principle ("garbage in, garbage out"), this concept holds for scientific models as well as for philosophical models. Clearly, the assumptions of a model have to be justified: We did, indeed, make sure that the crucial independence assumption (i.e., that the common cause variable screens off the other two variables) holds. We also discussed the possibility of including other variables (such as the difficulty of the problem in the NAA example) that may influence the conclusion. Second, by doing so, the Bayesian approach makes the reasoning transparent. It forces the modeler to put all assumptions on the table and it shows what can be concluded from a certain set of assumptions. The Bayesian approach provides a reasoning tool that may point to a hole in a putative argument.

2. Confirmation is an absolute, not a relative notion, and indirect confirmation cannot provide it.

   Rovelli (this volume, Chapter 7) claims that the common-sense understanding of confirmation means "very strong evidence, sufficient to accept a belief as reliable" and then goes on to claim that "only empirical evidence can grant 'confirmation' in the common sense."

   In response, we note that Rovelli conflates "confirmation" and "acceptance." While confirmation is a relation between theory and evidence, acceptance is not. According to BCT, a theory is confirmed if its probability increases after the evidence holds. In this case, the evidence is *one* reason in favor of the theory and confirmation is, therefore, an *epistemic notion*. For example, a theory is confirmed if its probability increases from 1% to 5%. Acceptance, by contrast, is *a pragmatic notion*. For example, one would not accept a theory if its probability is only 5%. Instead, the threshold for acceptance should be greater than 50%; that is, it must be more likely that the theory is true than that it is false.[9] Rovelli (this volume), Chapter 7 would set the threshold for acceptance even as high as 95%. If one sticks to this (Bayesian) use of the terms "confirmation" and

---

[9] Interestingly, this intuitive way of explicating acceptance is not without problems. For a discussion, see Leitgeb (2017).

"acceptance," then there is no problem with the claim that indirect evidence can confirm a theory (even if it does not lead to a high posterior probability).

3. The evidence provided by indirect confirmation is negligible.

Related to the previous issue, indirect evidence is always much less effective than direct empirical evidence. Hence, we should not take indirect evidence seriously.

In response, one may agree with the claim that direct evidence is typically better than indirect evidence. The observation of a new phenomenon that can be accounted for using string theory will provide us with much more confirmation than the NAA. However, if there is no such evidence (yet), then the search for indirect evidence for the various candidate theories is a means to better navigate in the epistemic landscape. Having discriminating indirect evidence, however small it is, in favor of string theory may provide a reason to keep working on the program and developing it, while having no evidence whatsoever does not support this claim (or justify the effort that is put into this research program).

4. BCT is subjective.

BCT assumes that scientists assign subjective degrees of belief to their hypotheses, which they subsequently update in the light of evidence. The resulting posterior probabilities are also subjective, which in turn implies that confirmation is subjective. As a result, BCT does not provide an objective account of the relation between theory and evidence.

In response, one may first point to the serious problems that more objective accounts such as HD confirmation or Popper's falsificationism face (see our discussion in Section 4.2). These problems led to the development of the Bayesian approach. Furthermore, Bayesians admit that there is an ineliminable subjective element in scientific reasoning and stress that the Bayesian approach makes it explicit. By putting on the table where subjective judgments come in (and where they do not), it becomes possible to assess the reasoning of a rational agent and to criticize it (if appropriate). Again, the Bayesian approach has the advantage of being transparent.[10]

4.7 Conclusion

The difficulty of experimentally probing certain energy regimes or reaching certain target systems have persuaded many physicists to seriously explore the possibility of alternative methods of theory assessment. At the same time, the stronger reliance on alternative methods has led to profound disapproval from members of the scientific community who question the viability of these methods. This makes

---

[10] Much more can be said on this and other objections to the Bayesian approach. See, for example, Howson and Urbach (2006).

it necessary to rigorously analyze the argumentative strategies employed when applying these methods, to identify the assumptions involved, and to investigate whether these methods can provide confirmation in the same way experiments can.

We have argued that standard qualitative accounts of confirmation (or corroboration) do not provide the tools for an analysis of these methods, as they are restricted to direct (deductive) consequences of the theories only. After having introduced Bayesian confirmation theory, we discussed how this powerful methodological framework provides the appropriate flexibility to rationally reconstruct various indirect ways of theory assessment. More specifically, we used the theory of Bayesian networks to demonstrate the possibility of indirectly confirming a scientific theory via a common cause structure. Here the evidence confirms the theory under certain conditions via a mediating common cause propositional variable. This methodology was then illustrated by two examples, the No Alternatives Argument and analogue experiments.

The crucial task in evaluating these indirect ways of theory assessment in the Bayesian framework is to identify a variable, if there is one at all, that plays the role of the common cause. This is a nontrivial task, and there is no real heuristic to find it. In the case of the No Alternatives Argument, the evidence being used does not probe the theory directly but rather addresses the space of theories as a whole, more specifically the number of alternatives available in that space. By assessing how constrained the theory space itself is, it can indirectly provide confirmation for a theory. In our second illustration, we argued that analogue experiments of black hole Hawking radiation can indirectly provide confirmation by probing directly a universality claim (which, in turn, probes the theory in question). In both cases, BCT provides the tools to reconstruct the argumentative strategies, to make transparent all assumptions involved, and to identify the normative consequences for the practicing scientists. The possibility of indirect confirmation, relying on nontrivial Bayesian network structures, opens up the possibility of analyzing other alternative ways of theory assessment.

Section 4.6 considered several possible objections to BCT and provided some replies. There is much more to say about the vices and virtues of indirect ways of assessing scientific theories within the Bayesian paradigm, but we hope to have shown that BCT is a good framework with which to start to critically and rationally assess the various methods, and a coherent framework for the various exciting new scientific methodologies.

Acknowledgment

We would like to thank our collaborators Richard Dawid, Jan Sprenger, Karim Thébault, and Eric Winsberg for useful discussions and feedback as well as the Alexander von Humboldt Foundation for financial support.

References

Barceló, C., S. Liberati, M. Visser, et al. (2011). Analogue Gravity. *Living Reviews in Relativity 14*(3).

Bovens, L., and S. Hartmann. (2004). *Bayesian Epistemology*. Oxford: Oxford University Press.

Dardashti, R., S. Hartmann, K. Thébault, and E. Winsberg. (2018). Hawking Radiation and Analogue Experiments: A Bayesian Analysis. Preprint available at https://arxiv.org/abs/1604.05932.

Dawid, R. (2013). *String Theory and the Scientific Method*. Cambridge: Cambridge University Press.

Dawid, R., S. Hartmann, and J. Sprenger. (2015). The No Alternatives Argument. *British Journal for the Philosophy of Science 66*(1), 213–34.

Diaconis, P., and S. L. Zabell. (1982). Updating Subjective Probability. *Journal of the American Statistical Association 77*(380), 822–30.

Earman, J. (1992). *Bayes or Bust? A Critical Examination of Bayesian Confirmation Theory*. Cambridge, MA: MIT Press.

Eva, B., and S. Hartmann. (2018). Bayesian Argumentation and the Value of Logical Validity. *Psychological Review* 125(5), 806–821.

Hájek, A., and S. Hartmann. (2010). Bayesian Epistemology. In: J. Dancy et al. (eds.), *A Companion to Epistemology*. Oxford: Blackwell, pp. 93–106.

Hartmann, S., and J. Sprenger. (2010). Bayesian Epistemology. In: S. Bernecker and D. Pritchard (eds.), *Routledge Companion to Epistemology*. London: Routledge, pp. 609–20.

Hempel, C. G. (1945). Studies in the Logic of Confirmation II. *Mind 54*(214), 97–121.

Howson, C. (2000). *Hume's Problem: Induction and the Justification of Belief*. Oxford: Clarendon Press.

Howson, C., and P. Urbach. (2006). *Scientific Reasoning: The Bayesian Approach*. London: Open Court Publishing.

Leitgeb, H. (2017). *The Stability of Belief: How Rational Belief Coheres with Probability*. Oxford: Oxford University Press.

Pettigrew, R. (2016). *Accuracy and the Laws of Credence*. Oxford: Oxford University Press.

Schurz, G. (1991). Relevant Deduction. *Erkenntnis 35*(1–3), 391–437.

Sprenger, J. (2011). Hypothetico-Deductive Confirmation. *Philosophy Compass 6*(7), 497–508.

Sprenger, J., and S. Hartmann. (2019). *Bayesian Philosophy of Science*. Oxford: Oxford University Press.

Steinhauer, J. (2016, August). Observation of Quantum Hawking Radiation and Its Entanglement in an Analogue Black Hole. *Nature Physics 12*(10), 959

Unruh, W. G., and R. Schützhold. (2005). Universality of the Hawking Effect. *Physical Review D 71*(2), 024028.

van Fraassen, B. (1980). *The Scientific Image*. Oxford: Oxford University Press.

Vineberg, S. (1997). Dutch Books, Dutch Strategies and What They Show About Rationality. *Philosophical Studies 86*(2), 185–201.

Woodward, J. (2005). *Making Things Happen: A Theory of Causal Explanation*. Oxford: Oxford University Press.

5

Philosophy of Science and the String Wars: A View from the Outside[1]

MASSIMO PIGLIUCCI

5.1 Introduction

Trouble, as explicitly hinted at in the title of a recent book by Lee Smolin, has been brewing for a while within the fundamental physics community. Ideas such as string theory and the multiverse have been both vehemently defended as sound science and widely criticized for being "not even wrong," in the title of another book, by Peter Woit. More recently, George Ellis and Joe Silk have written a prominent op-ed piece in *Nature*, inviting their colleagues to defend the very integrity of physics. To that invitation, cosmologist Sean Carroll has responded that physics doesn't need "the falsifiability police," referring to the famous (and often misunderstood) concept introduced by philosopher Karl Popper to demarcate science from pseudoscience. The debate is not just "for the heart and soul" of physics, but has spilled into social media, newspapers, and radio. What is at stake is the public credibility of physics in particular and of science more generally – especially in an era of widespread science denial (of evolution and anthropogenic climate change, for instance) and rampant pseudoscience (e.g., the antivaccine movement). Since philosophers of science have been invoked by both sides of the dispute, it may be useful to take a look at the "physics wars" from a more detached philosophical perspective, in my case informed also by my former career as an evolutionary biologist. That field's recent history has peculiar similarities with what is currently going on in fundamental physics, in terms of strong internal disputes and of perception by a significant portion of the general public.

5.2 Prologue: Einstein Versus Freud and Marx

The general theory of relativity is sound science; "theories" of psychoanalysis, as well as Marxist accounts of the unfolding of historical events, are pseudoscience.

---

[1] A shorter version of this essay appeared in *Aeon* magazine. Partially reproduced here with permission.

This was the conclusion reached a number of decades ago by Karl Popper, one of the most influential philosophers of science. Popper was interested in what he called the "demarcation problem," or how to make sense of the difference between science and non-science, and in particular science and pseudoscience. He thought long and hard about this challenge and proposed a simple criterion: falsifiability. For a notion to be considered scientific it would have to be shown that, at least in principle, it could be demonstrated to be false, if it were, in fact, false.

Popper was particularly impressed by Einstein's theory because it had recently been spectacularly confirmed during the 1919 total eclipse of the Sun, so he proposed it as a paradigmatic example of good science. Here is how he later differentiated between Einstein on one side, and Freud, Adler, and Marx on the other side (Popper, 1962, p, 37):

> Einstein's theory of gravitation clearly satisfied the criterion of falsifiability. Even if our measuring instruments at the time did not allow us to pronounce on the results of the tests with complete assurance, there was clearly a possibility of refuting the theory. The Marxist theory of history, in spite of the serious efforts of some of its founders and followers, ultimately adopted [a] soothsaying practice. In some of its earlier formulations ... their predictions were testable, and in fact falsified. Yet instead of accepting the refutations the followers of Marx re-interpreted both the theory and the evidence in order to make them agree. In this way they rescued the theory from refutation ... They thus gave a "conventionalist twist" to the theory; and by this stratagem they destroyed its much advertised claim to scientific status. The two psycho-analytic theories [Freud's and Adler's] were in a different class. They were simply non-testable, irrefutable. There was no conceivable human behavior which could contradict them. ... I personally do not doubt that much of what they say is of considerable importance, and may well play its part one day in a psychological science which is testable. But it does mean that those "clinical observations" which analysts naively believe confirm their theory cannot do this any more than the daily confirmations which astrologers find in their practice.

As it turns out, Popper's high regard for the crucial experiment of 1919 may have been a bit optimistic: When we look at the historical details, we discover that the earlier formulation of Einstein's theory actually contained a mathematical error that predicted twice as much bending of light by large gravitational masses like the Sun – the very thing that was tested during the eclipse. And if the theory had been tested in 1914 (as was originally planned), it would have been (apparently) falsified. Moreover, some sociologists of science (Collins & Pinch, 2012) have suggested that there were significant errors in the 1919 observations, and that one of the leading astronomers who conducted the test, Arthur Eddington, may actually have cherry picked his data to make them look like the cleanest possible confirmation of Einstein (though this is actually disputed: Ball, 2007). Life, and science, are complicated.

This is all well and good, but why should something written near the beginning of the last century by a philosopher – however prominent – be of interest to physicists today? Because of the increasingly harsh controversy surrounding string theory (Becker et al., 2006; Woit, 2007). (For the purposes of this essay I will use the generic term "string theory" to indicate a broad family of related theories or theoretical frameworks that are connected with the concept of cosmic strings.) The theory is something that the fundamental physics community has been playing around with for a few decades now, in their pursuit of what Steven Weinberg (2011) grandly called "a theory of everything." String theory isn't really a theory of everything, and in fact, technically it isn't even a theory, not if by that name one means a mature and empirically confirmed conceptual construction, such as the theory of evolution (Gould, 2002) or the theory of continental drift (Frankel, 2012). Rather, string theory is better described as a general framework – the most mathematically sophisticated one available at the moment – to resolve a fundamental problem in modern physics: the fact that general relativity and quantum mechanics, two of the most highly successful scientific theories of all time, give us sharply contrasting predictions when applied to certain problems, such as the physics of black holes (Frolov & Novikov, 2012), or a description of the singularity that gave origin to the universe at the moment of the Big Bang (Silk, 2000).

I take it that physicists agree that this means that either or both theories are therefore wrong or incomplete. String theory is one attempt at reconciling the two by subsuming them into a broader theoretical framework. There is only one problem: While some in the fundamental physics community confidently argue that string theory is not only a very promising scientific theory, but pretty much "the only game in town" (Greene, 2010), others scornfully respond that it isn't even good science (Smolin, 2007; Woit, 2007), since it doesn't make contact with the empirical evidence. Vibrating superstrings; multiple, folded, dimensions of space-time; and other features of the theory are (so far) impossible to test experimentally, and they are alleged by critics of the theory to be the mathematical equivalent of metaphysical speculation (which I take to be a really bad word in physics circles). Surprisingly, at least for me as a philosopher of science, the ongoing, increasingly public and acerbic diatribe often centers on the ideas of one Karl Popper. What, exactly, is going on?

5.3 The Munich Workshop: A Turning Point?

I had a front-row seat at one round of such, shall we say, frank discussions when I was invited to Munich to participate in a workshop on the status of fundamental physics, and particularly on what some refer to as "the string wars," an event that led to the production of this very book. The organizer, Richard Dawid, is a philosopher of science with a strong background in theoretical physics. He is also a proponent of a highly speculative, if innovative, type of epistemology that supports the efforts of string theorists and aims at shielding them from the accusation of engaging in flights of mathematical fancy decoupled from any real science. My role at the workshop (and in this volume) was to make sure that participants – an eclectic mix of scientists and philosophers, with a Nobel winner thrown in the mix – were clear on something I teach in my introductory course in philosophy of science: what exactly Popper said and why, given that some of those physicists had hurled accusations at each other, either loudly invoking falsification as the ultimate arbiter of the ongoing dispute or equally strongly advocating the ejection of the very idea of falsification from scientific practice.

In the months preceding the workshop, a number of high-profile players in the field had been using all sorts of means – from manifesto-type articles in the prestigious *Nature* journal (Ellis & Silk, 2014) to Twitter (Castelvecchi, 2015) – to pursue a no-holds-barred public relations campaign to wrestle, or retain, control of the soul of contemporary fundamental physics. Let me give you a taste of the exchange, to set the mood: "The fear is that it would become difficult to separate such 'science' from New Age thinking, or science fiction," said George Ellis, chastising the pro-strings party; to which Sabine Hossenfelder added: "Post-empirical science is an oxymoron" (referring to a term that Dawid actually does not use, preferring instead "non-empirical confirmation"). Peter Galison made crystal clear what the stakes are when he wrote: "This is a debate about the nature of physical knowledge." On the other side, cosmologist Sean Carroll tweeted: "My real problem with the falsifiability police is: we don't get to demand ahead of time what kind of theory correctly describes the world," adding "[Falsifiability is] just a simple motto that non-philosophically-trained scientists have latched onto." Finally (but there is more, much more, out there), Leonard Susskind mockingly introduced the neologism "Popperazzi" to label an extremely naive (in his view) way of thinking about how science works.

This surprisingly blunt – and very public – talk from prestigious academics is what happens when scientists help themselves to, or conversely categorically reject, philosophy of science, especially when they get their philosophy second-hand, so to speak. It is even more surprising because lately we have seen a string of high-level dismissals of both philosophy in general and philosophy of science in particular by a number of scientists, usually physicists. Richard Feynman may or may not have said "Philosophy of science is about as useful to scientists as ornithology is to birds," but Steven Weinberg (2011, p. 166) definitely did write that "The insights of philosophers have occasionally benefited physicists, but generally in a negative fashion – by protecting them from the preconceptions of other philosophers." And Lawrence Krauss (Andersen, 2012) is on record with this little gem: "Philosophy is a field that, unfortunately, reminds me of that old Woody Allen joke, 'those that can't do, teach, and those that can't teach, teach gym.' And the worst part of philosophy is the philosophy of science." Finally, though the list could easily be much longer, science popularizer Neil deGrasse Tyson said: "My concern here is that the philosophers believe they are actually asking deep questions about nature. And to the scientist it's, what are you doing? Why are you concerning yourself with the meaning of meaning?" (extended commentary and relevant background in Pigliucci, 2014).

To say the least, these anti-intellectual comments are rooted in a profound ignorance of the field these prominent scientists are so casually pontificating about. Then again, the situation just sketched reflects a rather novel, and by no means universal, attitude among some physicists. Compare the previously cited contemptuousness to what Einstein wrote to his friend Robert Thorton in 1944 on the same subject:

I fully agree with you about the significance and educational value of methodology as well as history and philosophy of science. So many people today – and even professional scientists – seem to me like somebody who has seen thousands of trees but has never seen a forest. A knowledge of the historic and philosophical background gives that kind of independence from prejudices of his generation from which most scientists are suffering. This independence created by philosophical insight is – in my opinion – the mark of distinction between a mere artisan or specialist and a real seeker after truth. (Einstein, 1944)

By Einstein's standard, then, there are a lot of artisans but comparatively few seekers of truth among contemporary physicists!

To put things in perspective, Einstein's opinion of philosophy may not have been representative even then. Certainly, modern string theorists are a small group within the physics community, and string theorists on Twitter are an ever smaller, possibly more voluble subset within that group. The philosophical noise they make is likely not representative of what physicists in general think and say, but it matters all the same precisely because they are so prominent: Those loud debates on social media and in the popular science outlets define how much of the public perceives physics, and even how many physicists perceive the big issues of their field.

That said, the publicly visible portion of the physics community nowadays seems split between people who are openly dismissive of philosophy and those who think they have got the pertinent philosophy right but their ideological opponents haven't. The controversy doesn't concern just the usually tiny academic pie, but public appreciation of and respect for both the humanities and the sciences, not to mention millions of dollars in research grants (for the physicists, not the philosophers). Time, therefore, to take a more serious look at the meaning of Popper's philosophy and why it is still very much relevant to science, when properly understood. Throughout the following discussion, keep in mind that anything that applies to the science/pseudoscience divide applies *mutatis mutandis* also to the distinction between science and non-science, as well as to the distinctions among different types of scientific approaches and disciplines.

5.4 What Popper Actually Said

As we have seen, Popper's message is deceptively simple, and – when repackaged in a tweet – has deceived many a smart commentator in underestimating the sophistication of the underlying philosophy. If one were to turn that philosophy into a bumper sticker slogan, it would read something like this: "If it ain't falsifiable, it ain't science, stop wasting your time and money."

But good philosophy doesn't lend itself to bumper sticker summaries, so one cannot stop there and pretend that there is nothing more to say. Moreover, Popper changed his mind throughout his career about a number of issues related to falsification and demarcation, as any thoughtful thinker would do when exposed to criticisms and counterexamples from his colleagues. For instance, Popper initially was skeptical of the role of verification in establishing scientific theories, thinking that it was far too easy to "verify" a notion if one were actively looking for confirmatory evidence. Sure enough, modern psychologists have a name for this tendency, common to laypeople as well as scientists: confirmation bias (Nickerson, 1998).

Nonetheless, Popper agreed that verification – especially of very daring and novel predictions – is part of a sound scientific approach. After all, the reason Einstein became a scientific celebrity overnight after the 1919 total eclipse is precisely because astronomers had verified the predictions of his theory all over the planet and found them in satisfactory agreement with the empirical data. For Popper, this did not mean that the theory of general relativity was "true," but only that it survived to fight another day. Indeed, nowadays we don't think the theory is (entirely) true, because of the previously mentioned conflicts, in certain domains, with quantum mechanics. Even so, it has withstood a very good number of sophisticated challenges over the intervening century, and its most recent confirmation came very recently, with the first detection of gravitational waves.

Popper also changed his mind about the potential, at the least, for a viable Marxist theory of history (and about the status of the Darwinian theory of evolution, concerning which he was initially skeptical, thinking – erroneously – that the idea was based on a tautology). He conceded that even the best scientific theories are often somewhat shielded from falsification because of their connection to ancillary hypotheses and background assumptions. When one tests Einstein's theory using telescopes and photographic plates directed at the Sun, one is really simultaneously putting to the test the focal theory, plus the theory of optics that goes into designing the telescopes, plus the assumptions behind the mathematical calculations needed to analyze the data, plus a lot of other things that scientists simply take for granted and assume to be true in the background, while their attention is trained on the main theory. But if something goes wrong and there is a mismatch between the theory of interest and the pertinent observations, this isn't enough to immediately rule out the theory, since a failure in one of the ancillary assumptions might be to blame instead. That is why scientific hypotheses need to be tested repeatedly and under a variety of conditions before we can be reasonably confident of the results.

In fact, and contra frequent modern misconceptions, especially on the part of scientists who may be only superficially familiar with Popper's writings, the idea of falsificationism is not averse to the possibility of rescuing a given hypothesis by way of introducing ad hoc hypotheses (germane to the case of the 1919 eclipse and its relevance to Einstein's theory of general relativity). Gruünbaum (1976) provides a detailed analysis of this aspect of Popper's account. That paper directly quotes Popper as elaborating on the relationship between falsification and ad-hocness in this fashion:

The problem [is] whether or not the introduction of a new hypothesis is to be regarded as a potential advance. ... The goodness of an auxiliary hypothesis would depend on whether it is independently testable, or perhaps merely rescuing a refuted theory. ... [In the case of the Lorentz–Fitzgerald contraction hypothesis] we have an excellent example of degrees of *ad-hocness* and one of the main theses of my book [*The Logic of Scientific Discovery*, 1959] – that degrees of *ad-hocness* are related (inversely) to degrees of testability and significance.

This makes it exceedingly clear that Popper himself did not think that ad-hocness was out of the question in scientific practice: It all depends on how it is deployed. So any simplistic invocation of falsificationism by a physicist to reject a theory just because its defenders are invoking what currently appear to be ad hoc hypotheses is not warranted, and it is certainly not grounded in anything Popper actually wrote.

Popper's initial work pretty much single-handedly put the demarcation problem on the map, prompting philosophers to work on the development of a philosophically sound account of what science is and is not. That lasted until 1983, when Larry Laudan (1983) published a highly influential paper entitled "The Demise of the Demarcation Problem," in which he argued that demarcation projects were actually a waste of time for philosophers, since – among other reasons – it is unlikely to the highest degree that anyone will ever be able to come up with small sets of necessary and jointly sufficient conditions to define "science," "pseudoscience," and the like. And without such sets, Laudan argued, the quest for any principled distinction between those activities is hopelessly quixotic.

"Necessary and jointly sufficient" is logical–philosophical jargon, but it is important to see what Laudan meant. He thought that Popper and others had been trying to provide precise definitions of science and pseudoscience, similar to the definitions used in elementary geometry: A triangle, for instance, is whatever geometrical figure has the internal sum of its angles equal to 180 degrees (in Euclidean space). Having that property is both necessary (because without it, the figure in question is not a triangle) and sufficient (because that's all we need to know to confirm that we are, indeed, dealing with a triangle). Laudan argued – correctly – that no such solution of that type will ever be found to the demarcation problem, simply because concepts such as "science" and "pseudoscience" are complex, multidimensional, and inherently fuzzy, not admitting of sharp boundaries. In a sense, physicists complaining about "the Popperazzi" are making the same charge as Laudan: Popper's criterion of falsification appears to be far too blunt an instrument not only to discriminate between science and pseudoscience (which ought to be relatively easy), but a fortiori to separate sound from unsound science within an advanced field like theoretical physics.

Yet Popper wasn't quite as naive as Laudan, Carroll, Susskind, and others make him out to be. Nor is the demarcation problem quite as hopeless as all that. That is why a number of authors – including myself and my longtime collaborator, Maarten Boudry – have more recently maintained that Laudan was too quick to dismiss the issue, and that perhaps Twitter isn't the best place for nuanced discussions in the philosophy of science (Pigliucci & Boudry, 2013).

The idea is that there are pathways forward in the study of demarcation that become available if one abandons the requirement for necessary and jointly sufficient conditions, which was never strictly enforced even by Popper. What, then, is the alternative? To treat science, pseudoscience, and their siblings as Wittgensteinian "family resemblance" concepts instead. Ludwig Wittgenstein was another highly influential twentieth-century philosopher, who hailed, like Popper, from Vienna, though the two could not have been more different in terms of socioeconomic background, temperament, and philosophical interests (Edmonds & Edinow, 2001).

Wittgenstein never wrote about philosophy of science, let alone fundamental physics (or even Marxist theories of history). But he was very much interested in language, its logic, and its uses. He pointed out that there are many concepts that we seem to be able to use effectively, yet are not amenable to the sort of clear definition that Laudan was seeking. His favorite example was the deceptively simple concept of "game." If you try to arrive at a definition of games of the kind that works for triangles, your effort will be endlessly frustrated (try it out – it makes for a nice parlor, ahem, game). Wittgenstein (1953) wrote:

How should we explain to someone what a game is? I imagine that we should describe games to him, and we might add: "This and similar things are called games." And do we know any more about it ourselves? Is it only other people whom we cannot tell exactly what a game is? ... But this is not ignorance. We do not know the boundaries because none have been drawn. ...We can draw a boundary for a special purpose. Does it take that to make the concept usable? Not at all!

The point is that in a lot of cases we don't discover preexisting boundaries, as if games and scientific disciplines were Platonic ideal forms that existed in a timeless metaphysical dimension. Instead, we make up boundaries for specific purposes and then we test whether the boundaries are actually useful for whatever purposes we drew them. In the case of the distinction between science and pseudoscience, we think there are important differences with practical consequences, so we try to draw tentative borders to highlight them. Surely one would give up too much, as either a scientist or a philosopher, if one were to reject the strongly intuitive idea that there is something fundamentally different between, say, astrology and astronomy. The question is where, approximately, the difference lies.

Similarly, many of the participants in the Munich workshop, and the "string wars" more generally, did feel that there is an important distinction between fundamental physics as it is commonly conceived and what string theorists are proposing. As we have seen, Richard Dawid (2013; see also: Schindler, 2014) objects to the (admittedly easily derisible) term "post-empirical science," preferring instead "non-empirical theory assessment." But whatever one calls it, Dawid is aware that he and his fellow travelers are proposing a major departure from the way we have done science since the time of Galileo. True, the Italian physicist himself largely engaged in theoretical arguments and thought experiments (he likely never did drop balls from the leaning tower of Pisa, for instance), but his ideas were certainly falsifiable and have been, over and over, subjected to experimental test (most spectacularly, if a bit theatrically, by astronaut David Scott on the Moon, during the Apollo 15 mission).

5.5 The Broader Perspective: A View from a Distance and One from the Side

From the point of view of a professional philosopher of science (a view from a distance, as it were), there is something else that is even more odd about the diatribe surrounding string theory, and specifically the invocation or dismissal of Karl Popper. Keep in mind that Popper died in 1994, and that the bulk of his work was done between the 1930s and the 1960s. As important as that work was for the development of philosophy of science, it's not like the field suddenly stopped at the publication of *Conjectures and Refutations* in 1962. On the contrary, philosophy of science was just beginning its golden age, with vibrant contributions by Popper's brilliant student, Imre Lakatos (1978), with the so-called historicist turn inaugurated by Thomas Kuhn (1962) and his idea of paradigm shifts, not to mention the radical methodological anarchism of Paul Feyerabend (1975). And things didn't stop there, either. Later work by – to name just a very few – Bas van Fraassen (1980) on anti-realist models of scientific understanding, Ian Hacking (1983) and Nancy Cartwright (1983) on the nature of physical laws, James Ladyman and Don Ross (2007) on the relationship between metaphysics and science, and the development of the whole Bayesian approach to scientific reasoning (Howson & Urbach, 1999), on which Dawid bases his proposal of non-empirical theory assessment, are all simply not on the radar of any physicist I've ever read or talked to. Nor is it reasonable to expect physicists, already busy with their own theoretical and empirical work, to be conversant in the advancements of philosophy of science over the past half-century. But if an academic invokes (or dismisses) contributions from another academic field, then she has a moral duty to either become sufficiently familiar with the technical literature of that field or invite scholars from that field to join the conversation. So far, the Munich workshop has been a lonely, partial exception to the general trend sketched here. We can only hope that it set a precedent that will soon be emulated.

Yet another angle from which the present discussion may usefully be approached is to remind ourselves that the current status of fundamental physics is not at all a unique case in the history of science, or even among contemporary sciences. What we might call a view from the side comes from the comparison between the string wars in physics and the equally heated debate surrounding the so-called extended evolutionary synthesis in biology (EES; see Pigliucci & Müller, 2010). Indeed, *Nature*, apparently increasingly inclined to embrace public scientific controversy, has published a pro-versus-con analysis of the issue (Laland et al., 2014), which of course has failed to settle anything at all. There is no point, within our context here, to enter into the details of the controversy, but to give the reader a sense of what is going on, keep in mind that biology has its own version of the Standard Model, referred to as the Modern Synthesis (MS). It is so called because it reconciled the original Darwinian ideas of common descent among all organisms on earth and natural selection as the process that produces adaptation, with the newly discovered (at the beginning of the twentieth century) phenomenon of Mendelian inheritance. That reconciliation gave rise to population genetic theory, the mathematical version of the theory of evolution, still taught today in biology textbooks.

The MS took shape over a period of several decades, from the 1920s through the 1950s, thanks to the pioneering works of Fisher (1930) and Sewall Wright (1942) in theoretical population genetics, Dobzhansky (1937) in experimental population genetics, Mayr (1942) in systematics and speciation, Simpson (1944) in paleontology, Stebbins (1950) in botany, and Huxley (1942) at the level of overarching theory. Following the consolidation of the theory, however, many new empirical discoveries (e.g., the explosion of molecular biology and genomics) as well as conceptual advances (concerning phenomena such as evolvability, phenotypic plasticity, developmental constraints and facilitation, niche construction, epigenetic inheritance, and multi-level selection, among others) have forced a debate about whether the MS should be further updated and expanded into the EES.

The reason the comparison between the MS/EES and the pro–against strings debates is apt has, obviously, nothing to do with the specific scientific content of the two disputes. Rather, the similarities are to be found along three lines:

1. From the outside, it is extremely difficult to arrive at a reasonably informed opinion, since many scientists, including a number of luminaries, can be found on either side of each debate.
2. Both sides have been waging a public relations war not only in the pages of *Nature* and similar journals, but also by way of decidedly nontraditional venues, such as books aimed at the broader public and on social media.
3. In both cases the stakes include the credibility of the field with the general public – in the case of evolutionary biology in terms of the notoriously already difficult acceptance of even basic evolutionary principles by large swaths of especially American public, and in the case of physics in terms of notions that have nothing directly to do with the specific scientific discussion, such as the credibility of the idea of anthropogenic climate change.

It is because of these new methods and broader impact that the scientists involved in these controversies should tread more carefully, mindful of the image of their own discipline, and of science in general, that they are implicitly projecting to the lay public, which, I remind everyone, is actually largely paying for the costs of our beloved inquiries.

Returning to fundamental physics in particular, the broader question appears to be: Are we on the verge of developing a whole new science, or is this going to be regarded by future historians as a dead end, temporarily stalling scientific progress? Alternatively, is it possible that fundamental physics is reaching an end not because we've figured out everything we wanted to figure out, but because we have come to the limits of what our brains and technologies can possibly do? These serious questions ought to be of interest not just to scientists and philosophers, but to the public at large. The high table of discussion should therefore feature an interdisciplinary range of scholars interested in the very nature of science and how it changes over time, including both historians and sociologists.

Meanwhile, what is weird about the string wars and the concomitant use and misuse of philosophy of science is that both scientists and philosophers have bigger targets to address together for the sake of society, if only they would stop squabbling and focus on what their joint intellectual forces may accomplish. Rather than laying into each other in the crude terms sketched in this chapter, they should work together not just to forge a better science, but to counter true pseudoscience

(Pigliucci & Boudry, 2013). Homeopaths and psychics, to mention a couple of obvious examples, keep making tons of money by fooling people, while damaging their physical and mental health. These are worthy targets of critical analysis and discourse, and it is the moral responsibility of public intellectuals or academics – be they scientist or philosopher – to do their best to improve as much as possible the very same society that affords them the luxury of discussing esoteric points of epistemology or fundamental physics.

References

Andersen, R. (2012, April 23). Has physics made philosophy and religion obsolete? *The Atlantic*, http://tinyurl.com/6too5dl (accessed on 13 December 2016).

Ball, P. (2007, September 7). Arthur Eddington was innocent! *Nature*, doi:10.1038/news070903-20 (accessed on 2 December 2016).

Becker, K., Becker, M., & Schwarz, J. H. (2006). *String Theory and M-Theory: A Modern Introduction*. Cambridge: Cambridge University Press.

Cartwright, N. (1983). *How the Laws of Physics Lie*. Oxford: Oxford University Press.

Castelvecchi, D. (2015) Feuding physicists turn to philosophy. *Nature* 528:446–7.

Collins, H. M., & Pinch, T. (2012). *The Golem: What You Should Know about Science*. Cambridge: Cambridge University Press.

Dawid, R. (2013) *String Theory and the Scientific Method*. Cambridge: Cambridge University Press.

Dobzhansky, T. (1937). *Genetics and the Origin of Species*. New York: Columbia University Press.

Edmonds, D., & Edinow, J. (2001). *Wittgenstein's Poker: The Story of a Ten-Minute Argument Between Two Great Philosophers*. New York: Ecco.

Einstein, A. (1944). *The Collected Papers of Albert Einstein* Einstein to Thornton, 7 December 1944, 61-574. Princeton, NJ: Princeton University Press.

Ellis, G., & Silk, J. (2014). Defend the integrity of physics. *Nature* 516:321–3.

Feyerabend, P. (1975). *Against Method*. New York: Verso Books.

Fisher, R. (1930). *The Genetical Theory of Natural Selection*. Oxford: Clarendon Press.

Frankel, H. R. (2012). *The Continental Drift Controversy: Wegener and the Early Debate*. Cambridge: Cambridge University Press.

Frolov, V., & Novikov, I. (2012). *Black Hole Physics: Basic Concepts and New Developments*. New York: Springer.

Gould, S. J. (2002). *The Structure of Evolutionary Theory*. Cambridge, MA: Harvard University Press.

Greene, B. (2010). *The Elegant Universe: Superstrings, Hidden Dimensions, and the Quest for the Ultimate Theory*. New York: W. W. Norton & Co.

Gruünbaum, A. (1976). Ad hoc auxiliary hypotheses and falsificationism. *British Journal for the Philosophy of Science* 27:329–62.

Hacking, I. (1983). *Representing and Intervening: Introductory Topics in the Philosophy of Natural Science*. Cambridge: Cambridge University Press.

Howson, C., & Urbach, P. (1999). *Scientific Reasoning: The Bayesian Approach*. Chicago: Open Court.

Huxley, J. (1942). *Evolution: The Modern Synthesis*. New York: Harper and Brothers.

Kuhn, T. S. (1962). *The Structure of Scientific Revolutions*. Chicago: University of Chicago Press.

Ladyman, J., & Ross, D. (2007). *Every Thing Must Go: Metaphysics Naturalized*. Oxford: Clarendon Press.

Lakatos, I. (1978). *The Methodology of Scientific Research Programmes: Philosophical Papers Volume 1*. Cambridge: Cambridge University Press.

Laland, K., et al. (2014). Does evolutionary theory need a rethink? *Nature* 514:161–4.

Laudan, L. (1983). The demise of the demarcation problem. *Boston Studies in the Philosophy of Science* 76:111–27.

Mayr, E. (1942). *Systematics and the Origin of Species, from the Viewpoint of a Zoologist*. Cambridge, MA: Harvard University Press.

Nickerson, R. S. (1998). Confirmation bias: A ubiquitous phenomenon in many guises. *Review of General Psychology* 2(2):175–220.

Pigliucci, M. (2014, May 12). Neil deGrasse Tyson and the value of philosophy. *Scientia Salon*, http://tinyurl.com/l6la5oj (accessed on 13 December 2016).

Pigliucci, M., & Boudry, M. (eds.). (2013). *Philosophy of Pseudoscience: Reconsidering the Demarcation Problem*. Chicago: University of Chicago Press.

Pigliucci, M., & Müller, G. (eds.). (2010). *Evolution, the Extended Synthesis*. Cambridge, MA: MIT Press.

Popper, K. (1962). *Conjectures and Refutations*. New York: Basic Books.

Schindler, S. (2014). A theory of everything. *Philosophy of Science* 83(3):453–8.

Silk, J. (2000). *The Big Bang*. New York: Macmillan.

Simpson, G. G. (1944). *Tempo and Mode in Evolution*. New York: Columbia University Press.

Smolin, L. (2007). *The Trouble with Physics: The Rise of String Theory, the Fall of a Science, and What Comes Next*. Boston: Houghton Mifflin Harcourt.

Stebbins, G. L. (1950). *Variation and Evolution in Plants*. New York: Columbia University Press.

van Fraassen, B. (1980). *The Scientific Image*. Oxford: Oxford University Press.

Weinberg, S. (2011). *Dreams of a Final Theory: The Scientist's Search for the Ultimate Laws of Nature*. New York: Knopf Doubleday.

Wittgenstein, L. (1953). *Philosophical Investigations*. Oxford: Blackwell.

Woit, P. (2007). *Not Even Wrong: The Failure of String Theory and the Search for Unity in Physical Law*. New York: Basic Books.

Wright, S. (1942). Statistical genetics and evolution. *Bulletin of the American Mathematical Society* 48:223–46.

Part II

Theory Assessment beyond Empirical Confirmation

6

The Significance of Non-Empirical Confirmation in Fundamental Physics

RICHARD DAWID

6.1 Introduction

Fundamental physics today faces the problem that empirical testing of its core hypotheses is very difficult to achieve and even more difficult to be made conclusive. This general situation is likely to persist for a long time. As a consequence, it may substantially change the perspective on theory assessment in the field.

During most of the twentieth century, fundamental physics was perceived as a scientific field in which theories typically could be empirically tested within a reasonable time frame. In high energy physics, collider experiments provided a steady and focused stream of empirical data that could be deployed for testing physical hypotheses. The process of confirming predictions of the standard model of particle physics, to give one important example, started very soon after it had been formulated in the late 1960s and early 1970s with the discovery of charm quarks in 1974; in the following years, it led to a wide range of empirical discoveries. In cosmology, the empirically confirmed principles of general relativity were clearly distinguished from the speculations of cosmological model building that generated a multitude of competing ideas and models without seriously insisting on the correctness of any of them at the given point.

In this general scientific climate, it was plausible to focus on empirical confirmation as the only reliable basis for assessing a theory's viability. It seemed to make little sense to pursue a detailed analysis of the degree of trust one was allowed to have in a theory on the basis of non-empirical evidence when conclusive empirical evidence that would decide the fate of the theory could normally be expected to be just a few years ahead. And even if some specific cases might have rendered such an analysis interesting, the general character and status of fundamental physics seemed well understood without it.

Today, the situation is very different. String theory has been playing the role of a well-established approach toward a universal theory of all interactions for more than three decades and is trusted to a high degree by many of its exponents in the absence of either empirical confirmation or even a full understanding of what the theory amounts to. Cosmic inflation is trusted by many theoreticians to a degree that in the eyes of many others goes substantially beyond what is merited by the supporting empirical data. Multiverse scenarios in the eyes of critics raise the question to what degree they can be endorsed as scientific hypotheses at all, given that their core empirical implications to a large extent seem not empirically testable in principle. What is at stake here is the understanding physicists have of the status of the theory they work on throughout their lifetimes. In the most far-reaching cases, the stakes are the status a given theory can acquire at all. The question as to how much credit can or should be given to non-empirical theory assessment therefore has turned from a fringe topic in physics into a question at the core of the field's self-definition.

In each individual case, the described question has to be answered based on a detailed physical analysis of the merits and problems of the theory under scrutiny. There is a sense in which the general issue of what can be accepted as a solid basis for theory assessment is nothing more than the sum of the individual physical assessments of theories and their specific claims. However, physicists' views of the scientific contexts they encounter solidify into a generalized background understanding of how theory assessment should work in a scientific way. This "philosophical background" then exerts an implicit influence on the individual physicist's theory evaluations. In periods when significant changes in the overall research context induce shifts with respect to the described philosophical background, the scientist's instrumentarium of theory assessment becomes less stable and more controversial. There are clear indications that fundamental physics today finds itself in such a situation. By elucidating general characteristics of the ongoing shifts and making them explicit, philosophical analysis can contribute to the process of developing an altered general understanding of theory assessment that is adequate to fundamental physics under the new circumstances.

In Dawid (2013), it was argued that a considerable degree of trust in an empirically unconfirmed theory could be generated based on 'non-empirical theory confirmation'. Non-empirical confirmation, like empirical confirmation, is based on observational evidence. This evidence, however, does not lie within the confirmed theory's intended domain. That is, it is not of the kind that can be predicted by the theory it confirms.[1]

The difference between empirical and non-empirical confirmation may be illustrated by looking at a specific example. If string theory ended up making specific

---

[1] I use the term "non-empirical" confirmation to emphasise the important distinction between testing a theory *empirically* by confronting its predictions with empirical data and understanding its chance of being viable based on observations that are entirely unrelated to the theory's predictions.

quantitative predictions, data in agreement with those predictions would lie within the theory's intended domain and therefore amount to empirical confirmation. In the absence of empirical confirmation, exponents of the theory may rely on different kinds of reasoning. For example, they may argue that the theory is supported by the striking difficulties encountered in coming up with promising alternatives. Those difficulties clearly cannot be predicted by string theory itself. The observation that those difficulties exist is a contingent observation about the research process of which the development of string theory is just one part. Therefore, this observation does not constitute evidence within string theory's intended domain. If one concludes, as I will, that the observation amounts to confirmation of string theory nevertheless, it can only be non-empirical confirmation. Much of this chapter is devoted to making the concept of non-empirical confirmation more precise than this rough sketch.

Dawid (2013) spells out three specific *non-empirical* lines of reasoning that play an important role in generating trust in string theory. It is argued that those strategies, while playing a particularly strong role in today's fundamental physics for a number of reasons, have always constituted an important element of physical reasoning whose significance had been neglected or underrated by philosophers of science as well as by many scientists.

The present chapter looks at the issue of non-empirical confirmation from a slightly different angle. Rather than discussing the conceptual details of individual arguments, it develops, step by step, the conceptual framework for taking non-empirical confirmation seriously at all. Sections 6.2–6.4 spell out what we can plausibly hope for when assessing non-empirical arguments in favour of a scientific theory. Section 6.5 then develops guidelines for identifying arguments of non-empirical confirmation that can fulfil those expectations. Section 6.6 demonstrates that the three arguments of non-empirical confirmation presented in Dawid (2013) are promising for the very reason that they do satisfy the conditions developed in Section 6.5. This fact, I argue, offers a plausible reason why the three presented arguments may be taken to be more powerful than other reasons for having trust in a theory that might come to mind. The three arguments themselves are only briefly rehearsed in Section 6.6. The reader interested in a more detailed presentation and discussion of those arguments may look into Dawid (2006); Dawid (2013); and Dawid, Hartmann, and Sprenger (2015).

6.2 Strategic or Epistemic?

It is uncontroversial that physicists assess theories even prior to empirical testing. But what kinds of questions do they ask when making those assessments? In the previous section, it was claimed that many exponents of string theory, cosmic inflation, and multiverse scenarios have generated a degree of trust in their theories that cannot be explained entirely – or in some cases not at all – by empirical confirmation of those theories. It was implicitly taken for granted that scientists aim at assessing the degree of trust they should have in a theory's truth or at least in its *viability* – that is, its capacity to adequately represent the empirical data that will be collected in the given research context in the future.

One might take one step back, however, and consider a more restrained, possibly less contestable point of view. In that view, the crucial question for the working scientist is simply whether it makes sense to work on a given theory. Endorsing a theory in this light might be perceived as the result of fairly pragmatic considerations that do not address the more ambitious question of whether a theory is likely to be viable.

While I concur that questions of research strategy are a main motivating force behind theory assessment, I do not think that what is at stake in non-empirical theory assessment can be reduced to the pragmatic issue of deciding upon research strategies.

Clearly, a scientist can have a number of rational reasons for working on a given scientific theory that are entirely unrelated to the question of the theory's viability. To begin with, personal strategic reasons may come into play. A young physicist might decide to work on the most popular theory because this seems most promising with respect to job perspectives; or she might decide to work on a nascent, less developed theory because that seems most promising with respect to assuming a leading role in a research field. Reasons can also appeal to what is strategically in the best interest of the scientific discipline. It may be most effective to first work out a theory that has already been proposed before starting a tedious search for possible alternatives. Among a spectrum of known approaches to solving a problem, it may be reasonable to first work out the approach that is easiest to develop. Even though none of these considerations addresses the question whether a given theory is likely to be viable, they can all make sense and do play a role in physics.

However, their relevance for deciding on the optimal research strategy in a scientific field is limited in two ways. First, one often finds lines of reasoning that pull in different directions and whose balance changes over time. For example, the idea that it is most effective to work on a theory that has already been developed may be countered by the idea that it is sometimes more productive to start from scratch rather than to be bound by worn-out ways of thinking. Second, and perhaps even more importantly, entirely strategic considerations of theory preference look plausible only if based on the assumption that all alternative hypotheses are equally promising. Any deviation from that assumption draws into question the implications of considerations that disregard epistemic issues.

The strongest reasons for working on a theory in this light are those that have an epistemic foundation suggesting that the theory in question is likely to be viable. To the extent that epistemic arguments can be developed, it is of high importance for the physicist to take them into account. The physicist herself may well treat those arguments pragmatically as a way of understanding whether there are good reasons to work on a given theory. From an operative professional perspective, framing what is at issue in that way is perfectly adequate. In the end, however, a strong commitment to working on a particular theory hinges on the question whether there exists a good reason for having trust in the theory's viability.

I want to point out a second reason for emphasising the epistemic element in non-empirical theory assessment. Reducing what is at stake in non-empirical theory assessment to the question of justifying work on a theory seems at variance with the main motivation for doing fundamental physics. Fundamental physics today clearly is not driven by perspectives of technological utilization, but rather by the quest for acquiring knowledge about the world. In fields aiming for technological utilization, it is plausible that non-empirical theory assessment is primarily motivated by the strategic question as to whether one should focus on a given research program. When medical researchers look for a cure for a disease, assessments of a specific approach may be motivated entirely by the need to decide whether one should work on that approach or on a different one. All that counts is the final result of having a cure that is sufficiently well tested to use. Assessing the truth of a hypothesis may be viewed as having little relevance as long as it does not open up a perspective of utilization in the foreseeable future.

It is important to distinguish that situation from the situation in which fundamental physics finds itself today. In a research field that is motivated primarily by the quest for knowledge, understanding the probability of a theory's viability is not just a pragmatic step on the way toward conclusive confirmation: It is also an epistemic goal in its own right. Knowing that there is a high probability that dinosaurs became extinct due to a comet impact constitutes a valuable element of my knowledge about the world even though that hypothesis has not been conclusively confirmed. It would remain valuable even if there was no hope of conclusive confirmation in the foreseeable future. Similarly, to the extent there are reasons to assume that string theory or the multiverse has a high chance of being viable, that constitutes an important element of knowledge about the universe even in the absence of conclusive empirical confirmation of those theories.

It may still make sense to ignore intermediate epistemic states between ignorance and conclusive knowledge in contexts where they last only for a brief period of time before the case is settled based on conclusive empirical evidence. In contemporary fundamental physics, the typical time scale for that intermediate state has grown beyond the length of a scientific career. In such a situation, ignoring the epistemic difference between knowing nothing about a unified description of nuclear interactions and gravity and, to the extent that can be established, knowing that string theory is likely to represent such a description amounts to a substantial misrepresentation of present day knowledge about fundamental physics.

6.3 Confirmation

6.3.1 Bayesian Confirmation Is Not Conclusive Confirmation

In light of the previous arguments, the crucial question is: Can there be strong reasons for believing that a theory is probably viable even if empirical confirmation of that theory has not emerged? The natural framework for discussing this question in the philosophy of of science is provided by Bayesian confirmation theory. Bayesian confirmation theory expresses confirmation in terms of the theory's probability, which is exactly the kind of perspective that allows for a specific analysis of the question stated previously. According to Bayesian confirmation theory, evidence E confirms a hypothesis H iff the posterior probability of H given evidence E is higher than the prior probability of H:

$$P(H|E) > P(H) \tag{6.1}$$

In Bayesian wording, we want to understand whether and to what degree non-empirical theory assessment amounts to theory confirmation: We are looking for *non-empirical theory confirmation.*

Ellis and Silk (2014) and Rovelli (this volume, Chapter 7) have asserted that the term *non-empirical confirmation* is misleading in suggesting that the given theory has been established as viable and is in no need of further empirical testing. It is important to emphasise that the Bayesian definition of confirmation does not imply anything of that kind. It is true, however, that the communication between theoretical physics and the philosophy of science on this issue is made difficult by an unfortunate mismatch between the ways the term "confirmation" is used in the two fields.

In statistics and in parts of experimental physics that engage in statistical data analysis, the use of the term "confirmation" is confined to Bayesian data analysis, where it denotes the increase of probability just described. In contemporary philosophy of science, confirmation is to a large extent understood in terms of Bayesian confirmation theory as well. While some important voices in the philosophy of science (see, for example, Achinstein, 2001) disagree with a Bayesian understanding of confirmation, the Bayesian approach is prevalent enough in the field that the term "confirmation" in the title of a contemporary paper in the philosophy of science without further qualification amounts to the announcement that the paper is written within the framework of Bayesian epistemology. Relying on a Bayesian definition of confirmation when discussing the issue thus provides the most effective framework for conveying to a philosophical audience what is being discussed.

In theoretical physics, however, the term "confirmation" is often used in accordance with its non-technical application, in the sense of what one might call "conclusive confirmation". The theory has been established to be viable in a given regime beyond reasonable doubt. Due to those two conflicting uses of the term "confirmation", the term "non-empirical confirmation" can indeed cause misunderstandings for persons who are not familiar with the Bayesian use of the term.

Let me state unequivocally: Non-empirical confirmation does not mean conclusive confirmation. Readers who for terminological reasons feel uncomfortable with the term "non-empirical confirmation" might, before their mental eye, replace it with the expression "non-empirical theory assessment" – albeit with the asterisk that "assessment" is to be understood in terms of attributing a probability of being viable to the theory in question.

6.3.2 Viability Rather than Truth

This leads to a second important specification of the way confirmation is used in the present context. The previous paragraph defined confirmation in terms of a theory's viability. The canonical formulation of Bayesian confirmation, however, is based on a theory's probability of truth. The fact that I deviate from this canonical approach is closely related to the role confirmation is going to play in this analysis and deserves closer attention.

Truth is a complicated concept. Whether a mature and empirically well-confirmed theory may be taken to be true or approximately true has been hotly debated in the philosophy of science for a long time. Things are particularly difficult in theoretical physics, where some theories are predictively highly successful even though they are known not to be true for conceptual reasons. Quantum field theories are predictively highly successful based on the first orders in a perturbative expansion that is strongly suspected not to converge and therefore difficult to appraise in terms of truth values. The standard model of particle physics is highly successful but known to constitute merely an effective theory to whatever more fundamental theory can account for the inclusion of gravity at the Planck scale.

Many Bayesian epistemologists are not troubled by the notorious difficulties related to calling scientific theories true for one reason: Bayesian confirmation theory relies on a differential concept of confirmation. Evidence confirms a hypothesis if it increases its truth probability. The verdict that it does so in a given scenario is invariant under the choice of priors for the hypothesis as long as one excludes the dogmatic priors zero and one. All doubts about the the truth of scientific theories can be relegated to specifying the priors, however. Therefore, even if one attributes a truth probability very close to zero to a scientific theory for philosophical reasons, one can still talk about the formal increase in a theory's truth probability based on new evidence.

While this line of reasoning works sufficiently well in contexts of empirical testing, it seems less than satisfactory in the context of investigating the role of non-empirical confirmation in fundamental physics. First, as noted earlier, the understanding that some predictively very successful theories in fundamental physics are, strictly speaking, not true is related to the understanding that they are inconsistent. If so, it seems adequate to attribute probability zero to their truth, which would bar any meaningful updating of truth probabilities.

Second, and more directly related to the agenda of this chapter, defining Bayesian confirmation in terms of truth probabilities has an important effect on the ways confirmation can be discussed. The preceding analysis stated the main reason why, according to the understanding of most Bayesian epistemologists, not even a rough consensus on absolute values for truth probabilities can be established empirically: Different philosophical views may suggest very different truth probabilities even after a theory is empirically very well confirmed.

This implies that a Bayesian analysis of confirmation in terms of truth probabilities will be convincing only to the extent it avoids any reference to absolute probabilities. Doing so can work with respect to empirical confirmation if one is willing to decouple the issue of the significance or conclusiveness of the evidence from the issue of the absolute probabilities of the theory. Whether a scientific claim can be called conclusively tested is then taken to be decided by the involved scientists based on specifying significance criteria for p-values within the framework of frequentist data analysis.[2]

Non-empirical theory confirmation, however, is not based on a solid setup of well-specified rounds of testing aimed at surpassing strict significance limits. In turn, the issue of the significance or conclusiveness of non-empirical evidence must be decided by assessing absolute values of probabilities.[3] In such a case, it is important to specify what is denoted by the probabilities in a way that is philosophically uncontroversial and can be linked to empirical data. Truth, for the reasons spelled out previously, fails to meet those conditions.

I therefore propose to understand confirmation as an observation-based increase of the probability that the theory is viable in a given regime. Viability here is defined as the agreement of the theory's predictions with all empirical data that can be

---

[2] Strictly speaking, this is philosophically satisfactory only once one has spelled out the connection between p-values and truth probabilities. I won't address that issue here, however.

[3] As always in Bayesian epistemology, we do not necessarily aim at extracting actual numbers. However, the aim must be to demonstrate that fairly high absolute probabilities can be made plausible in the process.

possibly collected within a given regime of empirical testing. Regimes of empirical testing are specified based on well-established background knowledge about a given research context. An example of specifying a regime of testing in the context of high energy physics would be specifying a certain energy scale up to which a theory is tested. The theory is viable within that regime if it correctly accounts for all possible empirical testing up to the specified energy scale.[4] The probability of a theory's viability, therefore, can be specified only with respect to a certain regime of testing.

6.4 Significant Non-Empirical Confirmation?

6.4.1 There Is No Plausible Claim of Conclusive Non-Empirical Confirmation

How strong should claims of non-empirical confirmation, understood in the sense spelled out previously, be taken to be? In Section 6.3.1, I emphasised that the Bayesian concept of confirmation does not *imply* conclusive confirmation. But could non-empirical evidence get so strong that we have to accept it as conclusive?

The conceptual status of non-empirical confirmation is somewhat reminiscent of the conceptual status of third-person testimony. There is an irreducible difference between observing something with one's own eyes and learning about it from a third person. The best way to check a report about an event is to directly observe an event of the same type. Still, a sufficiently dense web of reliable testimony for a given fact can provide a basis for taking that fact for granted. (Science crucially relies on this is process). Whether a sufficiently reliable web of testimony can be attained is a contingent fact about the world or, more specifically, a given context of inquiry.

In a similar way, there exists an irreducible difference between empirical and non-empirical confirmation. The best way to check whether non-empirical confirmation works is to control it based on empirical testing. Whether one is willing to accept non-empirical evidence as conclusive depends on the reliability of this kind of confirmation in the past. Let us assume, for a moment, that we live in a world where a given strategy of non-empirical confirmation is omnipresent and has a 100% success rate over many generations. In such a world, it would be just as irrational not to rely on non-empirical evidence of that kind as it is in our world to refuse to step on bridges because of the Humean problem of induction. Understanding the strength as well as the limitations of non-empirical confirmation is itself a matter of observing the world.

---

[4] To account for all possible empirical testing, a theory must not make predictions that disagree with empirical data. Not all empirical data need to be predicted by the theory, however. On the presented definition, a theory can be empirically viable even if some or many aspects of the phenomena represented by the theory remain unexplained.

The strength that non-empirical confirmation can achieve in principle needs to be clearly distinguished from the strength that one attributes to non-empirical confirmation in actual physics. No case of conclusive non-empirical confirmationhas ever occurred in any field of science up to this day. Nor can it be expected, given our present understanding of the scientific process, that non-empirical confirmation will lead to the conclusive confirmation of a hypothesis in physics in the future. A number of factors that will emerge later in the discussion reduce the reliability of non-empirical confirmation in a way that keeps it far from being conclusive even under optimal circumstances. Strong empirical testing therefore must be expected to remain the only path toward conclusive confirmation of a theory in fundamental physics.

6.4.2 But One Needs to Establish Significant Non-Empirical Confirmation

Nothing in this chapter aims to suggest that non-empirical confirmation can be conclusive in our world. Still, it is important to claim something stronger than the mere point that non-empirical evidence amounts to confirmation in a Bayesian sense at all. Though by no means trivial, the latter claim would be too weak to be interesting. For the Bayesian, even a minimal increase in probability amounts to confirmation. Therefore, the mere fact that non-empirical evidence can confirm a theory does not establish that it can play a relevant role in scientific reasoning. To support the latter claim, one needs to make plausible that non-empirical confirmation is not just marginal but significant in the sense that it can lead to substantial probabilities for a theory's viability even when starting with low initial expectations.[5] It has already been pointed out that the need to look at absolute probabilities to establish significant confirmation is one crucial reason for defining confirmation in terms of a theory's viability rather than its truth.

The distinction between marginal confirmation and significant confirmation will play an important role in our analysis once the specific strategies of non-empirical confirmation have been spelled out. It will be argued that each of the individual strategies of non-empirical confirmation in isolation does constitute confirmation but cannot be established to be significant. As we will see, significant confirmation can be made plausible only by considering two or even three arguments of non-empirical confirmation in conjunction.

---

[5] In the context of a Bayesian formalization, posterior probabilities are always a function of the subjective priors. In any specific case, a high posterior probability hinges on a prior that is not too low. In the formalized model, the statement that some evidence provides significant confirmation therefore does not guarantee a high posterior, but merely implies that high posteriors are reached based on the given evidence from a wide range of priors.

6.5 Zooming in on Non-Empirical Confirmation

Let us recapitulate what we are looking for: We want to identify strategies of attaining significant non-empirical theory confirmation. That is, we are looking for kinds of observations that lie beyond a theory's intended domain (which means that they are not of the kind that can be predicted by the theory) but can nevertheless substantially increase the theory's probability of being viable.

A wide range of arguments may be taken to increase a physicist's trust in a theory: (a) One may point to a theory's elegance, beauty, or simplicity and argue that a theory that shows those qualities to a high degree must be expected to be viable. (b) One might cite one's gut feeling (or the gut feeling of some very prominent physicist) and argue that a very strong gut feeling about the theory's viability indicates that the theory is probably viable. (c) A recent suggestion has been that one might trust a theory based on its mathematical fertility (Rickles, 2013).

Arguments of the described kinds clearly do influence scientists' expectations regarding a theory's viability. There is no reason to dogmatically deny that they can indeed be helpful in certain contexts. However, the listed lines of reasoning all heavily depend on subjective expectations regarding the properties that a viable or true theory should be expected to have. Due to this strong subjective element, they don't provide a convincing basis for genuinely scientific strategies of theory assessment. I will briefly address suggestion (c) a little later. At this point, I want to focus on suggestions (a) and (b), which share two substantial flaws.

First, there are no clear parameters for measuring a theory's elegance, beauty, or simplicity, let alone for measuring the intensity of a scientist's gut feeling. Any attempt to infer a theory's viability from those qualities inevitably looks hopelessly vague and subjective.

Second, even if we had a way of making the attribution of elegance or similar qualities more precise, the way the argument is set up confines it to evaluating characteristics of the theory itself and drawing inferences from that evaluation. At no point does the argument reach out to an observation about the actual world. This limited scope of the argument creates a serious problem. Inferring a theory's viability from internal features such as elegance, beauty, and the like. must rely on the understanding that the given features are conducive to that theory's viability. If no observations about the external world are involved in the argument, the claim about the connection between the features in question and the theory's viability cannot be treated as empirically testable, but rather must be accepted as a dogmatic posit. An inference from those features to the theory's viability on this basis cannot be accepted as scientific reasoning.

In a nutshell, one might characterize the status of the discussed arguments in the following way: They are epistemic in nature because they address a theory's prospects of being viable. Nevertheless, the vagueness and subjectivity of the criteria applied and the lack of explicit connections to observations about the external world imply that the arguments' significance cannot be established based on a concise and scientifically legitimate line of reasoning. An individual scientist may act based on the suspicion that the way she understands and deploys the arguments guides her toward theories with better prospects of success. This may be sufficient for justifying the use of those arguments when deciding which way to go in the absence of stronger arguments. It does not allow for an intersubjective justification of substantial trust in a given theory, however. The described arguments don't amount to substantial scientific theory confirmation.

Adherents to a canonical understanding of theory confirmation would assert that this is how far one can get in the absence of empirical confirmation. By presenting the case for non-empirical theory confirmation, I contest that view. I assert that specific arguments of non-empirical confirmation are substantially stronger than the lines of reasoning analysed so far. They are stronger because their argumentative structure lies closer to the rationale of empirical confirmation in important respects. The next question we need to address is: Which conditions should be fulfilled by a promising candidate for substantial non-empirical confirmation? In the following discussion, I introduce three such conditions.

6.5.1 Observations about the World

The first criterion for promising non-empirical theory confirmation can be extracted directly from the previous discussion. We need to distinguish between the system made up of the scientists and the theories they have developed, on the one hand, and the world beyond that system, on the other hand. To amount to a scientifically legitimate kind of reasoning, non-empirical confirmation must be based on observations about the *external* world beyond the system made up of scientists and their theories.

With respect to a given theory, observations about that *external* world can be divided into two parts. First, there is the intended domain of the theory, consisting of observations of the kind that can be predicted by the given theory. Second, there are observations beyond that intended domain. Observations in the intended domain can be confronted with the theory's predictions and on that basis can provide empirical confirmation or dis-confirmation. Observations about the external world that lie outside a given theory's intended domain by definition don't contribute to empirical testing of that theory but may nevertheless be relevant for assessing the theory's viability. They are candidates for non-empirical confirmation.

6.5.2 A Predictive Meta-level Hypothesis

The second condition to be met by a convincing argument of non-empirical confirmation is related to the mechanism of confirmation itself. To speak of confirmation, we need a clear argument why the confirming evidence increases the probability of the viability of H. In the case of empirical confirmation, this probability increase can be immediately understood. For the sake of simplicity, I spell it out for the case of a deterministic theory where data E are strictly implied by theory H. In other words, we have $P(E|T) = 1$. The core argument remains unchanged in the case of probabilistic predictions. Using $P(E|T) = 1$ and the total probability $P(E) = P(T)P(E|T) + P(\neg T)P(E|\neg T)$, we can write the Bayes formula as

$$P(T|E) = \frac{P(E|T)P(T)}{P(T)P(E|T) + P(\neg T)P(E|\neg T)} \tag{6.2}$$

$$= \frac{P(T)}{P(T) + P(\neg T)P(E|\neg T)}. \tag{6.3}$$

If we exclude $P(T) = 1$ and $P(E|\neg T) = 1$, which are the trivial cases that we are sure either about the viability of H or about the outcome of the experiment already before measurement, we get $P(T|E) > P(T)$, which means that E confirms H. In effect, H gets confirmed by E because measuring E sets to zero the probabilities of potential (conceived or unconceived) alternative theories that are in conflict with E.

Non-empirical confirmation is based on data F that lie beyond the theory's intended domain. Therefore, the confirmation mechanism at work in the case of empirical confirmation is absent. How can we make plausible that non-empirical evidence F nevertheless increases the probability of the viability of theory H? The most straightforward strategy is to find a construction that structurally resembles the one that applies in empirical confirmation. If a hypothesis Y can be found that is predicted by F, then F confirms Y. If the viability of Y is positively correlated with the viability of H, the increase of the probability of Y that is induced by F feeds down to H. F then amounts to non-empirical confirmation of H.

Y must not be expected to be a genuine scientific theory that can be empirically tested in a scientific sense. The way in which F confirms Y will be more vague – which is one core reason why non-empirical confirmation should not be expected to be conclusive. However, it shall be argued, a vague relation between Y and F can be sufficient for generating quite substantial trust in a scientific theory H if F is sufficiently well specified.

We are looking for a meta-level mechanism that resembles empirical confirmation and feeds down to the ground-level hypothesis. An argument of that form will likely be the closest one can get to empirical confirmation in the absence of empirical data in the theory's intended domain.

It is instructive to compare the proposed approach to claim (c) mentioned at the beginning of this section. Rickles (2013) suggests that a theory's mathematical fertility – that is, its tendency to lead to the development of new mathematics – increases the theory's probability of being viable. The mathematical fertility of a given physical theory may be understood in terms of a contingent observation F_M

about the research process and therefore satisfies condition 5.1. However, Rickles's claim does not involve a meta-level hypothesis Y that predicts F_M and is positively correlated with a given theory's viability. Rather, the argument relies on the further observation F_{FM} that observation F_M has often been made with regard to viable theories in the past.

To see that Rickles's argument is inconclusive, think of the following point. One might suspect that mathematical fertility is correlated with a theory's complexity: A complex world most likely requires complex mathematics, which in turn may be taken to increase the prospects of mathematical fertility. This view would, to some extent, provide an explanation of the mathematical fertility of our viable theories if we are ready to acknowledge that we live in a complex world. It would also suggest that the mathematically most fertile theories are about worlds even more complex than ours. Indeed, thinking about a world of very many spatial dimensions was mathematically fertile for a nineteenth-century mathematician for the very reason that it moved beyond what seemed physically realistic. F_{FM} in this light does not imply a non-empirical confirmation value of F_M, because mathematical fertility might be found to an even higher degree in worlds other than ours. A meta-level hypothesis Y that meets the conditions stated earlier in this subsection avoids this problem.

6.5.3 Viability Rather than Truth (Again)

A third consideration is important to emphasise. In recent years, a question debated in some parts of fundamental physics has been whether a physical theory can be called scientific if it does not make any empirical predictions even in its fully developed form. While thinking about non-empirical confirmation might open up new perspectives on this debate, one would not want to set up non-empirical confirmation in a way that implies a positive answer to this highly contentious question. As pointed out earlier in this section, it is crucial for establishing the plausibility of non-empirical confirmation to exploit all available connections to empirical data. The more non-empirical confirmation gets decoupled from observation, the more questionable its implications become. In this spirit, the goal of making the strongest possible case for non-empirical confirmation suggests confining the analysis to theories that do have empirical implications. We do want to account for theories such as string theory that are not sufficiently understood at this point to support extracting specific empirical predictions. But we want non-empirical confirmation to be applicable only to the extent such theories are expected to reveal specific empirical implications once they are sufficiently understood.

Confining our analysis to empirically relevant theories leads us back to the choice between addressing a theory's truth and its viability, which was discussed in Section 6.3.2. It offers one more strong reason for preferring the concept of viability over the concept of truth: If we understood confirmation in terms of an increase of truth probability, it would be difficult to exclude cases of theories that are predictively empty. A theory's truth may be taken to be distinct from its empirical adequacy, which means that a non-empirical argument could in principle increase a theory's probability of truth without changing that theory's probability of being empirically adequate. This opens up the possibility of confirming predictively empty theories.

Empirical viability, to the contrary, has been defined in terms of the theory's empirical predictions. A theory is viable in a given regime if its predictions agree with all data that can in principle be collected in that regime. If the theory is predictively empty in that regime, its viability in that regime is trivial and non-empirical evidence plays no role in establishing it. Non-empirical confirmation defined in terms of viability therefore is applicable only to theories that are predictive in principle.

We have now specified three conditions that should be fulfilled by non-empirical confirmation of a theory H to make it as similar to empirical confirmation as possible:

1. The observations on which non-empirical confirmation is based should be about the external world rather than merely about the system of scientists and their theories.
2. It should be possible to construe an argument of non-empirical confirmation based on "soft" empirical confirmation of a meta-level hypothesis Y by well-specified non-empirical evidence F, which, in conjunction with a positive correlation between Y and the viability of H, establishes that F confirms H.
3. Non-empirical confirmation should be applicable only to empirically predictive theories, which is guaranteed by defining confirmation in terms of the theory's viability.

An argument of non-empirical confirmation that meets those three conditions has a plausible path toward being significant.

6.6 A Specific Realization

We are now in a position to connect to the three core strategies of non-empirical confirmation presented in Dawid (2013). As will be shown in this section, all of those strategies satisfy the three conditions specified in the previous section. Let me restate the three arguments of non-empirical confirmation.

NAA: The no-alternatives argument. Scientists have looked intensely and for a considerable time for alternatives to a known theory H that can solve a given scientific problem but haven't found any. This observation is taken as an indication of the viability of theory H.

MIA: The meta-inductive argument from success in the research field. Theories in the research field that satisfy a given set of conditions have shown a tendency of being viable in the past. This observation is taken to increase the probability that a new theory H that satisfies those conditions is also viable.

UEA: The argument of unexpected explanatory interconnections. Theory H was developed to solve a specific problem. Once H was developed, physicists found that H also provides explanations with respect to a range of problems whose solution was not the initial aim of developing the theory. This observation is taken as an indication of the theory's viability.

Since the arguments are formulated as inferences to a higher probability of the theory's viability, by definition they satisfy condition (3). It is also straightforward to see that condition (1) is satisfied, since for all three arguments the respective observation about the external world has been spelled out in the first sentence. This leaves us with the task of understanding that the arguments are of the form required by condition (2). Since Dawid (2013) provides an extensive analysis of this point, it will suffice in the present chapter to give a brief sketch of the line of reasoning. The reader interested in a more thorough discussion is referred to Dawid (2013), Chapters 2.2–3.2.

The meta-level hypothesis Y that seems most adequate in all three presented arguments of non-empirical confirmation is a statement on local limitations to scientific underdetermination. To understand that statement, we need to clarify the terminology.

Scientific underdetermination (called transient underdetermination in a slightly different context by Lawrence Sklar [1975] and Kyle Stanford [2006]), measures how many alternative theories that are not empirically fully equivalent to each other (that is, they could be distinguished empirically in the long run) can account for a given empirical data set.

Local scientific underdetermination only accounts for theories that could be distinguished within a given "experimental horizon" – that is, a specified class of experiments. For example, local scientific underdetermination up to a given energy scale in high energy physics accounts for the spectrum of possible theories that may have different empirical implications up to that energy scale. The concept of local scientific underdetermination therefore directly relates to a theory's viability within a given experimental horizon as introduced in Section 6.3: Local scientific underdetermination with respect to a given empirical horizon counts those theories that can have different empirical viability conditions with respect to that empirical horizon.

Limitations to scientific underdetermination denote constraints on scientific underdetermination of some kind. Such limitations are a natural candidate for *Y* for a simple reason: Any trust in predictions of an *empirically confirmed* scientific theory must be based on implicit assumptions of limitations to scientific underdetermination.

If scientific underdetermination were entirely unlimited, one would expect that any imaginable continuation of the observed regularity patterns could be accounted for by a plausible scientific theory. There would be no basis for expecting that, among the unconstrained multitude of such theories, precisely the one that scientists happened to have developed would be the empirically viable choice. Only a rather radical rejection of the idea of unlimited underdetermination can lead physicists toward trusting the predictions provided by their theory. In other words, any degree of trust in a theory's viability in a fairly straightforward way translates into the meta-level hypothesis that local scientific underdetermination is strongly limited.

A very high degree of confidence in a theory actually translates into the understanding that, within the considered empirical horizon, there probably is no alternative scientific theory at all that can account for the available data in a satisfactory way but makes different predictions within the empirical horizon. If such unconceived alternatives exist, how could we rule out that one of those, rather than the known theory, was the viable one?

The question remains: How can scientists infer hypotheses about the spectrum of possible alternatives? Obviously, such meta-level hypotheses cannot be supported by the empirical data that support the scientific theory under scrutiny: Those data would support any unconceived alternative as well. Support of the meta-level hypothesis therefore needs to come from observations beyond the theory's intended domain – observations of the very kind that amount to non-empirical confirmation. As stated in Section 6.5.2, a confirmatory value of *F* for *Y* can be established by showing that *Y* predicts *F* (makes *F* more probable).

Looking at the three kinds of observations *F* listed previously, we find that they all do confirm a hypothesis *Y* on limitations to local scientific underdetermination on those grounds.

Most straightforwardly, the NAA observation that scientists have not found alternative theories is predicted by a strong hypothesis *Y*. In the most extreme case, if no alternatives to theory *H* exist, scientists cannot find any. (For a general Bayesian formalization of NAA and a proof that it amounts to confirmation, see Dawid, Hartmann, and Sprenger [2015]).

The MIA observation that theories comparable to H tend to be predictively successful once tested also is predicted by a Y hypothesis on the given ensemble of theories: If in the cases considered there usually are no or very few possible alternatives to the theories under scrutiny, one would expect a fairly high percentage of those theories to be viable.

The UEA case is a little less straightforward but eventually allows for the same conclusions. The basic reasoning is the following. Let us imagine a set of problems that need to be solved in a given context. Let us first assume that the number of theories that can be developed in that context is much higher than the number of problems. In this case, there is no reason to expect that a theory developed to solve one specific problem will be capable of solving other problems as well. Next, let us assume that the number of theories that can be developed in the given context is much smaller than the number of problems. In that case, one must expect that each possible theory will, on average, solve a substantial number of problems and therefore will show UEA. In the extreme case where only one theory can be developed, that theory must be expected to solve all problems. Therefore, Y does predict UEA.

We can conclude that NAA, MIA, and UEA do rely on one common meta-level hypothesis Y that provides the basis for establishing their character as non-empirical confirmation. The three arguments thus satisfy all three conditions set out in Section 6.5 for promising arguments of non-empirical confirmation. More-over, the fact that all three arguments are related to the same meta-level hypothe-sis shows that they constitute a coherent set of arguments that may work well in conjunction.

The last point becomes particularly important once one aims at understanding the arguments' actual significance. What has been established up to this point is that NAA, MIA, and UEA each amounts to non-empirical confirmation. As pointed out in Section 6.4, however, non-empirical confirmation can be taken to be scientifically relevant only if it can be shown to be significant.

The problem is that none of the three arguments in isolation can be established as significant confirmation. NAA faces the problem that hypothesis Y, asserting strong limitations to local scientific underdetermination, is not the only hypothesis that can explain the observation F_{NAA} that no alternatives have been found. One might also propose hypothesis U as an explanation, asserting that scientists just have not been diligent or clever enough to find the alternatives that exist. The significance of NAA therefore depends on the prior probabilities one attributes to Y and U. An observer who attributes a very low prior to Y and a substantial prior to U will take F_{NAA} to significantly increase the probability of U and find only a fairly insignificant increase in the probability of Y.

MIA faces a different problem: Observing a tendency of predictive success for a specified set of theories does not lend strong support to claims about the prospects of a new theory as long as there is no good reason to assume that the new theory shares the success prospects of the ones previously considered.

UEA may seem a little more autarkic than the other two arguments but also remains questionable in isolation. Unexpected explanatory interconnections may be due to deeper underlying connections between the problems addressed by the theory that are not related to the specific theory itself. If so, they don't provide good arguments for trusting a specific theory.

The fact that the three arguments rely on the same Y comes to the rescue in this situation. It provides the basis for strengthening one argument by adding another argument. This can eventually generate significant non-empirical confirmation of a theory based on two or three arguments in conjunction.

For example, MIA offers exactly what is missing in NAA: It discriminates between support for hypotheses Y and U. While the observation of a tendency of predictive success in a field can be explained by a hypothesis Y on limitations to underdetermination, it clearly cannot be explained by a hypothesis U that states the limited capabilities of the involved scientists.

Conversely, NAA can provide what is missing in MIA in isolation: Being itself related to Y, NAA provides a plausible criterion for specifying the ensemble of theories one should consider when looking for a tendency of predictive success. When other theories that allow for a NAA argument show a tendency of predictive success, that provides a plausible basis for expecting predictive success for a new theory where NAA also applies. Thus, NAA and MIA in conjunction can generate significant non-empirical confirmation.

In a case like string theory, the question whether the new theory (string theory) is in the same category as prior theories (like the standard model of high energy physics) that are the prime examples of previous predictive success is particularly tricky because the new theory is so much more complex and insufficiently understood. One might fear that scientists' capability of assessing the spectrum of possible alternatives may be substantially reduced in this more difficult scientific environment, which would render the application of MIA questionable even in the presence of the NAA. In such a situation, UEA assumes a crucial role. It can establish that the given research context is understood well enough to extract the observed instances of unexpected explanatory interconnections. Therefore, apart from providing an argument for limitations to scientific underdetermination in its own right, UEA strengthens the case for the applicability of MIA. A case like string theory thus requires use of all three arguments of non-empirical confirmation in conjunction to generate significant non-empirical confirmation.

6.7 Conclusion

The emerging picture of non-empirical confirmation is the following. While there is no reasonable basis for expecting that non-empirical confirmation can become *conclusive* in our world, it seems plausible that non-empirical confirmation can be *significant* in certain contexts.

The basis for significant non-empirical confirmation lies in an argumentative structure that is closely related to the structure of empirical confirmation in a number of respects.

- Confirmation is confined to empirically relevant statements by defining it in terms of viability rather than truth.

- Non-empirical confirmation is based on an observation F about the external world beyond the system of the scientists and their theories.

- The connection between successful prediction and confirmation that characterizes empirical confirmation is, in the case of non-empirical confirmation, reflected by the fact that non-empirical evidence F is predicted by a meta-level hypothesis Y.

- Non-empirical confirmation is directly connected to empirical confirmation based on MIA: Significant non-empirical confirmation arises only if empirical confirmation exists somewhere else in the research field.

- Arguments of non-empirical confirmation generate a web of interconnected reasoning. The individual arguments can strengthen each other in an effective way because they can all be construed in terms of the same meta-level hypothesis Y, the hypothesis of strong limitations to local scientific underdetermination. The very same hypothesis Y plays a crucial role for understanding the relevance of *empirical* confirmation.

All these points in conjunction can make non-empirical confirmation significant. This does not imply that non-empirical evidence for a theory should be dogmatically constrained to evidence that can be framed in precisely the suggested way. Nevertheless, the list of characteristics shared by NAA, MIA, and UEA demonstrates that those arguments have fairly non-trivial conceptual merits. There are plausible reasons why those arguments in particular can become significant and play a leading role in supporting empirically unconfirmed theories.

How strong the web of non-empirical confirmation should be taken to be in a given case is, of course, a matter of detailed scientific analysis of a physical theory. In some examples in recent fundamental physics, very strong and nearly unanimous trust in a hypothesis emerged based on non-empirical confirmation. A particularly strong example is the trust in the viability of the Higgs mechanism before the discovery of a Higgs boson in 2012.[6] String theory constitutes an example where a large number of the theory's experts have come to the conclusion that non-empirical confirmation is significant though not conclusive. Others doubt the significance of non-empirical confirmation in this case. Cosmic inflation is a case where supporting data for the theory exist but are not conclusive. Non-empirical confirmation arguably plays a substantial role in increasing trust in this theory in the eyes of many of its exponents.

Acknowledgments

This chapter has profited substantially from very helpful comments from the audience of the "Why Trust a Theory?" workshop in Munich and in particular from discussions with Peter Achinstein, Radin Dardashti, Daniele Oriti, Carlo Rovelli, and Karim Thébault.

References

Achinstein, P. (2001), *The Book of Evidence*. Oxford: Oxford University Press.
Dawid, R. (2006). "Underdetermination and Theory Succession from the Perspective of String Theory." *Philosophy of Science* 73(3): 298–322.
Dawid, R. (2013). *String Theory and the Scientific Method*. Cambridge: Cambridge University Press.
Dawid, R., S. Hartmann, & Jan Sprenger. (2015). "The No Alternatives Argument." *British Journal for the Philosophy of Science* 66(1): 213–34.
Ellis, G., & J. Silk. (2014). "Defend the Integrity of Physics." *Nature* 516: 321–3.
Rickles, D. (2013). "Mirror Symmetry and Other Miracles in Superstring Theory." *Foundations of Physics* 43(1): 54–80.
Rovelli, C. (2017). "The Dangers of Non-Empirical Confirmation." This volume.
Sklar, L. (1975). "Methodological Conservativism." *Philosophical Review* 84: 384.
Stanford, K. (2006). *Exceeding Our Grasp: Science, History and the Problem of Unconceived Alternatives*. Oxford: Oxford University Press.

---

[6] That trust was invested in the general characteristics of the Higgs mechanism. The specific details of the actual Higgs model – whether the Higgs boson is fundamental or has constituents; whether it is of a standard model type, supersymmetric, or something else – remain open and in part have not been clarified as yet.

7

The Dangers of Non-Empirical Confirmation

CARLO ROVELLI

Scientists have always relied on non-empirical arguments to trust theories. They choose, develop, and trust theories *before* finding empirical evidence. The entire history of science stands as witness to this. Kepler trusted Copernicus's theory *before* its predictions surpassed Ptolemy's; Einstein trusted general relativity theory *before* the detection of the bending of light from the Sun. They had non-empirical arguments, which proved good.

According to a popular version of the Popper–Kuhn account of the scientific activity, theories are generated at random, sort of fished out from the blue sky, and then judged only on empirical ground. This account is unrealistic: Theorists do not develop theories at random. They use powerful theoretical, non-empirical, motivations for creating, choosing, and developing theories. If these did not exist, the formidable historical success of theoretical physics would be incomprehensible. To evaluate theories, they routinely employ a vaste array of non-empirical arguments, increasing or decreasing their confidence in this or that theoretical idea, *before* the hard test of empirical confirmation (on this, see chapter VIII of [1]). This is the context of a "preliminary appraisal" of theories, or "weak" evaluation procedures [2].

In the book *String Theory and the Scientific Method* [3], Richard Dawid describes some of these non-empirical arguments that motivate theoretical physicists' confidence in a theory, taking string theory as his case study. This may imply that the use of non-empirical arguments is somewhat of a novelty in scientific practice. It is not.

But the theorist's "preliminary appraisal" of theories is in the context of discovery, which is quite another matter than the hard empirical testing of a theory, which is in the context of justification. Fogging the distinction is a mistake.

Dawid uses a Bayesian paradigm to describe how scientists evaluate theories. Bayesian confirmation theory employs the verb "confirm" in a technical sense, which is vastly different from its common usage by lay people and scientists.

In Bayesian theory, "confirmation" indicates any evidence in favour of a thesis, *however weak.*

In Bayesian parlance, for instance, seeing a Chinese person in Piccadilly Circus "confirms" the theory that the majority of Londoners are Chinese. Nobody says so outside Bayesian theory. For lay people and scientists alike, "confirmation" means something else: It means "very strong evidence, sufficient to accept a belief as reliable."

This unfortunate ambiguity has played a role in the reaction to Dawid's work. Some scientists appreciated his recognition of their theoretical reasons for defending a theory, but some string theorists went further: They were all too happy that string theory, which lacks "confirmation" (in the standard sense), was promoted by Dawid to have plenty of "confirmation" (in Bayesian sense), raising sharp contrary reactions [4].

Unfortunately, Dawid himself has done little to dispel this ambiguity. This generates a problem for his views, for the following reason.

Bayesian confirmation theory allows us to talk about the spectrum of intermediate degrees of credence between theories that are "confirmed," in the common sense of the word, or "established," and theories that are still "speculative" or "tentative." But doing so obfuscates precisely the divide that *does* exist in science between a confirmed theory and a tentative one. We trust the existence of the Higgs particle, which is today the weakest of the confirmed theories, with a 5-sigma reliability – namely, a Bayesian degree of confidence of 99.9999%. In their domains of validity, classical electrodynamics and Newtonian mechanics are considered even more reliable: We routinely entrust our lives to them. No sensible person would entrust her life to a prediction of string theory.

The distinction is there and is clear. A philosophy of science blind to this distinction is a bad philosophy of science. It is so important that phrasing it in terms of a higher or lower Bayesian degree of belief obfuscates the point: In science we *do* have theories that are "confirmed" or "established," which means that they are extremely reliable in their domain. Then we have other theories that perhaps enjoy the confidence of some scientists but are considered tentative: We wouldn't entrust to them even our life *savings.*

The distinction between reliable theories and speculative theories may not always be perfectly sharp, but is an essential ingredient of science. As Thoreau puts it: "To know that we know what we know, and to know that we do not know what we do not know, that is true knowledge" [5]. The very existence of reliable theories is what makes science valuable to society. Losing this from sight is not understanding why science matters.

Why is this relevant for non-empirical confirmation? Because non-empirical evidence is emphatically insufficient to increase the confidence of a theory to the point where we can consider it established – that is, to move it from "maybe" to "reliable."

The reason only *empirical* evidence can grant "confirmation", in the common sense of the word, is crucial and important: We all tend to be blinded by our beliefs. We pile up non-empirical arguments in support these. The historical success of science is grounded in the readiness to give up beloved beliefs when empirical evidence is against them. We create theories with our intelligence, use non-empirical arguments to grow confidence in them, but then *ask nature* if they are right or wrong. They are often wrong. Witness the recent surprise of many theorists in not finding the low-energy super-symmetric particles they expected.

As T. H. Huxley put it: "The great tragedy of Science is the slaying of a beautiful hypothesis by an ugly fact" [6]. Tragedy, yes, but an incredibly healthy one, because this is the very source of science reliability: checking non-empirical arguments against the proof of reality.

Perhaps no science illustrates this better than medicine. The immense success of Western medicine is largely (one is tempted to say "almost solely") based on a single idea: checking statistically the efficacy of the remedies used. Through the application of this simple idea, the life expectancy of us all has more than doubled in a few centuries. This feat has been achieved by simply *not* trusting non-empirical arguments.

Dawid's merit is to have emphasised and analysed some of the non-empirical argument that scientists use in the "preliminary appraisal" of theories. His weakness is to have obfuscated the crucial distinction between this and validation: the process where a theory becomes reliable, gets accepted by the entire scientific community, and potentially becomes useful to society. The problem with Dawid is that he fails to say that, for this process, only *empirical* evidence is convincing.

String theory, Dawid's case study, illustrates well the risk of over-relying on non-empirical confirmation and the need of empirical validation. As Dawid notices, non-empirical arguments support the credence of strings. These arguments are valuable, but too weak to grant reliability. I mention one: String theorists commonly claim that string theory has no alternatives ("the only game in town"). This is the first of Dawid's non-empirical arguments. But as any scientist knows very well, any "no alternatives" argument holds only under a number of assumptions, and these might turn out to be false. In fact, not only *do* alternatives to string theory exist in the real world, but these alternatives are themselves considered credible by their supporters precisely because they themselves have "no alternatives" under a *different* set of assumptions! As a theory of quantum gravity, an alternative to string theory is loop quantum gravity, considered the "only game in town" by those who embrace it, under *their* set of assumptions. Any theory, whether physically correct or incorrect, has "no alternatives" under suitable assumptions; the problem is that these assumptions may be wrong. Here we see clearly the weakness of non-empirical arguments. In science, we learn something solid when something challenges our assumptions, not when we hold on to them at any cost.

String theory is a living proof of the dangers of excessive reliance on non-empirical arguments. It raised great expectations thirty years ago, promising to compute all the parameters of the Standard Model from first principles, to derive from first principles its symmetry group $SU(3) \times SU(2) \times U(1)$ and the existence of its three families of elementary particles, to predict the sign and the value of the cosmological constant, to predict novel observable physics, to understand the ultimate fate of black holes, and to offer a unique, well-founded unified theory of everything. Nothing of this has come true. String theorists, instead, have predicted a negative cosmological constant, deviations from Newton's $1/r^2$ law at sub-millimeters scale, black holes at the European Organization for Nuclear Research (CERN), low-energy super-symmetric particles, and more. All this was false. Still, Joe Polchinski, a prominent string theorist, writes [7] that he evaluates the Bayesian probability of string to be correct at 98.5% (!). This is clearly nonsense.

From a Popperian perspective, these failures do not falsify the theory, because the theory is so flexible that it can be adjusted to escape failed predictions. But from a Bayesian perspective, each of these failures *decreases* the credibility of the theory. In Bayesian theory, this is an unavoidable consequence of the fact that a positive result would have increased credibility. The recent failure of the prediction of supersymmetric particles at the Large Hadron Collider (LHC) is the most fragrant example. By Bayesian standards, it lowers the degree of belief in string theory dramatically. In fact, in Bayesian parlance, not having found low-energy symmetry where it would be expected is an empirical *confirmation* that the theory is *wrong*.

In other words, one cannot have the cake and eat it, too. Being Bayesian on positive hints and Popperian when things go badly wrong is not rational: It is being blind.

Scientists who have devoted their life to a theory have difficulty in letting it go; they hang on to non-empirical arguments to save their beliefs, in the face of empirical results that Bayes confirmation theory counts as negative. This is human. A philosophy that takes this as an exemplar scientific attitude is a bad philosophy of science.

References

[1] C. Rovelli. *The First Scientist: Anaximander and His Legacy*. Chicago: Westholme, 2011.
[2] K. Schaffner. *Discovery and Explanation in Biology and Medicine*. Chicago: University of Chicago Press, 1993.
[3] R. Dawid. *String Theory and the Scientific Method*. Cambridge: Cambridge University Press, 2013.

[4] G. Ellis and J. Silk. "Scientific method: Defend the integrity of physics." *Nature* **516** (2014): 321323.

[5] H. D. Thoreau. *Walden*. London: Penguin Classics, 1854.

[6] T. H. Huxley. "Biogenesis and abiogenesis (1870)". In *Collected Essays*, ch. 8, p. 229. New York: Harper, 1959.

[7] J. Polchinski. "Why trust a theory?" arXiv:1601.06145.

8

No Alternative to Proliferation

DANIELE ORITI

8.1 Introduction

In the following, we present some reflections on the role of non-empirical theory assessment (NETA) in fundamental physics. We discuss the role, virtues, and dangers of NETA at a more general level, and then offer our critical appraisal of the specific criteria for carrying out this assessment proposed by Dawid [1].[1]

Before we proceed, it is useful to clarify the perspective and the limitations of our contribution.

First, we will be mostly concerned with issues of methodology of science, and only indirectly touch on the strictly related problems in epistemology. Also, while we will comment on the psychological and sociological aspects of scientific dynamics, arguing that they should be taken into account when discussing its methodology, we will not attempt any detailed analysis of them, instead relying mostly on insights from the literature and from personal experience. In fact, and this is probably the most important cautionary note, the perspective adopted in this chapter is (inevitably) that of a theoretical physicist working in quantum gravity, not that of a professional philosopher of science.

This should explain, if not justify, some possible naiveté (hopefully, limited) in addressing the complex topic of NETA, but also somehow reassure the reader of our genuine interest in the same topic. Indeed, in quantum gravity the issue of NETA takes centre stage, due to the disproportion between observational constraints and theoretical constructions. The variety of the theoretical landscape of different quantum gravity approaches forces any quantum gravity theorist to confront the issue

---

[1] Throughout the text, we use 'non-empirical assessment' to include any form of evaluation of a theory that does not amount to a *direct* confrontation of the same theory with the results of experiments or observations. It may also include aesthetic criteria or simple continuity with previous historical developments, but of course any such evaluation would have some strength only to the extent in which it can be supported by rational arguments. It will not be our purpose in this chapter to propose or analyse specific criteria for NETA, beside the ones proposed by Dawid.

of selecting one's own scientific path without empirical guidance, resting entirely on (tentative) NETA. At the same time, like any other working scientists, we are subject to and witnesses of the cognitive and sociological biases that shape the community itself as well as the very development of our theories. They are even more prominent in the case of quantum gravity, due again to the lack of experimental constraints, which may explain our sensitivity to these aspects of NETA. However, despite the fact that some work on NETA, and Dawid's [1] in particular, concerns one specific approach to quantum gravity (i.e., string theory), we will refrain from commenting on string theory itself or other approaches to quantum gravity. We will try instead to keep our opinions on specific quantum gravity approaches in the background, and tackle the issue of NETA from a purely philosophical point of view.[2]

Our perspective as quantum gravity theorists is also one main reason why we will approach the topic from a somewhat practical perspective. We interpret the question 'What constitutes good non-empirical theory assessment criteria, if any?' to have both descriptive and prescriptive aspects, and to refer to the scientific practice rather than the realm of pure philosophical reasoning. Indeed, we are interested in which non-empirical assessment practices can be argued to be *fruitfully* applied by scientists to achieve progress, thus with an inevitable *prescriptive* aspect, implicitly suggesting what scientists 'could be doing' to achieve progress. We do not shy away from this aspect of the discussion, exactly because we approach the methodological issue from the scientist's point of view (it is not philosophers' task to tell scientists what to do, but it is certainly scientists' duty to try to do it better, while listening carefully to the insights of the philosophers). At the same time, we know very well that any such prescription can only be tentative, and that there is no such thing as *the* scientific method, intended as a simple set of rules scientists follow or should follow, or as the defining feature of science that demarcates it sharply from other forms of knowledge.

Since our take on the issue of NETA is necessarily influenced by our general view of the nature of science, it is probably also useful to devote a few words to making this perspective explicit. We share a view of science as an adaptive system [2], whose progress could be described in terms very close to the evolutionary ones used in biology, in which scientists (and scientific communities) 'evolve' their theoretical (conceptual, mathematical) tools to fit in a constantly changing empirical

---

[2] Having said this, we also believe that some work on the subject, in particular Dawid's, is based on an overly generous evaluation of string theory and its achievements as a theory of quantum gravity (without diminishing in any way the justified scientific appeal and the many results of the framework), and a consequent dubious application of NETA criteria to it (God will forgive him for this). To provide a proper justification of this opinion would require a much more careful analysis that would bring us outside the scope of this chapter, so we leave it for a future one.

world of phenomena. This analogy could be phrased in a more impersonal manner, speaking instead of theories themselves evolving through selective pressure in the environmental niches defined by empirical facts. In this view, empirical and non-empirical theory assessment constitute such necessary selective pressure, ultimately driving progress. However, this more abstract way of describing the process of scientific evolution is at risk of neglecting the very human (historical and cultural) components of the same evolution, thereby missing much of what is really going on.[3] For this reason, another useful analogy would be thinking in terms of a 'marketplace of ideas' about the world – an admittedly very vague image, but one that gives a good intuition of the dynamical, very human, but at the same time very constrained nature of theory development, assessment, and spreading. Thus, without denying or even diminishing the objective nature of scientific understanding [9],[4] it helps to keep in mind that this understanding is the product of human minds first, and human communities second. In fact, we believe that another good way to characterize the scientific way of proceeding is as a constant struggle to understand the world while overcoming (or last, keeping on a leash) our many cognitive [5] and sociological [6, 7] biases,[5] which often mislead us. Given the strength and range of such biases, one would even be tempted to *define* the scientific method as the set of tools we have developed over time to make it easier to prove ourselves wrong (as individuals and as communities). Admittedly, this is a very weak but, we believe, apt version of methodological falsificationism. If theories are to be considered as constantly under siege from both empirical and non-empirical selection pressure, it goes without saying that the *temporary* and *dynamical* character of scientific explanations should remain at centre stage as well, and any contribution to scientific methodology should be founded on it [3]. Finally, it should be clear that the only notion of *truth* – in particular, the only notion of scientific truth – that this vision of science allows as meaningful is a partial, temporary, and approximate one. We believe that any attempt at theory assessment that is motivated by the attempt to establish something more than a truth of this type, or implicitly assumes that this is even possible, is actually betraying the very nature of science.

This should suffice as a sketch of our broader point of view on science, which informs our approach to the issue of NETA. In particular, it explains our insistence, in the following discussion, on the importance of *theory proliferation*, raised to the

---

[3] We take this to be a well-understood lesson of modern philosophy of science, from Kuhn, Lakatos, Feyerabend, Laudan, and many others (for a summary, see [4])

[4] We believe that the analysis of social aspects enriches our understanding of scientific objectivity, without undermining it, as in the often distorted picture of science offered by some social constructivists.

[5] By the latter, we mean both the subset of the former that have to do with belonging to a group, and the collective effects of group behaviour that may end up being misleading with respect to the goal of a rational understanding of the natural world; they include institutional factors as well as non-institutionalised aspects of group behaviour; see also the more sociological discussion in [8].

level of a methodological principle, the *principle of proliferation* (PoP).[6] Given the central role it will play in our reflections, let us make more explicit what we mean by this.

The PoP can be loosely phrased as follows: 'Construct as many alternatives as possible to the current (dominant) theoretical framework, and use the set of all such alternative theories (including the currently favoured one) as the object of (empirical) testing, not any given theory in isolation'. The first prescription applies at *any* moment in the evolution of a scientific domain; it does not refer only to moments of crisis following the discovery of some observational anomaly, or some conflict between experiments and the currently accepted theory, or other internal difficulty with it. Also, it clearly applies both to the context of theoretical discovery and to the context of theory justification, assuming we want to retain this distinction. The other very important point to stress is that the PoP urges us to actively *construct* and *develop* alternatives to any existing theory, no matter how well supported, and does not merely state the need for *tolerance* of any such alternative. To be used as prescribed, these alternative theories have to be sufficiently developed and coherent, and should be able to replace the given theory, if shown to fit experience (or other assessment criteria) better than that theory. This suggests that we devote as much effort as possible to the *development* of full-fledged alternative *theories*, rather than just their invention as plausible alternative hypotheses.

The PoP, in various forms, has been discussed and argued for (but also against) by several authors [10], and, most vocally, Feyerabend [11]. As Feyerabend has stressed, however, it can actually be traced back to J. S. Mill and understood as part of his philosophical liberalism [12].

The arguments for the PoP in philosophy of science refer, for obvious reasons, to the empirical assessment of theories. Indeed, it has been argued that the adoption of the PoP leads to an increase of testability, for various reasons. First, it improves the understanding of each element in the set, by clarifying what would be denied of the 'established' theory if one alternative was shown to be correct, and by elucidating the proper meaning of the established theory itself in relation to its alternatives. Second, an alternative theory may suggest a new empirical test of a given theory, which would otherwise be left unimagined (even if it was fully within the conceptual capabilities of the given theory). Moreover, alternative theories provide means to magnify discrepancies of a given theory with observations – for example, by

---

[6] By this, we do not mean that the PoP, or any of the arguments we will present in this chapter, follows necessarily from our general view of science, or that it has to be understood in conjunction with that perspective. All of these arguments have their own intrinsic strengths and weaknesses. We digress briefly to sketch our more general perspective of science simply for the sake of transparency ('this is where we come from and this is our set of background prejudices') and completeness ('you may understand better our arguments if you frame them within this broader picture').

asserting such discrepancies as central facts, when they were actually only negligible possibilities in the given theory. Finally, alternative theories allow us to evade the psychological constraints of a given theory, which may prevent us from noticing its weaknesses.

Notice that the PoP also discourages the elimination of older 'disconfirmed' or simply disfavoured theories, because they can always win back support, and because their content enriches the given 'established' theory. Of course, the PoP does not imply that a scientist cannot hold tenaciously to a given theory she finds promising or convincing, and that is furthermore currently corroborated by observations or non-empirical assessment criteria. On the contrary, the principle of tenacity (PoT) is the natural balance of the PoP at the level of scientific communities, and maybe even at the level of individual scientists. However, the PoP goes together necessarily with the PoT, while the converse is not true: One can hold a PoT without logically holding also to the PoP; the result, however, is dogmatism, which hampers progress. Overall, the PoP suggests a picture of progress in knowledge as an ever increasing landscape of alternatives, rather than a convergence toward an ideal view, with each alternative forcing the others into greater articulation, and constantly competing, and all together enriching our mental capabilities.[7] At the very least, even theories that have long receded into the background provide a 'measure of development' for the more contemporary theories. Why this view is in line with the picture of science as an adaptive system (it is a call to protect and enrich 'biodiversity') or as a marketplace of ideas should be obvious, as it should be obvious why the PoP would be regarded as precious by anyone concerned with the cognitive and sociological biases hampering our search for a solid understanding of the world. Indeed, the PoP has been advocated by authors holding a naturalistic perspective of science and an 'evolutionary' approach to scientific progress and methodology, while being at the same time supported by results in cognitive science and psychology [15].

The arguments made here refer mainly to empirical assessment, but the usefulness of the PoP go well beyond it. In fact, the original arguments given by J. S. Mill in support of the PoP were not directly focused on empirical tests (so much that some authors have denied that Mill was referring to science at all [13]). They are as follows: (1) A rejected theory may still be true (we are not infallible); (2) even when false, it may contain a portion of truth, and only the collision of different views allows the emergence of such truth; (3) a theory that is fully true but not contested will be held as a matter of prejudice or dogma, without a full comprehension of its rational basis; and (4) the full meaning of a theory, even

---

[7] The dynamical combination of these two principles also forms the basis of Feyerabend's general view of methodology and rationality [14].

if true, can be comprehended only by contrasting it with other views. As argued by Feyerabend and others [11], (1) and (2) are supported in history of science: A theory may win competition by chance, greater attention devoted to it, and so on, before competitors have had time to show their strengths or because they are simply temporarily out of steam, only to win it back later. For their part, (3) and (4) receive support based on what happens when a given theory takes centre stage: The risk of a decrease in rationality and understanding, since it no longer needs to produce good arguments or evidence in its support, means that the theory may become part of the general education and academic discourse without having been fully understood and without having necessarily solved its basic problems, debates having been held on its own terms that create additional troubles to opponents or critics, and so on. The favoured theory can, of course, point to its many positive results, but the important point is to evaluate such results, and this should be done in comparison with some (equally) well-developed alternative. Proliferation is then not just an epistemological expression of liberalism, but the necessary ingredient for a rational enquiry into the nature of things. Therefore, also on the basis of Mill's original, more general arguments, we will argue that the PoP is central and absolutely necessary to NETA.

8.2 The Need, Purpose, and Risks of Non-Empirical Theory Assessment

Let us now reflect on the issue of NETA at a rather general level, before turning to a criticism of Dawid's analysis in the following section.

We share the view of several authors that the methodological issue of identifying good (empirical as well as non-empirical) theory assessment criteria is intertwined with the more epistemological issue of scientific underdetermination [16, 17], in the sense that criteria for theory assessment and selection are limitations to scientific underdetermination – that is, to the number of allowed alternative theories that are compatible with the same set of observations that corroborate the given one.[8] Such limitations to scientific underdetermination are necessary to have any confidence in our theories. This is simply the core of a fallibilist and critical epistemology, since it amounts to the fact that we can have (limited, provisional) confidence only in theories that we have critically assessed and that survived some form of (natural) selection. However, we do not think that limitations to scientific underdetermination can or should be strong enough to reduce the drift in theory space toward a deterministic path, in which each step forward is the only allowed one. In other

---

[8] This is only loosely speaking, because to be fully compatible with all existing observations may be too strong a requirement. After all, what counts as a corroborating observation is partly theory-dependent (for a summary of work on this point, see [18]), and one never truly eliminates theories that fail to be corroborated by an observation.

words, we hold that scientific underdetermination is unavoidable. In line with the broader perspective on science outlined in the previous section, we are naturally led to embrace such underdetermination as something to exploit as much as possible by creativity and ingenuity (which is the essence of the PoP). After all, it is exactly this underdetermination that makes scientific progress possible in the first place, since it is in the environment created by scientific underdetermination that new theories are found, ready to replace previous ones, and a wider environment increases the chances that we will find in it the conceptual and theoretical tools we need to explain the world better than we currently do. At the same time, as working scientists we are mainly interested in the scientific underdetermination of *existing* alternatives to a given theory, including poorly developed ones or less empirically corroborated ones. We believe that such alternatives have to be constructed and developed explicitly, at least to some extent, if they have to be taken as reason to be cautious in trusting a given theory, but also if they have to provide the comparative measure of competitive success for the same theory. As a result, we are not convinced that there is any advantage in going beyond local, and transient, underdetermination [1, 16]. We do not underestimate how difficult it is, in practice, to construct functioning theories. On the contrary, precisely because we are well aware of how difficult it is, we are less concerned with the existence of merely logically possible alternatives to any given theory, which we basically take for granted. In fact, the PoP stems from the resulting urge to devote as much intellectual energy as possible to this difficult task of constructing theoretical alternatives that can be used for a meaningful assessment of existing theories. We also stress that scientific underdetermination, the related competition among theories, and the theory assessment on which it is based should be understood relative to their temporal, dynamical nature and embedded in the historical (and very human) scientific development; it is an abstraction to consider them in purely logical and conceptual terms, as if theories were born complete and competing in some fictitious atemporal (and asocial) platonic arena of ideas. It may be a useful abstraction, in some limited case, but it is also a potentially misleading one, especially when it comes to methodological issues. This implies that when we write 'theories', we really intend *research programmes* [3], and that the aim of 'theory assessment' is to evaluate as objectively as possible their progressing or degenerating status – not any logical, atemporal truth content or their abstract likelihood of being empirically confirmed.

The preceding argument applies to both empirical and non-empirical theory assessment. Let us now focus on the latter.

NETA constitutes a necessary component of the work of scientists – one that is routinely performed, even if sometimes in a rather unconscious manner (like most methodological decisions and most conceptual analysis, especially when observational constraints and mathematical consistency conditions are already stringent).

So, it is certainly true that a picture of theory assessment that includes only empirical criteria (what Dawid calls 'canonical theory assessment') is too limited to really describe how science proceeds.[9] Therefore, more philosophical work aimed at strengthening its rational basis is certainly needed. It should also not be controversial at all that any non-empirical assessment criterion we may identify is and will always remain subordinate in its scope and weaker in its impact than any form of empirical theory assessment, save in peculiar circumstances (e.g., trivial or routine experimental tests in highly controlled and well-tested conditions can sometimes carry less heuristic weight compared to surprising theoretical discoveries; even in this case, though, as we will argue, it is tricky to turn the appreciation of such heuristic value into a proper assessment of its 'likelihood' as a theory of the world). Such a criterion remains subordinate for the simple reason that scientific theories are attempts at understanding the natural world, and therefore fitting the empirical data as well as possible is part of their defining goal. The criterion is weaker because the empirical constraints are for some rather mysterious reason especially recalcitrant, although not entirely immune, to being subjected to our sociological and cognitive biases; thus they have proved much more effective in applying selective pressure against our best theories than has any non-empirical criterion. Thus, while the role of NETA is stronger nowadays and in specific scientific subfields, we do not think that this amounts to any radical change of paradigm for scientific methodology.

We agree with scholars who emphasise the importance of non-empirical assessment, on the basis that this should not be confined to the realm of 'theory construction' but rather should be properly understood in the context of 'theory justification' – that is, not as part of the heuristics of discovery, but as a genuine form of theory assessment, alongside (but subordinate to) empirical assessment. However, one main reason for objecting to this confinement is that the very distinction between the context of discovery and the context of justification appears very blurred when the dynamical, historical, and very human character of scientific progress is taken into account. This is a standard result of the modern philosophy of science [19] as well as the daily experience of practicing scientists: Theories under development are constantly assessed in terms of their results and their promise, and the selective pressure they encounter is not applied only on

---

[9] We do not think this is really a picture that is held by anyone who has reflected on methodological issues or who has direct experience with the daily work of scientific communities. Therefore, it is at best an abstraction of a more refined understanding, and attacking this view as such risks attacking a straw man. We believe that it is certainly a straw man at the descriptive level: The influence of assessment criteria beyond direct empirical observations is well documented by historians of science and by practicing scientists. It may not be a straw man at the normative level, though, since one could hold the view that only empirical assessment should matter and other considerations should be rigorously excluded. In our opinion, this view would be very weak if the exclusion is pretended to be absolute and not just a matter of preferential status reserved for empirical considerations, exactly because it is very poorly supported by both history and practice of science. Indeed, this view may be so weak that attacking it might not be worth the academic effort, so to speak.

the 'end result' of theory development, assuming there is one at all, but leads to a constant redirecting and modifying of the same theories, including (albeit more rarely) their core assumptions. This is one more reason to talk about research programmes, rather than theories, as the true object of theory assessment. Given this blurred distinction, however, proponents of NETA criteria should be careful in not making the opposite mistake, by assuming that such criteria belong to the context of justification only and are simply applied to the results of theory construction, again as if the assessment was taking place in an abstract world of platonic ideas. Non-empirical assessment criteria *guide* theory construction and influence significantly the establishment of any partial consensus around a given theory. Thus, such a consensus, whenever present, cannot be taken as the *basis* for NETA, but rather is the *result* of it and is influenced by a number of other factors that are not directly classifiable as rational theory assessment (e.g., sociological factors).

The need for a more detailed analysis of NETA criteria is all the more pressing in research areas such as quantum gravity [21], cosmology, and fundamental physics more generally, where observational constraints are fewer and weaker, where much work necessarily must rely on theoretical considerations alone, and where the underdetermination of theory by data is more noticeable and inevitable due to the difficulties in acquiring such data in the first place. Having said this, one should add some important cautionary remarks, since fundamental physics offers also a vivid example of the true scope and risks of NETA.

There are several puzzling aspects of current fundamental physics and cosmology [22], due to observations that are anomalous with respect to the accepted theoretical frameworks. Dark matter and dark energy are two examples, and we have plenty of data about them. Might they be used to constrain, for example, quantum gravity theories? The answer, of course, depends on whether they refer to physical phenomena that can be influenced by (if they do not directly originate from) the fundamental quantum properties of spacetime. But this is an issue that is up to the various quantum gravity theories to settle; that is, it is a *theoretical* issue, and different quantum gravity formalisms [21] will have a different perspective on it (one can cite several examples of heuristic arguments or even concrete simplified models that suggest a fundamental quantum gravity explanation of such observational puzzles). It is affected by what a given quantum gravity formalism says about issues such as separation of scales, the validity of the continuum approximation for spacetime, and the regime in which effective field theory applies, to name a few considerations. Moreover, a very active area of research focuses on 'quantum gravity phenomenology' [23], where the observational consequences of various quantum gravity scenarios are explored within a variety of contexts. Some of these effective field theory models incorporate putative quantum gravity effects, usually based on violation or deformation of spacetime symmetries, and/or on the introduction of minimal lengths or other quantum gravity scales. In this category one should also probably include quantum cosmology models and other symmetry-reduced or otherwise simplified models of quantum black holes, all of which produce tentative observational signatures of the underlying quantum gravity structures. One can attribute the failure of such simplified models, and thus of empirical data, to constrain strongly the more fundamental theories to the need to improve our observations, of course. Nevertheless, the current inability of fundamental frameworks to connect to these effective simplified scenarios also plays a major role in their failure (see [24] for some recent work in this direction, in the context of cosmology).

In the end, how much we really need to rely on non-empirical assessment criteria is a theory-dependent fact. If the goal remains to maximise the testability and adherence to the empirical reality of our theories, then our non-empirical assessment criteria – that is, the way we select and shape our fundamental theories, indirectly driving their evolution in one direction or another – should be such that they encourage connection with observations. Then, we have to be wary of any criterion that not only is of a purely non-empirical type (which may be a necessity), but also makes it more difficult to move toward later empirical tests. On the contrary, we should remember that the relation between theories and observations is not a passive one, in which theories sit there waiting to be tested by future observations. Rather, theoretical developments themselves, and the non-empirical criteria we adopt to constrain them, may facilitate or delay (or make impossible) such future testing. In fact, one could even raise this suggestion to an assessment criterion in itself: The community should favour (devote more resources and efforts to, regard as more promising, trust more as a viable description of nature) a theory or a theoretical development of an existing framework that brings it closer to empirical testing, or that points to new phenomena amenable to empirical testing (rather than in the opposite direction). Another criterion in the same spirit would amount to a negative assessment of any theory that fails to connect, even approximately or tentatively, to observations (i.e., fails to make predictions or to suggest testing observations of some of its basic tenets) after many attempts or after a long enough time. Obviously, the lack of such a criterion would imply that the theory is not viable, but it would amount to a non-empirical assessment of insufficient promise to be shown as viable in future observations, simply because it will probably fail to connect to them in a solid way. From this point of view, the PoP acquires an immediate important status, exactly because it makes it easier to identify possible ways of testing theories and encourages their (collective) confrontation with observations. The more empirical assessment criteria recede in strength as the PoP takes centre stage and should provide the main basis for theory development and for NETA.

Facilitating or, conversely, making more difficult to move closer to empirical assessment is only one way in which NETA could be both useful and dangerous. There are more aspects to be considered in this regard. First, good non-empirical assessment criteria should fully embrace fallibilism regarding scientific theories; they should aim to define some form of 'non-empirical falsificationism' and avoid any attempt to resurrect at the non-empirical level the dead corpse of verificationism. Now, it is true that, on the one hand, we never truly falsify theories [3], so that any such attempt should simply be understood as applying additional (stronger or weaker) selective pressure on proposed or existing theories. It is also true that, on the other hand, any strategy aiming at verifying (falsifying) a theory can be a small step toward falsifying (verifying) it, when it fails. However, the subtle methodological and psychological difference between the two approaches should not be underestimated. In other words, the point of NETA (as of empirical theory assessment) is not to give us reasons to "trust" a theory, but rather to provide weapons to challenge our trust in it. It is far easier to be fossilised in our trust of a theory, especially absent experimental tests, than to challenge it. The danger is not so much the reliance on non-empirical assessment criteria, but the weakening of the skeptical attitude toward our own theories that should accompany them (and the empirical ones).

Second, we believe that reliable NETA criteria should be justified by a careful analysis of the history of science. Such analysis should complement any given rational argument in their favour. This does not mean, of course, that if a criterion has been used in the past, that fact is *in itself* an argument (let alone a sufficient one) in support of its rational validity. It means, though, that detailed case studies as well as broader surveys of scientific practices and many key historical turning points can allow us to infer which NETA criterion has proved useful in the past, on a rather regular basis, and may potentially be useful in the present and in the future. The history of science can also offer a sort of 'empirical testing ground' for the assessment criteria we identify, in the sense that detailed cases studies and broad surveys can be used to 'falsify' our methodological claims, or at least their universal claims. The key words, however, are 'detailed' and 'broad'. Absent such detailed and broad confrontations of proposed NETA criteria within the history of science, any attempted inference of the former from the latter risks being convenient cherry picking. To put it differently, in our analysis of historical scientific developments, just as in our analysis of the current scientific situation in one or another theoretical domain, it is crucial to distinguish rational assessment, referring to the identification of rational criteria for NETA that have been used, from post-hoc rationalization of historical developments that were independent of such supposed criteria. In looking back at the history of science, we should pay due respect to it, and avoid the risk of doing the opposite – that is, understanding history on the basis of present decisions.

If we have to bring together all these main aspects of what should be the basis of good non-empirical theory assessment, it seems to us that the PoP should be regarded as the foundation of it, just as it has been argued to be crucial for empirical theory assessment, and even more so, due to the peculiarities and dangers of non-empirical variant. This foundational role stems from the fact that NETA, just like the empirical version, is always comparative, or at least it certainly works best when it is comparative. It is a drastic simplification to picture the process of empirical testing of a theory as if it occurred in theoretical isolation from competing theories. This is even more true for the confrontation of theoretical frameworks with our favourite non-empirical assessment criteria. Moreover, the very existence of competing theories enhances the strength of our non-empirical assessment criteria. We believe one can see the immediate non-empirical counterpart of the standard arguments in favour of the PoP, described earlier, as a basis for the conviction we have just stated.

In brief, the PoP is naturally the engine of a methodology of NETA that is subordinate to empirical assessment and aimed at promoting it, fuelled by skepticism toward the theories it is applied to, aimed at challenging rather than corroborating them, fully embedded within the historical scientific development and intended to contribute to its dynamics, rather than fictitiously placed in the abstract realm of logical possibilities.

8.3 Dawid's NETA Criteria: A (Non-Empirical) Critical Assessment

The general view on science we have outlined so far, and the resulting focus on some specific aspects of NETA that we deem crucial, as discussed in the previous section, inform our appraisal of the specific NETA criteria that have been proposed by Dawid [1].

These are the unexpected explanatory coherence argument (UEA: 'If a theory that was developed to explain the phenomenon X is unexpectedly found to *also* explain the independent phenomenon Y, then this theory is more likely to be true') argument,[10] the no-alternatives argument (NAA: 'If no alternative to a given theory is found to explain X, then this theory is more likely to be true') [25], and the meta-inductive argument (MIA: 'If a theory, which is supported by UEA and NAA [or other non-empirical assessment criteria], has been developed within a research program/tradition in which the same non-empirical criteria have proved successful in selecting empirically corroborated theories, then this theory is more likely to be true').

---

[10] We admittedly paraphrase the arguments in a rather rough manner, but not incorrect or unfair. Obviously, 'true' in this context means 'observationally viable'.

While we find Dawid's application of these criteria to string theory extremely generous and not really solid, we share a basic appreciation of all of them, in that they are indeed often employed in the practice of scientists. In particular, the UEA and the NAA are often cited by practicing scientists in support of their favourite theories, while the MIA is only implicitly employed, if at all. However, the NAA is rarely stated in such strong form by cautious practicing scientists, but rather mentioned in the form of a weaker 'best among the alternatives' argument or the disappointed 'only remaining option' argument. The only community that asserts the NAA in such strong form is the string theory community, and, to be fair, only some subsets of it. We will discuss the NAA argument in more detail later on, explaining why we fail to see its usefulness, why we see instead its counterproductive aspects, and why we think that the PoP represents instead the only useful core of the chain of reasoning leading to the NAA. Our conclusion, in this respect, will then be that the NAA plays no useful role, and that on the contrary we should recognize the full importance of the PoP, in a sensible non-empirical (part of) scientific methodology. The latter is, of course, not an assessment criterion in itself, but should be understood as the necessary precondition for theory assessment, and a sincere recognition of its necessity also runs contrary 'in spirit' to the adoption of a NAA.

The UEA refers to a property of many successful theories, and unexpected explanatory power of theories under development is unanimously considered a sign of their promise. We emphasise that, to have any value at all, the UEA applies only *after* the successful explanation of phenomenon X is achieved.[11] We also emphasise that the UEA works in support of the given theory *in comparison* to its competitors, not in an absolute sense. Indeed, it is a bit more difficult to move from promise to likely observational viability, as Dawid does, and it is not a step that can be taken on the sole basis of the UEA; in fact, it is not a step that scientists take, in general. Dawid explains this point well, mentioning several possible explanations of the unexpected explanatory power of a given theory that do not imply its own observational viability. In particular, we agree with his conclusion that the UEA can be used as a non-empirical support of a given theory only in conjunction with another criterion. However, while he identifies this additional criterion to be used together with the UEA in the NAA, we point out that in practice the only way to rule out the alternative explanations for the unexpected explanatory power of a given theory is to proliferate the theoretical approaches to the explanation of the same (theoretical or observational) phenomena that are connected by the theory in

---

[11] For example, one sometimes hears the argument that string theory is a likely explanation of gravity and fundamental interactions, on the basis of its unexpected and beautiful applications in mathematics. This could be an instance of the UEA only after we have agreed that string theory is, in fact, a good explanation of gravity and other interactions; otherwise, the whole argument is a non-sequitur.

question. Thus it is the PoP that truly works in conjunction with the UEA to provide stronger non-empirical support to a given theory. Consider, as an example, the issue of determining whether the unexpected explanatory power of a given theory, with respect to some phenomenon Y, is really due to some underlying more general principle that is shared by other theories, such it does not really imply much about the viability of the theory being assessed. It should be obvious that the only way to do so is to imagine, construct, and develop alternative theories that incorporate the more general principle, but differ significantly from (or are even incompatible with) the theory being assessed, and then to determine whether if they have the same explanatory power concerning the phenomenon Y. The extent to which this is possible will give an indication of the strength of the theory in question as an explanation of Y, and it is determined by the non-empirical assessment criteria that will constrain the construction and development of alternative theories to it (assuming that they rely on the same empirical basis). This point should become even clearer in the following discussion, after our consideration of the NAA, its problems, and its relation with the PoP.

The MIA has a different type of problem. In its core, it is basically a formalization of the sensible suggestion to look carefully at the history of science, in particular the history closest to the theory being assessed, to determine whether any specific NETA criterion has proved useful in the past. If it has, present theories supported by it are naturally deemed more promising.[12] The main problem with turning this sensible suggestion into an assessment prescription is that the strength of the prescription is proportional to the depth and accuracy of the historical analysis and to the strength of its conclusions. Also, the type of historical analysis needed to run the meta-induction, if one wants to do it properly, is extremely difficult and complex. Moreover, as with any induction, it is proportional to the number of case studies that have been analysed. One would need to show convincingly that (a) the NETA criteria that one aims to 'test' were indeed primarily responsible for the support of a given theory (it is not enough to show that they contributed to that support, among many other factors, or to show that this is non-trivial at all); and (b) they have an historical track record of success in picking up the right theory – that is, they did so regularly, in a very high number of cases and, most important, they only rarely failed. It is not enough to show that they did so on a few occasions: It is the success *rate* that matters, not the existence of many instances of success. Obviously, these are indeed very difficult tasks for professional historians, let alone philosophers of science or scientists wishing to pick up useful assessment

---

[12] Again, the shift from 'more promising' to 'more likely to be empirically confirmed' is implicit in the practice of scientists, but subtle and very hard to justify at the conceptual level.

criteria for their theories (or, more likely, confirm their prejudices on the validity of their own criteria).

Another smaller criticism that can be put forward regarding the MIA concerns the reference to a given 'research tradition'. Several traditions live side by side in any scientific domain, which have complex inter-relations, and there are many potentially different ways of pushing forward each of them (depending on which aspect of a given tradition plays a central role in the new theory). Thus, even when a specific set of assessment criteria are found to be successful within a given tradition, it may not be obvious to what extent the new theory that we wish to assess by the same criteria is truly 'in the same tradition' (so that one can run the MIA) any more than alternative theories. Moreover, anchoring the potential success of a given theory to past successes of a given research tradition runs the risk of being a truism, since most theories are further developments of successful paradigms, inheriting most of their methodology, basically by definition (bar the case of true scientific revolutions). If it reduces to this level, the MIA is not very useful, because it could apply to any of the features of the given tradition, rather than specifically to any particular methodological criterion. The attempt to apply the MIA to the NAA runs into even more serious troubles, in our opinion, coupling the problems with historical induction applied to it to intrinsic problems of the NAA itself.

The more serious problems of Dawid's criteria, in fact, relate to the NAA.

The first problem with the NAA is that it is simply never true that there is only a single theory or model or hypothesis that is able to explain, in principle, a given phenomenon, if one intends, by this, that the whole community of scientists working on explaining that phenomenon was able to come up with only a single such hypothesis or model, after long enough trying. We know this for a fact in the context of quantum gravity research. Likewise, we believe that a careful enough analysis of the historical situation of any specific subfield of science, concerned with the explanation of any specific phenomenon, would show that, at the time, there were several competing ideas for such explanation. Some are more developed, some less developed; some have a larger number of followers, others have less consensus around them; some are developed within mainstream research directions, some are put forward by somewhat lateral sectors of the community; some are more in line with accepted paradigms, some based on more heretical (and thus risky) suggestions. Recall that the NAA pretends to infer the viability of a theory from the *absence* of alternatives. Thus, one should first ascertain with some confidence this absence, if the argument is to be made at all (even assuming that it is valid, which we will dispute). We argue that this is never, historically, the case.

Does this mean that we contest the fact that sometimes a given community of scholars can be *left* with no serious alternative to a given theory for explaining a given phenomenon? Of course not. It is basically the definition of accepted science that the (large majority of the) community of scholars has ended up agreeing on a given explanation for a phenomenon, at the exclusion of all others. However, this is (and has always been) the result of empirical assessment, performed either as a final way of eliminating (actually, disfavouring) alternatives to a given theory or as a way of further corroborating a theory that had already won large support so as to finally convince the remaining skeptics. In both cases, the empirical assessment was, ultimately, a comparative assessment. Therefore, it was first made possible, and then made stronger, by a previous phase of intense theory proliferation. This is the basis of the PoP. The question is whether an entirely non-empirical counterpart to this situation is possible. Can we have a situation in which 'there is no alternative' to a given theory before any empirical testing has been performed?

We can see only two ways in which this can be achieved. Neither of them, we argue, allows us to infer anything, per se, about the observational viability of the theory that is found 'without alternatives'.

The first way is by matter of definitions. One can exclude one or more alternatives to a given theory because they do not fit a set of previously chosen desiderata or preconditions, often amounting to adherence to the established paradigm. There is nothing wrong with this approach, of course. The tricky part is which specific desiderata should be chosen to define a 'successful' explanation of a phenomenon, because this amounts to a specific hypothesis about which aspects of the existing paradigm should be maintained despite the inability to account for some characteristic of the world[13] and which other aspect should be understood as bearing the responsability for this inability. These desiderata can be of very different types. They can include specific mathematical ingredients that one believes should be part of the sought for theory, such as manifest background independence in the case of quantum gravity, or physical requirements we would like to be realised by our theory, such as that it should describe fundamental interactions, including gravity, in a unified manner as manifestation of a single physical entity, rather than simply in a unified mathematical language. The desiderata can also include requirements on the level of mathematical rigor that a compelling physical theory should demonstrate, or on the number of additional hypotheses it can be allowed to make to fit existing observations (since a newly proposed theory, based on some new hypothesis that stands in contrast to a current paradigm, may be at first stated in contrast to some part of the observational basis of the same paradigm) – that is, requirements on how 'speculative' or 'radical'the theory is allowed to be. Also, it is often the case that the very definition of the phenomenon to be explained is itself influenced by a number

---

[13] Here we take this inability for granted, for the sake of the argument. However, one may simply be dissatisfied with the specific way in which the existing theory describes the world, and consider alternatives to it just for this reason.

of assumptions and preconceptions, carried over from the existing theories. Because some of these assumptions and preconceptions are necessarily reconsidered and dropped when developing a new theoretical framework, our very definition of the phenomenon to be explained by the new framework will often be revised in due process or, at the latest, once a new theoretical framework is adopted to describe it. All this is normal, but it shows that one can always put enough restrictions on the *definition* of what we mean by a candidate explanation of a given phenomenon to be left with a single such explanation. But these are often restrictions to both our imagination and the paths we decide to explore. It is very dangerous to apply too many of them, if we are really interested in finding the best possible explanation of a given aspect of the natural world.

Conversely, assuming we are in a situation where a single candidate explanation of a given phenomenon is available, it is often enough to lift some of our preconceived assumptions about the phenomenon itself, or how a good explanation of it should look like, to come up with other tentative alternative explanations to be explored further. Both the PoP and the usual scientific practice encourage us to be careful in damping this further exploration by trying to decide in advance how it has to be carried on. If this is the way in which a 'no-alternative situation' is realised, we do not have anything that helps establish the viability of the survivor theory; on the contrary, we have a problem. We should go back a few steps, and let the PoP come to help.

The second way in which a 'no-alternative' situation can be established is by actual comparison of alternatives, followed by elimination of some of them as not viable or not promising. This can take place in two cases, distinguished by the stage of development of one or more of the alternatives being compared.

The first case is when the comparison includes theories that are severely under-developed or in a very early stage of development. These may range from solid and vast theoretical frameworks that have several key open issues in their foundational aspects or in their physical implications[14] to theories that are barely above the status of hypotheses or templates of actual scientific theories or models.[15] It should be clear that the distinction between theories in this category, and their placement across the mentioned range, is a matter of degree and not entirely objective. This is the most common situation in practice, if one includes in the comparison theories in their very early stages of development ('very young research programmes'), and even theories barely beyond the level of interesting speculations. The former can be obtained from the existing paradigm or accepted partial theories by simply modifying some aspect of them or by pushing some of their features a little

---

[14] Personally, we would include in this category most current approaches to quantum gravity.
[15] Here we would include all the quantum gravity approaches that do not fit in the previous category.

further; the latter can be generated quite easily with a little imagination, and in fact abound in active communities facing some interesting open issue. This is also the case where comparison is the most difficult. How can we compare a large albeit incomplete, research programme that has most likely already obtained a number of partial results and therefore attracted the interest of a large subset of the community, with a young and small research direction that is, as expected, much less developed and with fewer partial results to its credit? How do we compare a theory that, perhaps because it corresponds to an extension of an existing paradigm, has obviously a more solid foundation, with a wilder speculation, presenting many more shaky aspects but maybe still a promising solution to the problem at hand? How can we make this comparison fair?

Notice again that we are talking about the comparative *dynamics* of theories, not an abstract side-by-side comparison in some logical space. This means, for example, that any partial success of a given theoretical framework can immediately suggest a modification of an alternative to it that allows it to achieve the same success, or the generation of a new hypothesis for the solution of the same problem, which is then added to the landscape to be developed further. In a case like this, how can we reach a 'no-alternative' situation?

We could decide to set a threshold on the level of development, below which we judge a theory 'too underdeveloped' to be compared with the others or to represent a genuine alternative to them. The problem is that any such threshold will be very ambiguous both quantitatively (the 'level of development' is not something that can be measured with any precision) and qualitatively (Which aspect counts as crucial for considering a theory 'well developed'? Its mathematical foundations? Its conceptual clarity? Is it worse to lack a clear connection to possible observations or to be unsure about the fundamental definition of the theory itself?). And again, is the *level* of development more important than the *rate* of development – that is, its progressive state as opposed to stagnant situation? Or is it more important to estimate the *efficiency of its development rate*, meaning its rate of development corrected by the amount of resources (time, human power, funding) that have been devoted to it?

In the end, it is not just unfair but logically incorrect to infer anything definite about the viability of a theory from the (possibly) true fact of it being the most developed candidate explanation of a given phenomenon. Such a move, far from being logically cogent, amounts to nothing more than noticing how difficult it is to construct a working explanation of a given phenomenon. Any practicing theoretician knows how difficult it is to come up with a compelling explanation of a phenomenon (especially if we are not talking about effective models but candidate fundamental theories). Thus, the correct attitude in the face of this difficulty is to refrain from drawing conclusions from the underdeveloped status of a hypothesis or a framework in the absence of observational support for any of the competing hypotheses on the table, and instead to devote more work to bringing them to the level that makes any comparison meaningful at all.[16]

The second case (which may be obtained after the application of some development threshold in the first case) is when we compare only well-developed theories with each other. Notice that this situation is may be fairer, but not much less ambiguous than the first case. In the absence of empirical tests, any such comparison involves judgements about which outstanding open issues of each theoretical framework are serious shortcomings, perhaps even preventing the theory from being considered as a solid candidate for the explanation of a phenomenon, and which issues can be instead looked at with some indulgence as temporary shortcomings in an otherwise compelling theory. In addition, some of the open issues of one theoretical framework may relate directly to aspects that are considered *defining features* of a successful explanation of the phenomenon of interest. For example, we may consider unification of all forces to be a defining feature of a successful theory of quantum gravity (to be clear, we do not agree with this definition). In turn, we would evaluate negatively a proposed theory of quantum gravity that, while complete in other aspects, has not made such unification a central feature, and thus is largely underdeveloped in this aspect, even when it incorporates several strategies for achieving such unification. As a consequence, often this second case is at risk of being an instance of the elimination of alternatives by definition that we discussed earlier This implies that this second case provides a very shaky ground for representing a starting point of the NAA.

Finally, let us notice that the more careful formulation of the NAA as 'despite a long search for alternatives to the candidate theory A, no alternative has been found', including explicitly a temporal aspect, is, in fact, a restatement of the first case, more often, or of the second case. Indeed, it is usually the shorter version of the more precise statement: 'despite a long search for alternatives to the candidate theory A, no alternative has been found to be equally compelling, in the sense of being equally developed and solid and of fitting all the assumed desiderata'.

In the end, on both historical grounds (by which we also mean the daily work of theoreticians) and conceptual grounds, we conclude that the very premise of the NAA is hard to justify. We identify the root of the difficulty in the attempt to run NETA on the logical grounds of abstract theories, as if they came to us fully formed and definite, neglecting the dynamical and permanent-work-in-progress nature of scientific theories.

---

[16] Let us stress that we are objecting to the inference from 'well developed' or even 'best developed' to 'likely true' or 'likely empirically viable'. We do appreciate the importance of reaching the former status, exactly because it is difficult to achieve that feat.

Indeed, the main problem with the NAA, even going beyond all the ambiguities involved in setting up its premise, is that it creates the following contradiction: Even if the premise ('there is currently no alternative to A') was in some sense correct, it is the *result* of some provisional NETA that has been already carried out. Therefore, as far as the analysis of NETA is concerned, the interesting issue is to understand how such assessment had been performed, not the result of it. The theory emerging 'without alternatives' as a result of such assessment will be only as compelling, as a scientific theory, as the (non-empirical) criteria used to arrive at it. The NAA therefore ends up portraying itself as a non-empirical assessment criterion when it is at best a restatement of the *result* of some unexplored (and not even formalised) set of non-empirical assessment criteria having been applied at an earlier stage. But then one has to conclude that the NAA is not, and cannot be, a non-empirical assessment criterion itself. Moreover, it is actually at risk of diverting the needed attention to the real (non-empirical) assessment criteria that led to its premise ('there is no alternative to theory A'), even assuming this premise is an accurate statement of fact (and, we emphasize again, we believe this is basically never the case in such crude terms).

Moreover, the premise of the NAA is inevitably a statement of fact about the *consensus* reached in a given research community (or a subset of it). As such, it risks being nothing more than a sociological observation, which should carry little heuristic power in itself. In particular, and for the reasons presented previously, it fails to provide a rational ground for that consensus, which is taken as the starting point of the argument, and has been achieved by other means (which can also be rather contingent [20]). At the same time, the application of the NAA inevitably feeds back on that consensus, reinforcing it. This leads to another problem, again resulting from the failure to take properly into account the dynamical and temporal aspects of theory assessment – that is, how the use of specific NETA criteria, and in particular the NAA, may feed back on theory development. We have already discussed, in the previous section, how important it is to ensure that our non-empirical assessment criteria are not only well grounded in the history of science and in the practice of scientists, but also encourage scientific progress rather than hamper it, favouring the progressive dynamics of research programmes rather their degeneration [3]. What do we have instead, in the case of the NAA? We have a consensus that is maybe well founded for other reasons and that has been achieved by the application of other non-empirical assessment criteria, which we do not explore. We introduce an additional argument that is not much more than a restatement of the obvious fact that the rational assessment criteria we employ to achieve consensus are designed to improve the likelihood that a given theory is correct. This has the simple effect of reinforcing the existing consensus.

In practice, the NAA works as a pure mechanism for strengthening the consensus, rather than as a tool for critically examining it. This is exactly the opposite of what a non-empirical assessment criterion should do – that is, to facilitate the progressive dynamics of research programmes. Its applications makes it more difficult, rather than simpler, to identify critical aspects of our favourite theories, or any prejudice that has unnecessarily informed their construction or earlier assessment. And it follows blindly along the flow of any existing sociological or cognitive bias that we may have been prone to in developing our currently favoured theory.

To appreciate this aspect in the case of NAA, let us consider one extreme situation, in which its premise is clearly evident. This is the situation in which a community is looking at a new problem – for example, explaining some new phenomenon – for the first time, and a first candidate explanation for it is proposed. This is usually the simplest extension of the current theoretical framework (which has been, by definition, mostly successful until now). Now suppose we apply the NAA. This is indeed one (and possibly the only) case in which its premise holds and it is not the result of any previous theory assessment. In turn, the candidate explanation is a bit more likely to be correct (than it would be if alternatives existed). In this case, one is actually assigning some value only to the adherence to an established paradigm; the NAA would simply reinforce the competitive advantage that follows from this assumption. This automatic reinforcement is exactly the opposite of what a good non-empirical assessment criterion should do. Of course, the inference from this process of any viability of the theory under consideration is questionable, to say the least.

The NAA appears extremely weak in its negative consequences as well. Continuing the same example, it is obvious that the situation in which a single candidate explanation for the new phenomenon exists is very short-lived. Usually, the relevant community comes up with a number of alternative explanations that more or less radically depart from the existing paradigm. The NAA would imply that as soon as such alternatives are produced, the viability of the first proposed theory is ipso facto decreased – not only before any observational test is performed, but also before any other non-empirical assessment criterion is applied to any of them. We do not find this convincing at all.

The immediate objection to this example, and the doubts it raises, is that the NAA is not supposed to be applied at such preliminary stage of theory development, and that its premise asks that no alternative to a given theory is found 'after an extensive search'. Granted. But we fall back on the previous objections. It is never the case that no alternative is 'found' in the sense that none is proposed and studied. If one takes this dynamical perspective on theory assessment seriously, one should realize that the main extra element that 'time' brings to the picture is that both the original theory and its proposed alternatives are developed further, criticized, and assessed. As a result, some of them may stop progressing fast enough or become less popular. The new ingredients are therefore the additional non-empirical assessment criteria that are employed to determine such progress or lack thereof, and the consequent increase or decrease of popularity. They would be the interesting object of enquiry, as we emphasised previously, but the NAA does not add anything to their work. On the contrary, its only effect would be to stifle the development and exploration of alternatives and the reinforcement of any temporary consensus that has been achieved. It would increase the chances of theoretical impasse rather than growth.

In this dynamical perspective on theory development and assessment, what should be the reaction to a situation in which a problem is left unsolved; a number of theoretical explanations is proposed; and all are non-empirically assessed, with one of them emerging as more convincing than the others? The only progressive strategy is to double the efforts to develop such theory to the stage in which it can be empirically tested, to develop its alternatives further to overcome their current difficulties and to offer a stronger challenge to the favourite theory, and to produce even more theoretical alternatives that can offer the same challenge. The strategy should be to create a landscape in which theories push one another forward, force each other to explore hidden assumptions and to solve outstanding issues, and rush toward connecting to the observations. In other words, the most productive strategy for NETA, in a situation like the one outlined here (a single alternative seems to be available to solve a given problem) is the one based on the PoP (and that resists any temptation to use the NAA to put our minds at peace), even more than in a situation in which observational tests abound. If non-empirical assessment criteria end up instead *limiting* the range of explored alternatives or the effort in their development, they make it harder to proceed to empirical assessment – that is, to confrontation with the real world.

A closer look shows that Dawid's criteria, and in particular the NAA (in the attempt to establish its premise), crucially rest on the PoP at multiple levels; if they have any strength at all, it is only within a proliferating environment.[17] However, we have also argued that they do not have much strength, and that a proper appreciation of the importance of the PoPreally undermines them, in particular the NAA. This is because whenever a first stage of proliferation leads to a situation where the NAA could be invoked (merely as a statement of fact, but without much heuristic value), the response should be 'more proliferation'. If this point is missed, as the enthusiasts of Dawid's criteria in the theoretical physics community risk doing, the

---

[17] This point is probably obvious to Dawid himself, but not to many of those who have been hailing these arguments in the theoretical physics community.

same criteria may end up instead undermining theory proliferation, NETA with it, and, in perspective, scientific progress; they may end up encouraging stasis, rather than moving forward toward better future theories.

We believe that it is important to stress these points, even if some of them may sound obvious to many practicing scientists and to many philosophers of science. Similar points have been raised by other philosophers of science interested in NETA, most notably P. K. Stanford [16, 27]. Stanford makes a number of points we agree with, which are directly relevant to the present discussion and undermine the significance of any NAA, but also of the MIA applied to it: '[W]e have, throughout the history of scientific inquiry and in virtually every scientific field, repeatedly held an epistemic position in which we could conceive of only one or a few theories that were well-confirmed by the available evidence, while the sequent history of inquiry has routinely (if not invariably) revealed further, radically distinct alternatives as well-confirmed by the previously available evidence as those we were inclined to accept on the strength of that evidence.'[16]. In other words, subsequent investigation always produces new evidence for past scientific underdetermination – that is, new theories that are equally valid on the basis of previous evidence – and always shows how previous non-empirical limitations to scientific underdetermination were too weak. This is an historical fact, and any reflection of NETA should start from it. To presume the end of this process of continuous discovery of past scientific underdetermination is in many ways tantamount to presuming the end of science. No wonder the NAA is proposed in conjunction with a final theory claim.[18] The previously stated consideration from the history of science, which was implicit in our previous discussion, gives a basis for the 'meta-inductive' support for the central role of the PoP in NETA, in the same spirit but in exactly the opposite direction of the MIA, when applied in conjunction to the NAA. In the words of Stanford [16], 'the scientific inquiry offers a straightforward inductive rationale for thinking that there typically are alternatives to our best theories equally well-confirmed by the evidence, even when we are unable to conceive of them at the time'. This point is raised by Stanford to a 'New Induction over the History of Science' (NIoHS) principle in epistemology/methodology of science (which he is also careful to distinguish it from Laudan's 'pessimistic induction' in the context of the debate about scientific realism [28]). It leads us to advocate for the following: The application of Dawid's MIA to concrete cases is difficult because, absent a truly detailed analysis of the scientific context and minute developments of each such concrete case, it is too

---

[18] We do not discuss final theory claims, since it is beyond the scope of this chapter, and too much tied to current issues in fundamental physics. We do not discuss it also because we believe that our criticisms of the NAA make the further implications for a final theory claim less interesting. However, it is important to keep in mind that the NAA leads naturally in the direction of a final theory claim, as Dawid correctly argues.

prone to cherry picking and to twisting the history to our purposes. Historical analysis shows that the logic 'I cannot conceive any plausible alternative to X; therefore the likelihood of not-X is low' is flawed. Far too many examples from the history of science can be cite when this 'no possible alternative can exists' conviction, however strongly held, was later shown incorrect [16].

There are two possible ways out of this conclusion, and against the NIoHS, both not very convincing to us. One could try to argue that some new theory also enables our imaginative capabilities to exhaust the space of serious possibilities. This is not impossible, but it would have to be based on a serious analysis of our cognitive capabilities as well as a deep philosophical (epistemological) analysis. As such, it cannot be internal to any specific scientific theory; that is, one cannot deduce this exhaustion of possible alternatives from internal aspects of a given theory,[19] and certainly the historical fact of absence of alternatives to a given scientific theory cannot be such argument. Another possible objection would counter that scientific communities are much more powerful in exploring possible alternatives than individual scientists, and that modern ones are even more effective than earlier ones; therefore it is not so implausible that they might manage to explore all possibilities and in the end find there is really no alternative to a given present theory. Once more, we believe instead that sociological developments have affected the nature and workings of scientific communities in a way that makes the exploration of theoretical alternatives less – not more – encouraged and effective, due to a general push for conservatism. Among the factors producing this result, one could cite professionalization, institutionalisation (with rigid machinery of science production, validation, and funding), and the rise of Big Science (which also favours risk aversion and conservatism, hierarchical organisation, and the like) [7]. To give just one example, we refer to Cushing's pyramid structure of scientific communities to explain convergence around one approach, which is particularly strong in the absence of empirical tests [29]. Therefore one cannot avoid adopting something like the NIoHS and, in conjunction with a cautious MIA, taking seriously its warning against the NAA.

In the end, we do not find the NAA, nor the MIA applied to it, very convincing as useful NETA criteria. On the contrary, we find it potentially damaging for the healthy progress of scientific communities. Based on the arguments presented in this chapter, we call instead for the PoP to be recognised as the key to NETA, and the true engine of scientific progress.

---

[19] This goes directly against the final theory claim based on the structure of string theory [26].

8.4 Conclusions: No Alternative to Proliferation

In this chapter, we presented some reflections on NETA, based on a general view of science as akin to an adaptive system in evolutionary biology, with empirical and non-empirical theory assessment playing the role of the selective pressure that ultimately drives 'progress' (in the non-teleological sense of 'evolution'). We also emphasixed the other key aspect that our perspective relies on: the very human nature of science, as the product of individuals and communities, which does not contradict but actually improves (when it functions well) its objective content. This emphasis on the human aspects implies paying very careful attention to the social and cognitive biases that are part of theory development and assessment, and which assessment criteria should help to fight. In addition, this general view suggests the main ingredients for our analysis of NETA: the importance of the dynamical aspects and feedback loops of theory construction and theory assessment, and the central role played by the proliferation of theories in achieving progress, expressed as the PoP for scientific methodology.

We tackled first the issue of NETA in general terms. We agreed on the link between theory assessment and scientific underdetermination, in which the first aims at reducing the second. However, we recalled that scientific underdetermination is also the humus of scientific progress, which should also be based on tools aiming at expanding the landscape of possible theories that generate the same underdetermination. This is a first general argument for the PoP. We argued that NETA deserves all possible attention, because it is routinely performed by scientists, so this philosophical analysis can lead to an improvement of both our understanding and our practice of science. We also emphasised that it is important to keep in mind the subordinate and weaker nature of NETA in comparison with the empirical version, in any attempt to identify good non-empirical assessment criteria.

To identify such *good* assessment criteria, we stressed that, since there is an important feedback loop between any form of theory assessment and theory development, our non-empirical theory assessment criteria should likewise be judged in terms of the feedback they produce. Also, since the relation between theories and observations is not a static and passive one, theoretical developments and the non-empirical criteria that constrain them may facilitate or delay future testing; therefore we have to be wary of any criterion that makes it more difficult to move toward later empirical tests. From this perspective, the PoP should be seen as the key aspect of non-empirical assessment (one of the main reasons why it was first proposed).

We also argued that good non-empirical assessment criteria aim at falsifying theories, rather than confirming them, just like the empirical criteria. Thus, NETA (like the empirical variant) should provide weapons to challenge our trust in our best theories, since it is far easier to become fossilised in our trust of a theory, especially absent experimental tests, than to challenge it. Again, the PoP naturally plays a role in keeping alive this moderate skepticism that lies at the heart of science.

Next, we turned to a critical analysis of the specific criteria for NETA proposed by Dawid. While we agreed with him that the UEA of a given theory is an important criterion that is routinely used, we emphasized that its strength lies in a simultaneous extensive use of the PoP to produce alternative theories and in the criteria that are then used to constrain or eliminate them, rather than, as Dawid argues, in the NAA. Thus our only objection to Dawid's UEA, beside an additional emphasis on the PoP, is really a critique of the NAA.

We argued that the NAA is both poorly supported and dangerous owing to its feedback on scientific progress. First, it is never really the case that there simply are no alternatives to a given theory, if we intend it as a pure statement of fact and in such absolute terms. Second, any temporary situation of no alternative to a given theory can be achieved in only two ways: (1) by restricting the terms of the problem (either what the phenomenon to be explained really entails or what the explaining theory is supposed to look like) or (2) by applying some (non-empirical) assessment criteria that disfavour the alternatives. These additional criteria may assess the stage of development of the alternative theories as well as their content, and weigh their inevitable shortcomings. This process is necessarily tentative, often ambiguous. Thus, even when it leads to establishing something like the premise of the NAA, it provides a very shaky ground for drawing any conclusion from it. Most importantly, if the premise of the NAA is the result of applying some non-empirical assessment criteria to a set of alternative theories, it is these criteria that we should identify and analyse. It is their strength that gives any credibility to the resulting theory found to be the only credibile one, and the NAA itself has no additional heuristic value. Third, the NAA rests on the consensus reached on a certain theory in a specific scientific community, but fails to provide any rationale for that consensus. Despite this, it becomes inevitably a tool to reinforce that consensus, and risks being nothing more. In other words, it does exactly the opposite of what a good NETA criterion should do – that is, to facilitate scientific progress. Once more, the crucial ingredients entering our analysis are the temporal and dynamical aspects of theory assessment and the consideration of how it informs theory development.

Our brief analysis, in the end, shows that the NAA crucially rests on the PoP at multiple levels, since its only chance to have any strength arises within a proliferating environment. However, our analysis also shows that the NAA does not have much strength, and that a proper appreciation of the importance of the PoP really undermines it entirely. We argue instead that whenever a first stage of proliferation leads to a situation in which the NAA could be invoked (merely as a statement of fact), the response should be 'more proliferation'. The failure to do so may end up undermining theory development, NETA, and scientific progress.

The importance of the PoP, the role it plays in NETA and the consequent undermining of the NAA find further support in Stanford's analysis based on the history of science, that goes exactly in the direction opposite to the NAA, even if it has the same spirit as the MIA suggested by Dawid. We also warned about the difficulties of applying the MIA to any NETA criteria, stressing how drawing conclusions from the history of science, as the MIA entails, requires a level of detail and broadness of the historical analysis that is hard to achieve. We conclude that, while we do not disagree with the MIA itself, a careful and broad look at the history of science is more likely to undermine the strength of the NAA rather than the opposite.

In the end, our conclusion is simple. In the search for a better understanding of the natural world, experiments and observations are our best ally; proliferation and fair competition between theories are their complement at the non-empirical level. Both feed the natural, reasonable, and reasoned healthy skepticism of scientists. So should do any NETA criteria. Especially when empirical constraints are scarce, the proper reaction should be more proliferation, and the careful use of those non-empirical assessment criteria that facilitate scientific progress by constraining, but also facilitating, such proliferation. Scientific progress requires an open and dynamical scientific community, and an open scientific community entails the highest degree of theoretical pluralism.

8.5 Acknowledgements

We thank S. De Haro, V. Lam, C. Revelli, C. Smeenk, K. Thébault, C. Wütrich, and especially R. Dawid, for numerous discussions and clarifications on the topics of this contribution, and R. Dardashti for a careful reading of the manuscript and many useful comments.

References

[1] R. Dawid. *String Theory and the Scientific Method.* Cambridge: Cambridge University Press, 2013.

[2] K. Popper. *Objective Knowledge: An Evolutionary Approach.* Oxford: Oxford University Press, 1979; D. L. Hull. *Science as a Process: An Evolutionary Account of the Social and Conceptual Development of Science.* Chicago: University of Chicago Press, 1988; R. Boyd & P. J. Richerson. *Culture and the Evolutionary Process.* Chicago: University of Chicago Press, 1985; R. Giere. Philosophy of science naturalised. *Philosophy of Science* 52(1985): 331–56; L. Laudan. *Progress and Its Problems.* Berkeley and Los Angeles: University of California Press, 1977.

[3] I. Lakatos. Falsificationism and the methodology of scientific research programmes. In I. Lakatos & A. Musgrave (eds.), *Criticism and the Growth of Knowledge*. Cambridge: Cambridge University Press, 1970, pp. 91–195; I. Lakatos. *The Methodology of Scientific Research Programmes (Philosophical Papers: Volume 1)*, J. Worrall and G. Currie (eds.). Cambridge: Cambridge University Press, 1978.

[4] C. Matheson & J. Dallmann. Historicist theories of scientific rationality. In E. N. Zalta (ed.), *The Stanford Encyclopedia of Philosophy* (Summer 2015 Edition). https://plato.stanford.edu/archives/sum2015/entries/rationality-historicist/; H. Longino. The social dimensions of scientific knowledge. In E. N. Zalta (ed.), *The Stanford Encyclopedia of Philosophy* (Spring 2016 Edition). https://plato.stanford.edu/archives/spr2016/entries/scientific-knowledge-social/.

[5] D. Kahneman. *Thinking, Fast and Slow*. New York: Farrar, Straus and Giroux, 2011; M. Piatelli-Palmarini. *Inevitable Illusions: How Mistakes of Reason Rule Our Minds*. New York: John Wiley and Sons, 1994; M. G. Haselton, D. Nettle, & P. W. Andrews. The evolution of cognitive bias. In D. M. Buss (ed.), *Handbook of Evolutionary Psychology*. Hoboken, NJ: Wiley, 2005, pp. 724–46; D. Kahneman, P. Slovic, & A. Tversky (Eds.). *Judgment Under Uncertainty: Heuristics and Biases*. Cambridge: Cambridge University Press, 1982.

[6] P. E. Smaldino & R. McElreath. The natural selection of bad science. *Royal Society Open Science* 3(2016): 160384; A. Clauset, S. Arbesman, & D. B. Larremore. Systematic inequality and hierarchy in faculty hiring networks. *Science Advances* 1(2015): e1400005.

[7] P. K. Stanford. Unconceived alternatives and conservatism in science: the impact of professionalization, peer-review, and Big Science. *Synthese* (2016): 1–18.

[8] L. Smolin. *The Trouble with Physics*. Boston: Houghton Mifflin, 2006.

[9] L. J. Daston & P. Galison. *Objectivity*. Cambridge, MA: MIT Press, 2007.

[10] K. Bschir. Feyerabend and Popper on theory proliferation and anomaly import: on the compatibility of theoretical pluralism and critical rationalism. *HOPOS: The Journal of the International Society for the History of Philosophy of Science*, 5(1), 24–55. G. Munevar. Historical antecedents to the philosophy of Paul Feyerabend. *Studies in History and Philosophy of Science* 57(2016): 9–16; M. Collodei. Was Feyerabend a Popperian? Methodological issues in the history of the philosophy of science. *Studies in history and philosophy of science* 57(2016): 27–56; P. Achinstein. Proliferation: is it a good thing? In J. Preston, G. Munévar, & D. Lamb (eds.), *The Worst Enemy of Science?: Essays in Memory of Paul Feyerabend*. Oxford: Oxford University Press, 2000.

[11] P. K. Feyerabend. How to be a good empiricist: a plea for tolerance in matters epistemological. In P. K. Feyerabend, *Knowledge, Science and Relativism: Philosophical Papers*, Vol. 3, ed. J. Preston Cambridge Cambridge University Press, 1999, pp. 78–103; P. K. Feyerabend. Problems of empiricism. In R. G. Colodny (Ed.), *Beyond the Edge of Certainty: Essays in Contemporary Science and Philosophy*. Englewood Cliffs, NJ: Prentice Hall, 1965, pp. 145–260; P. K. Feyerabend. Reply to criticism: comments on Smart, Sellars and Putnam. In R. S. Cohen & M. W. Wartofsky (eds.), *Proceedings of the Boston Colloquium for the Philosophy of Science, 1962–64: In Honor of Philipp Frank*. New York, NY: Humanities Press, 1965, pp. 223–61; P. K. Feyerabend. *Realism, Rationalism and Scientific Method: Philosophical Papers*, Vol. 1. Cambridge: Cambridge University Press, 1991; P. K. Feyerabend. *Knowledge, Science and Relativism: Philosophical Papers*, Vol. 3, ed. J. Preston. Cambridge: Cambridge University Press, 1999.

[12] J. S. Mill. On liberty: essays on politics and society. In J. M. Robson (ed.), *Collected Works of John Stuart Mill* (Vol. 18). Toronto: University of Toronto Press, 1977.

[13] S. Jacobs. Misunderstanding John Stuart Mill on science: Paul Feyerabend's bad influence. *Social Science Journal* 40(2003): 201–12; K. Staley. Logic, liberty and anarchy: Mill and Feyerabend on scientific method. *Social Science Journal* 36, no. 4 (1999): 603–14.

[14] R. Farrell. *Feyerabend and Scientific Values: Tightrope-Walking Rationality*. Dordrecht: Kluwer Academic Publishers, 2003.

[15] P. M. Churchland. To transform the phenomena: Feyerabend, proliferation, and recurrent neural networks. *Proceedings of the Philosophy Science Association* 64(1997): 408–20; M. Goldman. Science and play. *Proceedings of the Philosophy of Science Association* 1(1982): 406–14; G. Munevar. *Radical Knowledge*. Indianapolis: Hackett Publishing Company, 1981; C. J. Preston. Pluralism and naturalism: why the proliferation of theories is good for the mind. *Philosophical Psychology* 28, no. 6 (2005): 715–35.

[16] P. K. Stanford. Refusing the devil's bargain: what kind of underdetermination should we take seriously? *Proceedings of the Philosophy of Science Association 68(2001): S1–12*; P. K. Stanford. *Exceeding Our Grasp: Science, History, and the Problem of Unconceived Alternatives*. Oxford: Oxford University Press, 2006.

[17] R. Dawid. Novel confirmation and the underdetermination of scientific theory building. 2013. http://philsci-archive.pitt.edu/9724/.

[18] J. Bogen. Theory and observation in science. In E. N. Zalta (ed.), *The Stanford Encyclopedia of Philosophy* (Summer 2017 Edition). https://plato.stanford.edu/archives/sum2017/entries/science-theory-observation/.

[19] H. Siegel. Justification, discovery and the naturalizing of epistemology. *Philosophy of Science* 47, no. 2 (1980): 297–321; P. Hoyningen-Huene. Context of discovery and context of justification. *Studies in History and Philosophy of Science* 18, no. 4 (1986): 501.

[20] J. T. Cushing. Historical contingency and theory selection in science. *Proceedings of the Biennial Meeting of the Philosophy of Science Association* (1992): 446–57, 1992.

[21] D. Oriti (ed.). *Approaches to Quantum Gravity: Towards a New Understanding of Space, Time and Matter*. Cambridge Cambridge University Press, 2009.

[22] D. Spergel. The dark side of cosmology: dark matter and dark energy. *Science* 347(2015): 1100–2; J. Silk. Challenges in cosmology from the Big Bang to dark energy, dark matter and galaxy formation, *JPS Conference Proceedings* 14(2017): 010101. arXiv:1611.09846 [astro-ph.CO].

[23] G. Amelino-Camelia. *Living Reviews in Relativity* 16, no. 5 (2013); S. Hossenfelder. *Living Reviews in Relativity* 16, no. 2 (2013)

[24] A. Barrau (ed.). Testing different approaches to quantum gravity with cosmology. *Special Issue of Comptes Rendus Physique*, (2017).

[25] R. Dawid, S. Hartmann, & J. Sprenger. The no alternative argument. *British Journal for the Philosophy of Science* 66, no. 1 (2015): 213–34.

[26] R. Dawid. Theory assessment and final theory claim in string theory. *Foundations of Physics* 43, no. 1 (2013): 81–100.

[27] P. K. Stanford. Scientific realism, the atomic theory, and the catch-all hypothesis: can we test fundamental theories against all serious alternatives? *British Journal for the Philosophy of Science* 60 (2009): 253–69.

[28] L. Laudan. A confutation of convergent realism. *Philosophy of Science* 48(1981): 19–48.

[29] J. T. Cushing. The justification and selection of scientific theories. *Synthese* 78(1989): 1–24.

9

Physics without Experiments?

RADIN DARDASHTI

9.1 Introduction

Consider the following three theories: cosmic inflation, supersymmetry, and string theory. Cosmic inflation provides explanations, for instance, for the large-scale homogeneity and isotropy of the universe, the flatness of the universe, and the absence of magnetic monopoles. Supersymmetry extends the symmetry of the Standard Model of particle physics by introducing a new supersymmetric partner for each particle we know. It solves many open problems, such as the infamous hierarchy problem, and provides possible candidates for dark matter. String theory is a proposed unified theory of all fundamental forces – that is, gravity, the electromagnetic force, and the strong and weak nuclear forces. It is a theory most relevant at very high energy scales, the so-called Planck scale, where one expects all fundamental forces to become unified. These three theories are exemplar theories in modern fundamental physics, and they all have two things in common. First, they either lack any empirical support (string theory and supersymmetry) or rely on scarce empirical support (inflation).[1] Second, they are, nevertheless, being defended by scientists for several decades despite the lack of empirical support.

But why do scientists trust their theories in the absence of empirical data? In a recent book, Richard Dawid (2013) addresses this question. The reason, he argues, is based on the idea that one can assess theories non-empirically by assessing the extent to which scientific underdetermination is limited, or to put it differently, the extent to which the space of theories is constrained. This shift in fundamental physics toward regimes where empirical data are hard to come by will necessarily have an impact on scientific practice. In this chapter, we consider what this change in practice amounts to and which problems the practicing scientist then faces.

---

[1] See Ijjas et al. (2017) for a recent discussion.

We start in section 9.2 by discussing scientific methodologies with and without experiments and their normative consquences. The normative consequences of the methodologies without experiments discussed are faced with several problems that we discuss in section 9.3. These problems provide us with additional normative implications for scientific practice, which we discuss in section 9.4. We conclude in section 9.5.

9.2 Scientific Methodologies and Their Normative Consequences

9.2.1 Scientific Methodology with Experiments

Scientific theories are supposed to at least provide us with empirically adequate accounts of the world. It is therefore not surprising that when we want to assess[2] them, we confront them with the world. The history of the scientific method has, therefore, been a history of finding the right methodology to assess theories based on empirical data.[3]

One line of reasoning followed an inductivist methodology, where one tries to identify reliable methods to infer inductively from observations (the empirical data) to generalisations (the theories). Other methodologies were not concerned with the origins of the theories; that is, it did not matter whether the inductive method was used in the development of the theory or whether the theory appeared in someone's dream. The method of assessment relied on the consequences of the theory only. Popper's falsificationism and the hypothetico-deductive account or Bayesian account of confirmation fall under this category.[4]

Most methodologies were not only (or at all) about providing descriptively accurate accounts of how scientists reason but also aimed at providing a normative component, describing how scientists ought to reason. That is, they did not only aim to provide the framework to rationally reconstruct how scientists obtain scientific knowledge but also sought to provide the rules and guidelines of how they should go about obtaining it.

Let us consider some examples. Newton formulated, within the inductivist tradition, certain rules that scientists (i.e., natural philosophers at that time) should follow. His Rule III, for instance, says:

Those qualities of bodies that cannot be intended and remitted [i.e., qualities that cannot be increased or diminished] and that belong to all bodies on which experiments can be made,

---

[2] I use "assess" when I do not want to consider a specific kind of assessment. I leave it open whether the assessment is via Popperian corroboration, H-D confirmation, Bayesian confirmation etc. See Dardashti and Hartmann [this volume].

[3] See Nola and Sankey (2014) for a textbook introduction or Andersen and Hepburn (2015) for an overview article on the scientific method.

[4] See Pigliucci or Dardashti and Hartmann [this volume] for some discussion on these.

should be taken as qualities of all bodies universally. [Quoted in (Harper, 2011, p. 272); my emphasis]

John Stuart Mill's inductive account was about finding regularities in the observations. He tried to identify methods by which to identify the causes.[5] He proposes In 'Systems of Logic', for example, his famous method of difference:

If an instance in which the phenomenon under investigation occurs, and an instance in which it does not occur, have every circumstance save one in common, that one occurring only in the former; the circumstance in which alone the two instances differ, is the effect, or cause, or a necessary part of the cause, of the phenomenon. (Mill, 1843, p. 483)

Karl Popper claims that "[e]very genuine test of a theory is an attempt to falsify it, or to refute it" and that "[c]onfirming evidence should not count except when it is the result of a genuine test of the theory" (Popper, 1989, p. 36). So rather than showing that a theory makes correct predictions, one should focus scientific practice on trying to refute the theory.

The details and the viability of these normative claims do not matter at this point. However, all have in common that, if taken seriously, they have direct implications for scientific practice. That is, they are not solely (or at all) descriptive accounts of how science generates new knowledge but rather normative accounts of how scientific practice should proceed to reliably produce scientific knowledge.

9.2.2 Scientific Methodology without Experiments

As mentioned in the introduction, the situation of fundamental physics has changed in the last two to three decades. For most theories beyond the standard model of particle physics and theories of quantum gravity, empirical data are either scarce or completely absent. Nevertheless, theories such as string theory, supersymmetry, and cosmic inflation have all been defended for decades now, although none of the previously developed methodologies seem to straightforwardly apply.[6] Thus, we need to answer two questions: (1) Are there new methodologies scientists can rely on? and (2) what are the consequences for scientific practice? Note that there might already be consequences for scientific practice, even if one does not want to rely on any of the new proposed methodologies, as the lack of empirical data by itself may require a change in scientific practice. In what follows we argue that the same change in scientific practice is warranted on the basis of a new proposed methodology, Dawid's non-empirical account of theory assessment, as well as due

---

[5] See, for instance, Wilson [2016, Sect. 5].

[6] See Dardashti and Hartmann [this volume] for a discussion of Bayesian confirmation theory and its flexibility to account for new proposed methodologies. See Carroll [this volume] for a criticism of considering falsificationism as the only method scientists rely on.

to the lack of new empirical data. Indeed, the lack of experiments by itself requires the practicing scientist to make the most out of the empirical evidence already available.

Dawid's Account of Non-Empirical Theory Assessment

An answer to question (1) was proposed by Richard Dawid (2013) in his book *String Theory and the Scientific Method,* in which he argues for the possibility of confirming theories even in the absence of direct empirical evidence. His approach is based on the idea that one can assess how constrained theory space is and that, in cases where one has (non-empirical) evidence that the theory space is strongly constrained, one can use this evidence to confirm the theory.[7] The rationale is that a more constrained theory space increases the probability of those remaining theories being empirically adequate.

Hence, one crucial element for Dawid's proposed assessment is the idea of a theory space. Theory space, however, is a very difficult construct to assess. How can we possibly have access to it so as to analyse how constrained it is? Dawid discusses three arguments that, in case one can establish them, provide non-empirical evidence for how constrained theory space is (Dawid, 2013, Sect. 3.1). One of the specific arguments is the No Alternatives Argument (NAA). Consider the following observation: '[T]he scientific community, despite considerable effort, has not yet found an alternative to theory T fulfilling certain constraints' (Dawid et al., 2015, p. 217). One can understand this observation about the status of the theory as providing non-empirical evidence for there not *being* any alternatives or at least a very limited number of them. The conjunction of all three arguments is then supposed to provide relatively strong evidence for how constrained theory space is. The details of the other two arguments do not concern us for the purpose of this chapter, as we are already able to identify an important normative consequence of the account.

The evidence in the case of the NAA, as well as in the other arguments, depends crucially on how well theory space has actually been explored. For, if scientists have not actively looked for alternatives, the observation that they have not found any alternatives does not probe the actual number of alternatives in theory space. However, scientists usually do not actively search for alternative theories. Thus, to assess theories non-empirically, scientists will have to change their focus in research and engage in an active search for alternatives. This provides a first hint at the normative implications of non-empirical theory assessment for scientific practice.

---

[7] We refer the reader for the details of Dawid's non-empirical theory assessment to Dawid (2013), Dawid [this volume], and section 4 of Dardashti and Hartmann [this volume].

Making the Most out of the Available Empirical Data

Even if scientists do not want to rely on non-empirical methods of theory assessment, it is still advantageous for scientists – or so we will argue – to actively explore theory space. In the situation where new empirical data are difficult to come by, it is crucial to make the most out of the already available empirical data – that is, to maximize the amount of information we may gain from the available data. The active search for alternative theories does exactly that. See also Smolin (2006) or Oriti [this volume] for further arguments in support of theory space exploration.

Let us assume we are interested in whether we can trust the predictions of some confirmed theory T. We have made a large set of observations E, which are in agreement with the predictions of theory T and therefore confirm it. T also makes the predictions G and H. We usually will have some confidence in these predictions of T, as it has so far been an empirically successful theory. So the previous empirical success warrants an increase in our trust regarding the novel predictions G and H of T. Now assume that for some reason we will not be able to conduct the experiment that would allow us to probe whether G and H are obtained. In this circumstance we cannot further assess the theory T based on empirical data. Now assume further that someone comes up with an alternative theory, say T', that happens to also predict the set of observations E and is therefore similarly confirmed by it. In addition, it predicts G but disagrees about H. Let us denote the predictions by $Predictions(T) = \{E, G, H, \ldots\}$ and $Predictions(T') = \{E, G, H', \ldots\}$. How will the existence of this additional theory impact our beliefs regarding the predictions G and H? The same available empirical data, E, confirm two competing theories, which agree with respect to one prediction, G, and disagree with respect to another prediction, H. If we have no reason to trust one theory more than the other, then the proposal of the competing theory T' should lead to an increase in our trust regarding the prediction G, while it leads to a decrease with respect to the prediction H. We learn through the proposal of the competing theory T' that we should not have taken the available empirical data E as providing as much support for prediction H as they do for G, as the same evidence confirms just as much a theory T' that does not predict H. Now imagine that theorists come up with theories $T', T'', \ldots$, all of which agree with respect to the prediction G and disagree with respect to H. It is reasonable to assume that we would slowly become more and more certain about G being a feature of the world we live in and more uncertain about H. In this way, the proposal of competing alternatives allows us to better assess the untested predictions of the theory.

This simple example illustrates how the exploration of theory space, which involves the search for alternative theories, allows us to make the most out of the empirical data we already have. Of course, in cases where we could just conduct the experiments necessary to test *G* and *H*, we would not need to rely on competing theories to assess the viability of *G* and *H*. In contrast, in cases where you do not have empirical data, whether you intend to assess theories non-empirically á la Dawid or simply want to learn as much as possible from the already available empirical data, exploring theory space may be the change in scientific practice necessary to make progress.

9.3 Problems of Theory Space Assessment

To explore theory space is, obviously, easier said than done. For that purpose, it is crucial to identify certain pitfalls that one may face. In this section, we address several problems that arise in determining a theory's status with regard to its position in theory space. The approach is to consider specific cases from the history of physics in which scientists have mistakenly constrained theory space. This will allow us to identify which elements constrain theory space and what the possible pitfalls are in using them prematurely. The three discussed problems are the theoretical problem, the structure problem, and the data problem. While we discuss them separately, they are actually intricately intertwined.

9.3.1 The Theoretical Problem

When physicists develop new theories, they necessarily rely on certain theoretical assumptions. Further, the problem they wish to solve may not be an empirical problem but a conceptual one. The difficulty in justifying these two theoretical components of theory development creates what I call *the theoretical problem of theory space assessment*. We will illustrate that problem with a case study from 1970s particle physics.

In 1974, Georgi and Glashow proposed a grand unified theory based on the mathematical group $SU(5)$. They provide a unification of all fundamental interactions of particle physics. They start their paper by claiming:

We present a series of hypotheses and speculations leading inescapably to the conclusion that $SU(5)$ is the gauge group of the world. [Georgi & Glashow, 1974, p. 438]

They then go on to develop the theory and end with the claim:

From simple beginnings we have constructed the unique simple theory. [Georgi & Glashow, 1974, p. 440]

So here we have a case in which scientists make explicit statements with regard to theory space. They say they have provided a theory that 'inescapably' gives rise to

Table 9.1 *The Standard Model gauge group and its particle content (first generation only) in their respective representations. Any future theory of particle physics should contain these. See, for example, Griffiths (2008).*

Names		$SU(3)_C \times SU(2)_W \times U(1)_Y$
Quarks	$Q = \begin{pmatrix} u \\ d \end{pmatrix}$	$(\mathbf{3}, \mathbf{2})_{+1/3}$
	u^c	$(\bar{\mathbf{3}}, \mathbf{1})_{-4/3}$
	d^c	$(\bar{\mathbf{3}}, \mathbf{1})_{+2/3}$
Leptons	$L = \begin{pmatrix} \nu_e \\ e \end{pmatrix}$	$(\mathbf{1}, \mathbf{2})_{-1}$
	e^c	$(\mathbf{1}, \mathbf{1})_{+2}$
	$[\, \nu_e^c$	$(\mathbf{1}, \mathbf{1})_0]$
Gauge bosons	G	$(\mathbf{8}, \mathbf{1})_0$
	$A^\pm, A^3$	$(\mathbf{1}, \mathbf{3})_0$
	B	$(\mathbf{1}, \mathbf{1})_0$
Higgs	ϕ	$(\mathbf{1}, \mathbf{2})_{+1}$

the 'gauge group of the *world*'[8] and it is a 'unique' theory. Whether they meant that statement as strongly as suggested by these quotes is not relevant for our purposes.[9] Our only interest is reconstructing the necessary constraints that would lead to the conclusion they draw.

Let us start with the gauge group of the standard model of particle physics from which Georgi and Glashow develop their $SU(5)$ theory. The gauge group of the standard model is $SU(3)_C \times SU(2)_W \times U(1)_Y$. The particle content in their respective representations of the gauge group is listed in Table 9.1. What is important is that any future theory should accommodate the particle content of the Standard Model and contain the gauge group as a subgroup. This is crucial to guarantee the empirical adequacy of any future theory with respect to the evidence that has already confirmed the Standard Model.

The Standard Model gauge group is a Lie group. Luckily, simple Lie groups have been completely classified. That is, we have a fixed set of possible groups to choose from. Let us consider several of the constraints on any future gauge group G (Georgi, 1999, pp. 231–4):

(1) Group G must be rank $\geq$ 4. This is needed to contain the four commuting generators of the Standard Model gauge group.

---

[8] My emphasis.
[9] It is even doubtful that they considered it in that strong sense, as Georgi himself went on to propose $SO(10)$ as a possible group for unification.

(2) Group G must have complex representations. Some of the particles, like u^c, are in the complex representation of $SU(3)_C$ and so need to be accommodated by any future group.

(3) Group G should be a simple[10] group. This ensures that the gauge couplings are related.

Restrictions (1)–(3) will put strong constraints on the set of possible groups that come into consideration. For instance, if we restrict the groups to simple groups, we rule out groups like $SO(7)$ (rank 3) or F_4 and $SO(13)$ (both have no complex representation), to mention a few. However, groups like $SU(5)$, $SU(6)$, $SO(10)$, and others remain. Thus, these constraints do not seem to lead us 'inescapably' to $SU(5)$ as the gauge group 'of the world'. The only way $SU(5)$ could be considered the unique group is by adding another constraint:

(4) Group G should be the simplest group satisfying constraints (1)–(3).

So, given constraints (1)–(4). we may say that theory space allows for only one possible group, namely $SU(5)$.

Let us start by evaluating the different constraints. It is obvious that the constraints differ in the strength of justification one can give for them. It seems restrictions (1) and (2) are reasonably supported theoretical constraints on G, as they represent minimal requirements to account for the successes of the Standard Model gauge group. In other words, they are empirically supported. By comparison, constraint (3) seems not to be necessary in a similar sense. Having a simple group has the nice feature that the three gauge couplings of the Standard Model are then related to each other after the symmetry is spontaneously broken. There is, however, nothing empirically requiring this to be the case. Slansky, for example, even says with regard to simple groups that the 'restriction is physically quite arbitrary' (Slansky, 1981, p. 14). A couple of months after Georgi and Glashow, Pati and Salam (1974) actually provided a unification based on a group, which was not simple. The last constraint we considered, that we pick the simplest out of the simple groups, lacks any empirical justification (except perhaps the non-empirical justification of Occam's razor). Now we see how one may have been led to think one has found a "unique" theory but this assessment will be only as strong as the constraints that led to the uniqueness. At least in this case, not all constraints are well justified.

There is a further problem. While we saw why we may need to worry about wrong constraints, we may also need to worry about the very problem the theory aims to solve. Consider for a moment that all the theoretical constraints presented earlier are empirically well confirmed (imagine this is possible) and based on these

---

[10] A group is defined to be simple if it does not contain any invariant subgroup.

constraints there are no alternatives. We would be led to believe that this theory is empirically adequate, as there simply are not any other theories. However, it may now be the case that the problem we are considering is not a genuine problem. Unlike empirical problems, conceptual problems such as unification rest on shaky grounds. In the preceding example one considers the problem in need of a solution to be the unification of all the Standard Model interactions. But how do we know this is really a problem in need of a solution? The danger is that we may find a theory that has no alternatives for a problem that is not really a problem. In this case we would have developed a theory toward the wrong 'direction' of theory space.

Physicists have the impulse to develop theories without rigorously trying to justify every step of the way all of the assumptions involved in theory construction. This is, of course, fine if one can then do experiments to test the theory. Georgi and Glashow's $SU(5)$ theory predicted, for instance, that the proton would decay. To date, there has not been any observation of proton decay and the bound put on the lifetime of the proton disagrees with the predictions of $SU(5)$. If, however, experiments are not possible, one needs to address two theoretical problems: (1) to carefully assess all the constraints used in the development and (2) to provide good reasons that the proposed problem is genuine. These are necessary ingredients for the reliability of the theory space assessment.

9.3.2 The Structure Problem

Consider some physical assumption. While one may phrase that physical assumption colloquially in words, one usually has a more precise formal representation of it in mind. When I, for example, talk of probabilities and certain features they should satisfy, I have in mind more formally the mathematical structure of Kolmogorovian probabilities, which satisfy certain specific mathematical axioms. That is, I link the physical concepts with specific mathematical structures, which represent them. In the previous example, we considered symmetries of some internal space, which are represented in terms of the mathematical structure of Lie groups. The conceptual problem of unification, together with the other constraints, such as the particle content and their representations, then translated into the attempt to unify these different Lie groups into one bigger Lie group. The assessment of theory space was explicitly determined by the classification of all simple Lie groups. That is, we could rule out certain points in theory space just by ruling out Lie groups that do not satisfy the constraints. But is this the right place for the constraining to take place? The possible non-uniqueness of the mathematical representation of the physical assumptions is what I call *the structure problem of theory space assessment*.

To illustrate the problem that arises with this approach, let us consider an example. The strongest kind of assessment of theory space comprises no-go theorems, also known as impossibility results. One proves that, given certain assumptions, there are no theories in theory space that are able to satisfy these theorems. In the 1960s, scientists aimed to find a unification of internal and external symmetries. Not only did they fail, but they also claimed that such a unification is impossible. Later, and for independent reasons, scientists were able to non-trivially unify internal and external symmetries – not by changing certain physical constraints, but rather by changing the mathematical structures used. The mathematical structure used in the impossibility results were Lie algebras. It is instructive to consider this case in more detail. Let us quickly remind ourselves of some definitions.[11]

A **Lie algebra** consists of a vector space L over a field ($\mathbb{R}$ or $\mathbb{C}$) with a composition rule, denoted by $\circ$, defined as follows:

$$\circ : L \times L \to L$$

If $v_1, v_2, v_3 \in L$, then the following properties define the Lie algebra:

1. $v_1 \circ v_2 \in L$ (closure)
2. $v_1 \circ (v_2 + v_3) = v_1 \circ v_2 + v_1 \circ v_3$ (linearity)
3. $v_1 \circ v_2 = -v_2 \circ v_1$ (antisymmetry)
4. $v_1 \circ (v_2 \circ v_3) + v_3 \circ (v_1 \circ v_2) + v_2 \circ (v_3 \circ v_1) = 0$ (Jacobi identity)

Both the internal and external symmetries were represented as Lie algebras. The aim, similar to that in the previous example, was to find a Lie algebra that would non-trivially combine them. However, by the mid-1950s, mathematicians had already found a more general structure. The idea is that one can circumvent the pure commutator structure apparent in the anti-symmetry claim by defining the vector space as the direct sum of two vector spaces with different composition properties. This is achieved in what one calls $\mathbb{Z}_2$-graded Lie algebras:

A $\mathbb{Z}_2$-**graded Lie algebra** consists of a vector space L which is a direct sum of two subspaces L_0 and L_1. That is, $L = L_0 \oplus L_1$ with a composition rule, denoted by $\circ$, defined as follows:

$$\circ : L \times L \to L$$

satisfying the following properties:

1. $v_i \circ v_j \in L_{i+j \bmod 2}$ (grading)
2. $v_i \circ v_j = -(-1)^{ij} v_j \circ v_i$ (supersymmetrisation)
3. $v_i \circ (v_j \circ v_l)(-1)^{il} + v_l \circ (v_i \circ v_j)(-1)^{lj} + v_j \circ (v_l \circ v_i)(-1)^{ji} = 0$ (general Jacobi identity)

with $v_i \in L_i$ (i=0,1).

---

[11] See Corwin et al. (1975) for a review article, and Kalka and Soff (1997) and Mueller-Kirsten and Wiedemann (1987) for elementary discussions.

This new structure has several interesting features. First, the grading means that the composition of two L_0 elements gives an L_0 element, the composition of an L_0 element with an L_1 element gives an L_1 element, and the composition of two L_1 elements gives an L_0 element. So L_1 by itself is not even an algebra, since it is not closed, while L_0 by itself is. Second, the so-called supersymmetrisation leads to a commutator composition for all cases except when two elements are taken from L_1. In those cases, the composition is given by an anti-commutator. From this, it follows that Lie algebras are special cases of graded Lie algebras in the case where L_1 is empty. This generalization of the mathematical structure simply allows one to do more. In the previous example, the non-trivial unification of internal and external symmetries is now possible with this new mathematical structure. This is problematic, because it seems that any assessment of theory space will depend on the choice of the mathematical structure within which the different theoretical and empirical constraints are represented. This *structure problem* of theory space assessment is difficult to assess and has crucial implications for the interpretation of no-go theorems.[12]

9.3.3 The Data Problem

A final and maybe less obvious problem is what I call the *data problem of theory space assessment*. There seems to be nothing more uncontroversial than the constraints that come from empirical data. But what one may consider as empirical data once those data are implemented explicitly into a theory is far from obvious. Take one apparently obvious example, the dimensionality of space. One may say it suffices to open the eyes to see that, as an empirical fact, space is three-dimensional. The three-dimensionality of space, though, may seem to be a strong empirical constraint on physical theories. In formulating a theory, it seems obvious to simply start with three space dimensions, as has done for many centuries. Even with the advent of the general theory of relativity, where space itself became a dynamic entity, $D = 3$ was not under discussion. However, starting in theory development with the strong empirical constraint of $D = 3$ may put too strong of a constraint on the theory space. Let us consider this example a little bit further, by considering cases from physics where extra dimensions were introduced.

Unsurprisingly, theorists did not consider higher dimensional theories for the sole purpose of having a higher dimensional theory, but instead had different motivations. For instance, Nordström (1914) and Kaluza (1921) were hoping to unify gravity with electromagnetism and realised that a theoretical option would be to

---

[12] See Dardashti (2018).

introduce an additional space dimension.[13] More recently, Arkani-Hamed et al. (1998) proposed a solution to the hierarchy problem, the unexplained difference in strength between the electroweak force and the gravitational force, by introducing curled-up extra dimensions. The idea is that the gravitational constant is a dimension-dependent quantity. In case we have more dimensions, we can change the gravitational strength, thereby bringing it closer to the electroweak scale. In case the radius of the curled-up dimension is chosen to be small, we might not have been able to observe it by any experiment. At the time of Arkani-Hamed et al.'s (1998) proposal, millimetre-sized extra dimensions were still empirically possible.

At this point one may think that one may not be able to rule out curled-up extra dimensions by observation, but one can at least be sure that there are no additional extended non-compactified extra dimensions. Randall and Sundrum (1999), just like Arkani-Hamed et al., tried to solve the hierarchy problem and showed that one does not need to require the extra dimensions to be curled up if one allows for warped extra dimensions. As long as the electroweak interactions are constrained to three-dimensional space, the additional dimensions could be extended. The gravitational interactions, unlike the electroweak interactions, extend into these additional dimensions, which explains why gravity is weaker. Again, at the time of the proposal, there was no empirical evidence to rule out these higher dimensional spaces. Thus, although one may have thought that the dimensionality of space is an empirical fact and in no need of any further justification, matters turned out to be much more complicated.

The three-dimensionality of space, a seemingly obvious empirical fact about nature, turned out to be an incredibly flexible element in theory development. When we observe the world around us, we see three space dimensions. But as our vision is restricted to only electromagnetic interactions with the world. As the resolution of the eye is approximately one arc minute,[14] there is plenty of leeway for theory development. Further experiments will, of course, put further limits on this leeway but will never rule them out irrefutably. It is important to realise that this is not a unique example, but rather a general feature of empirical data that are used in theory development. The empirical data used by scientists in theory development never consist of protocol statements in the Carnapian sense but extrapolations thereof. Whether we consider the three-dimensionality of space or the homogeneity and isotropicity of the universe, none of these are irrefutable empirical statements.

---

[13] Since Einstein had not yet developed his theory of general relativity in 1914, Nordström used his own empirically inferior scalar theory of gravity.

[14] One arc minute corresponds to a resolution of about a millimetre at a distance of 30 cm.

9.4 The Normative Impact on Scientific Practice

In science, we find both reliable and unreliable methods of theory development and assessment. When scientists and philosophers of science develop scientific methodologies, they want to identify the reliable methods. Once you consider your methodology reliable, you are imposing normative rules on scientific practice to guarantee that the scientists follow the reliable method as closely as possible.

As we argued, in cases where we cannot rely on empirical data to assess theories, a shift toward an assessment of theory space is a way to make progress, whether or not you want to rely on Dawid's non-empirical theory assessment. We have considered several examples from physics to illustrate different ways one can mistakenly constrain theory space. In these cases, one considered theory space to be more constrained than what was warranted by the available empirical evidence. If we want to assess theories based on non-empirical theory assessment, we need to address these problems because they threaten the reliability of our assessment. The success of non-empirical theory assessment will therefore strongly depend on how well we can remedy these problems. However, there does not seem to be a straightforward way to address these issues. In turn, the aim of this chapter and formulating these problems should be understood in the sense of this adage: Recognising a problem is the first step toward solving it.

We saw that when we are developing a theory, we need to make certain assumptions to get started. Based on empirical, theoretical, and mathematical assumptions, we develop a theory able to solve some problem. Whether these are well justified does not matter as long as empirical evidence is available that can be used to test the theory. If no empirical evidence is available, however, it is crucial to consider the legitimacy of each constraint and assumption. An assessment of theory space to a large extent depends on the legitimacy of the assumptions involved in theory development and the constraints they impose. The theoretical, structure, and data problems of non-empirical theory assessment suggest that it will be very difficult to assess theory space. The question of how to address these problems needs to be addressed case by case. Doing so will require a careful analysis of how the constraints act on theory space in each case and the development of possible strategies of inductive justification of the constraints involved.

To assess theories non-empirically, a conscious shift of focus in scientific practice is needed. The most obvious change of perspective in scientific practice will be the focus on the context of discovery. That is, rather than coming up with new theories and then considering their empirical consequences, the focus should be on justifying the very assumptions that led to the theory in the first place. Let us now consider each of the problems in turn and discuss some possible ways to address them.

Theoretical Problem

It is important to consider the scientific problem one wishes to solve carefully. If, for instance, I have some fine-tuned element in my theory, I may want to try to solve it, but one should also recognise the possibility that the theory does not solve the problem at the next energy scale. Assuming and comparing only theories at the next energy scale that solve the problem may constrain theory space too strongly. This, of course, makes it necessary to analyse the very question of what constitutes a genuine scientific problem – a highly non-trivial problem. Moreover, one has to be aware of imposing strong theoretical constraints, even when they are wellconfirmed. Consider, for instance, the theoretical principle of CPT symmetry or Lorentz invariance. Lorentz invariance has been a successful ingredient of many well-confirmed theories in physics, from classical electrodynamics via special relativity up to the Standard Model of particle physics. This may provide good reasons to consider it a meta-principle that needs to be required of all theories, but doing so may again prematurely constrain theory space. Therefore, an approach has been to test violations of the Lorentz symmetry, such as by considering preferred frame effects, by extending the Standard Model of particle physics with Lorentz-symmetry breaking operators, or by using doubly special relativity, where it is shown that only a subgroup of the Lorentz group is needed to account for all the standard predictions [Mattingly, 2005; Liberati, 2013]. These methods explicitly address the problem of constraining theory space based on unwarranted theoretical constraints. In this case, an extensive set of experiments has tested various possible violations of Lorentz invariance within the different proposed frameworks [Kostelecký & Russell, 2009]. Thus, if the theory one is developing is at an energy scale where the empirical evidence does not show a violation of Lorentz invariance, requiring the Lorentz symmetry – and thereby constraining theory space – is empirically justified. In contrast, if one develops a theory several orders of magnitude beyond the empirical bounds, theory space is not yet constrained by the data, and one may prematurely constrain theory space by requiring it. These methods may potentially be effectively used to test the viability of theoretical constraints. Scientists, in the absence of empirical data to support their theories, should focus on these scientific practices to epistemically justify their guiding principles.[15]

Structure Problem

The structure problem seems less accessible. The problem is that scientists in practice do not recognise it as a constraint. When trying to solve some problem, one uses, understandably, the mathematical structures one has always used. They were

---

[15] See Crowther and Rickles (2014) for a special issue focused on the principles guiding theories of quantum gravity.

successful in the past; why should they not be just as successful in the future? However, if we recognise that we cannot solve a problem by using that structure, we may look for an alternative structure that allows us to solve the problem. If one finds a structure, one stops looking, as there is no need to continue the search because one has solved the problem. Once one has solved the problem with a specific mathematical structure, there is no incentive in continuing the search for other mathematical structures that may also do the job. In this case one may be led to think, based on the well-supported theoretical constraints, that theory space is constrained. However, now there is an incentive to consider alternative structures. If I do not have empirical data to test my theory, I may want to rely on theory space assessment. So if I would like to argue in favour of my theory, I need to actively pursue the possibility of alternative mathematical structures. As we mentioned, the mathematical structure of graded Lie algebras was developed in the mid-1950s. If O'Raifeartaigh or Coleman and Mandula had actively searched for alternative mathematical structures, even within the existing mathematical literature, they would have recognised their unwarranted constraint on theory space.

Data Problem

The data problem can be addressed similarly to how the problem of theoretical constraints can be addressed – namely, by not extrapolating the available data to energies, where we have no evidence for it, but rather by approaching the question head on by testing how far one can extrapolate. Take, for instance, the number of particle generations in the Standard Model. Three generations have been observed, amounting to 12 matter particles. Could there be any more particles, a fourth generation, we have not yet observed? Requiring future theories to have only three generations may put too strong of a constraint on the theory space. However, researchers have recently combined experimental data on Higgs searches from the particle accelerators LHC and Tevatron to conclude that a fourth generation of the Standard Model can be excluded with $5\,\sigma$. Similarly, take the example of the speed of light. It has been measured using many different methods, from astronomical measurements to interferometry. Thus, its value is well tested. In turn, for any future theory, one may require the speed of light to have that fixed constant value. But do we really have good reason to believe that it is actually constant? Some physicists have suggested the possibility of theories with a varying speed of light (Magueijo, 2003), though the viability of these theories remains a source of controversy. As Ellis and Uzan (2005, p. 12) point out in arguments against these variable speed of light theories: 'The emphasis must be put on what can be measured', which leads to the constraint that 'only the variation of dimensionless quantities makes sense'. So we see that a seemingly simple concept like the speed of light is actually 'complex and has many facets. These different facets have to be distinguished if we wish to construct a theory in which the speed of light is allowed to vary' (p. 12). These examples illustrate that the topic of how data can constrain theory space is a highly non-trivial matter that needs much more careful analysis.

There is, obviously, much more one can say about each of these problems. Here, we just illustrate how the different problems could, in principle, be addressed. While each of these problems needs to be discussed in much more detail, we can already recognise some more general features for scientific methodology. Most importantly, addressing these problems requires a careful reorientation in scientific practice. When I cannot test the novel predictions of the developed theory, I have no other choice but to assess carefully the elements that led me to it. This then may allow a confirmatory assessment to be performed in terms of theory space assessment. Alternatively, one can just wait and hope that further progress in technology will soon catch up with the energy scale, where the novel empirical predictions are being made. The trouble with this second possibility is that many of those theories where non-empirical theory assessment is necessary – that is, theories of quantum gravity – are by construct theories at the Planck scale. It is unclear whether in the foreseeable future experiments will be able to probe these scales.

9.5 Conclusion

In modern fundamental physics, we have many instances where we are lacking empirical evidence. For most of the twentieth century, it did not matter how theories were developed. After all, once we developed the theory, we could just test it by conducting experiments. Scientific practice itself was not affected so much by scientific methodology. As long as experiments could be done, there was enough external guidance that the developed methodologies did not need to affect the practice much. But now, if experiments are lacking, a whole new approach of theory development and assessment needs to be implemented.[16]

We have discussed how the lack of empirical data leads to two possible strategies: Either one relies on Dawid's non-empirical method of theory assessment or, more conservatively, one tries to make the most out of the available empirical data.[17] Independent of which strategy one chooses, scientific practice should shift, or so we have argued, toward an assessment of theory space (see also Oriti's contribution in this volume, Chapter 8). Any evaluation of a theory then relies on how reliable

---

[16] Another promising route, which is not discussed here, is the use of the often quite complicated relation of empirically unconfirmed predictions of empirically more or less confirmed theories, such as black hole entropy (see Wüthrich [this volume]) and Hawking radiation (see Thébault [this volume]), to test currently empirically inaccessible theories.

[17] Of course, there can be further possible non-empirical strategies, none of which has been significantly developed so far.

the assessment of theory space as a whole is. We have considered three problems of theory space assessment that any practicing scientist may face. Each of these problems may lead to an unreliable assessment of the extent to which theory space is considered to be constrained.

One may argue that these problems just show that it is impossible to assess theories reliably based on non-empirical methods. This might very well be the way it turns out, but it does not have to. Scientists have not aimed at systematically exploring the theory space. If they would, there is still the possibility that they might find many alternatives that are currently not excluded; in those cases there will be no non-empirical support for any particular theory. That is, in those cases, the available empirical evidence does not suffice to constrain the theory space significantly. But it also may be the case that a wide range of theory space can be excluded, which suggests that there really cannot be any alternatives.

Let us end on an important point. Nothing in this chapter has suggested in any way that one should move away from trying to empirically test these scientific theories (see the chapters by Quevedo and Kane in this volume, Chapters 21 and 22). The claim is that, if there are no empirical data to test these theories, an exploration of the theory space provides a method to make the most out of the available empirical data.

Funding

I would like to acknowledge the support of the Alexander von Humboldt Foundation. This work was partly performed under a collaborative agreement between the University of Illinois at Chicago and the University of Geneva and made possible by grant number 56314 from the John Templeton Foundation. The contents of this chapter are solely the responsibility of the authors and do not necessarily represent the official views of the John Templeton Foundation.

Acknowledgements

I would like to thank Mike Cuffaro, Richard Dawid, Stephan Hartmann, Arthur Merin, Karim Thébault, and Christian Wüthrich for various comments, discussions and suggestions that greatly helped in the development of this chapter.

References

Andersen, Hanne, & Brian Hepburn. (2015, Winter). Scientific method. In Edward N. Zalta, ed., *The Stanford Encyclopedia of Philosophy*.

Arkani-Hamed, Nima, Savas Dimopoulos, & Gia Dvali. (1998). The hierarchy problem and new dimensions at a millimeter. *Physics Letters B*, 429(3):263–72.

Corwin, L., Yu Ne'eman, & S. Sternberg. (1975). Graded Lie algebras in mathematics and physics (Bose–Fermi symmetry). *Reviews of Modern Physics*, 47(3):573.

Crowther, K., & Rickles, D. (2014). Introduction: Principles of quantum gravity. Studies in the History and Philosophy of Modern Physics, 46, 135–141.

Dardashti, Radin. (2018). What can we learn from no-go theorems? (Unpublished manuscript).

Dawid, Richard. (2013). *String Theory and the Scientific Method.* Cambridge: Cambridge University Press.

Dawid, Richard, Stephan Hartmann, & Jan Sprenger. (2015). The no alternatives argument. *British Journal for the Philosophy of Science*, 66(1):213–34.

Ellis, George F. R., & Jean-Philippe Uzan. (2005). c is the speed of light, isn't it? *American Journal of Physics*, 73(3):240–47.

Georgi, Howard. (1999). *Lie Algebras in Particle Physics: From Isospin to Unified Theories.* Vol. 54 of *Frontiers in Physics*. Boulder, CO: Westview Press.

Georgi, Howard, & Sheldon L. Glashow. (1974). Unity of all elementary-particle forces. *Physical Review Letters*, 32(8):438.

Griffiths, David. (2008). *Introduction to Elementary Particles.* New York: John Wiley & Sons.

Harper, William L. (2011). *Isaac Newton's Scientific Method: Turning Data into Evidence about Gravity and Cosmology.* Oxford: Oxford University Press.

Ijjas, Anna, Paul J. Steinhardt, & Abraham Loeb. (2017). Cosmic inflation theory faces challenges. *Scientific American*, 316(2): 32–9.

Kalka, Harald, & Gerhard Soff. (1997). *Supersymmetrie.* Wiesbaden: Teubner Studienbücher Physik. Vieweg+ Teubner Verlag.

Kaluza, Theodor. (1921). Zum unitätsproblem der physik. *Sitzungsberichte der Königlich Preußischen Akademie der Wissenschaften (Berlin)*, 966–72.

Liberati, Stefano. (2013). Tests of Lorentz invariance: A 2013 update. *Classical and Quantum Gravity*, 30(13):133001.

Magueijo, Joao. (2003). New varying speed of light theories. *Reports on Progress in Physics*, 66(11):2025.

Mattingly, David. (2005). Modern tests of Lorentz invariance. *Living Reviews in Relativity*, 8(5):2003.

Mill, John Stuart. (1843). *A System of Logic, Ratiocinative and Inductive: Being a Connected View of the Principles of Evidence and the Methods of Scientific Investigation*, Vols. 7 & 8. Toronto: University of Toronto Press.

Mueller-Kirsten, H., & Armin Wiedemann. (1987). *Supersymmetry - An Introduction with Conceptual and Calculational Details.* Teaneck, NJ: World Scientific Pub. Co.

Nola, Robert, & Howard Sankey. (2014). *Theories of Scientific Method: An Introduction.* New York: Routledge.

Nordström, Gunnar. (1914). Über die Möglichkeit, das Elektromagnetische Feld und das Gravitationsfeld zu vereinigen. *Physikalische Zeitschrift*, 15:504.

Pati, Jogesh C., & Abdus Salam. (1974). Lepton number as the fourth 'color'. *Physical Review D*, 10(1):275.

Popper, Karl. (1989). *Conjectures and Refutations: The Growth of Scientific Knowledge.* London: Routledge.

Randall, Lisa, & Raman Sundrum. (1999). Large mass hierarchy from a small extra dimension. *Physical Review Letters*, 83(17):3370.

Kostelecký, V. A., & Russell, N. (2011). Data tables for Lorentz and C P T violation. *Reviews of Modern Physics, 83*(1), 11.

Slansky, Richard. (1981). Group theory for unified model building. *Physics Reports*, 79(1): 1–128.

Smolin, Lee. (2006). *The Trouble with Physics*. Boston: Houghton Mifflin.

Wilson, Fred. (2016, Spring). John Stuart Mill. In Edward N. Zalta, ed., *The Stanford Encyclopedia of Philosophy*.

10

Scientific Methodology: A View from Early String Theory

ELENA CASTELLANI

10.1 Introduction

This chapter is devoted to addressing the question of whether the methodology followed in building/assessing string theory can be considered scientific – in the same sense, say, that the methodology followed in building/assessing the Standard Model of particle physics is scientific – by focussing on the 'founding' period of the theory. More precisely, its aim is to argue for a positive answer to this question in the light of a historical analysis of the early developments of the string theoretical framework.

The chapter's main claim is a simple one: There is no real change of scientific status in the way of proceeding and reasoning in fundamental physical research. Looking at the developments of quantum field theory and string theory since their very beginning, one sees the very same strategies at work in both theory building and theory assessment. Indeed, as the history of string theory clearly shows (see Cappelli et al., 2012), the methodology characterising the theoretical process leading to the string idea and its successive developments is not significantly different from the one characterising many fundamental developments in theoretical physics that have been crowned with successful empirical confirmation afterward (sometimes after a considerable number of years, as exemplified by the story of the Higgs particle).

Of course, theory assessment based on empirical support is problematic in the case of theoretical developments that concern regimes (energy/length scales) away from the current possibility of empirical access, such as those related to the string theoretical framework. This happens for reasons that can be both technical and theoretical – the technology is not advanced enough or the theory is not developed enough. In such cases, non-empirical criteria naturally play a major role in theory assessment (which does not mean renouncing the pursuit of empirical confirmation).

The stance adopted here is that the real issue at stake, when discussing the status of string theory, is not whether non-empirical theory assessment could or should substitute for the traditional way of confirming physical theories by confrontation with empirical data. To understand the debate in such terms, I think, is misleading and fruitless.[1] The genuine question is how to obtain a reasonable balance between empirical and non-empirical criteria for theory assessment, given the particular physical context considered. From the methodological point of view, the only real difference is how this balance is achieved and justified, depending on the specific scenarios one is dealing with. Thus, in the case considered here, the main question is: What are the admissible typology and extent of non-empirical support that can reasonably motivate pursuing the chosen line of research?

Criteria and strategies for this kind of support to theory assessment are of various types. The 'non-empirical confirmation arguments' individuated by Dawid (2013) – the 'no alternatives argument', the 'meta-inductive argument', and the 'unexpected-explanation argument' – work at a general epistemological 'meta-level'. By comparison, at the 'object level' of scientific practice, the philosophical discussion has traditionally focused on the so-called theoretical virtues – that is, virtues such as simplicity, coherence, and elegance, to mention some of the main ones.

Here, we will focus on a further type of criterium influential at the level of scientific practice. This is the criterium grounded on what can be called the 'convergence argument' – namely, the argument for non-empirical support based on the fact that some of the crucial new results are obtained in alternative, independent ways, and even from different starting points.[2] This sort of argument has been particularly effective in providing motivation for accepting apparently very unphysical features emerging from theoretical developments. A paradigmatic example is given by the story of the discovery of extra dimensions (22 extra space dimensions) in the framework of 'early string theory', an expression used to indicate the origin and first developments of string theory, from Veneziano's 1968 formulation of his famous scattering amplitude to the so-called first string revolution in 1984.[3]

In what follows, after some preliminary remarks about the nature and merit of the methodological debate concerning what is currently known as string theory, we focus on early string theory and its significance for the issue at stake. More precisely, the focus is on the first phase of this 'founding era' of the theory. The

---

[1] On the point that a physical theory must receive, sooner or later, an empirical confirmation – in whatever way – there can be no real disagreement.

[2] This criterion could be seen to have some affinity with the famous 'consilience of inductions' introduced, as one of the confirmation criteria (beside predictivity and coherence), by William Whewell in his 1840 *The Philosophy of Inductive Sciences, Founded upon Their History*. This point is examined in another paper in preparation, in collaboration with Radin Dardashti and Richard Dawid.

[3] See Cappelli et al. (2012) for detailed historical and, in many cases, first-hand reconstructions of these early developments of string theory.

aim is to show, by looking in some detail at the extra dimensions case, that the rationale guiding both the building and assessment methodology in cases like this one – where features far from experimental testing possibility are introduced and become a basic ingredient of the whole theoretical framework – is not significantly different from the methodology followed in many successful theoretical advances in physics, later empirically confirmed.

10.2 Scientific Methodology and String Theory: The Issue

Traditionally, the general methodological debate in the philosophy of science regards the following points:

- The modalities and strategies followed in building scientific theories; that is, questions about scientific heuristic or 'discovery'
- The modalities and strategies followed in assessing/confirming scientific theories (on the grounds of both empirical and extra-empirical support); that is, questions about 'justification'
- The inter-relations between theory building and theory assessing, and the influence of 'external' aspects (psychological, sociological, economical, political) in the scientific enterprise

The 2015 Munich Workshop on 'Why Trust a Theory?', on which this book is based, was substantially concerned with the second point (with some attention also being paid to the influence of sociological aspects on research strategies). In fact, the specific question was whether the methodology followed in assessing string theory (and other theories that are currently in a similar situation in regard to their empirical support) could be said *scientific* – which is not the same thing, of course, as asking whether string theory is *true*.

Indeed, methodological queries regarding string theory are generally concerned with justification, not discovery. There is a sort of implicit consensus that the problem doesn't lie specifically in the theory building phase. What is considered controversial is the scientific status of the results of such a building activity, because of the lack of empirical confirmation.

However, assuming a clear-cut distinction between the two modalities of scientific activity – model/theory building and assessment strategies– is not unproblematic, as is well known in philosophy of science.[4] Moreover, in the current debate on the status of string theory, the actual disagreement is about the use of non-empirical criteria in theory assessment, not about the modalities to be followed

---

[4] A detailed revisitation of this distinction (its merits and its flaws), also in the light of its history, is provided in the collective volume edited by Schickore and F. Steinle (2006).

for an empirical confirmation. The real issue, in fact, is the nature and legitimacy of the non-empirical criteria, to be considered in the light of the role they play in the intertwining of discovery/assessing strategies.

As mentioned in Section 10.1, two sorts of criteria for extra-empirical support to scientific theories can be distinguished:

- General meta-criteria, implying the confrontation of the theory considered with other theories. Best examples of such 'external' criteria are the three 'non-empirical confirmation arguments' individuated by Dawid (2013), much debated in this volume.
- More specific 'internal' criteria, based on those theoretical virtues or 'values' that are typically discussed in philosophy of science with respect to the issues of theory underdetermination and theory choice. An inventory of such values traditionally includes consistency, unifying power/generality, fertility, and explanatory power, besides the already mentioned simplicity, coherence, and elegance (or beauty).[5]

As said, other types of internal criteria can be individuated. Here, we consider the role of one of them, the *convergence argument*, by discussing a specific case study: the discovery of extra dimensions in the context of early string theory.

10.3 A Case Study from Early String Theory

Early string theory (EST) represents a particularly fruitful case for discussing the modalities of theory building and theory assessment in fundamental physics. Indeed, EST's history sheds light on the origin of many ideas (e.g., string, duality, supersymmetry, extra dimensions) and mathematical techniques that have become basic ingredients in this research field. Moreover, by highlighting the rationale of a scientific progress characterised by a close interplay of mathematically driven creativity and physical constraints, the historical analysis of these developments provides novel data for discussing the building/assessing issue and, in particular, the role and characteristics of evidential support in the construction process of a scientific theory.

Here, we focus on the first phase of this founding era of string theory: the *dual theory of strong interactions*, which flourished in the years 1968–1973.[6] During

---

[5] See, for example, Ivanova and Paternotte (2017), for an overview of the discussion of such kinds of virtues in evaluating theories.

[6] On the distinction between two phases of EST – a first phase lasting until the end of 1973, when EST was falsified as a theory of strong interactions, and a second phase lasting approximately a decade (1974–1984), when the theory was re-interpreted as a unified quantum theory of all fundamental interactions – see Castellani's contribution in Cappelli et al. (2012, Chapter 4, Section 1). More detailed descriptions are contained in Part 1 of the same text.

this period, following Veneziano's 1968 discovery of the 'dual amplitude' for the scattering of four mesons, a very intense model-building activity ensued, spanning from the first two models for the scattering of N particles – the generalised Veneziano model, known as the *dual resonance model* (DRM), and the *Shapiro–Virasoro model*[7] – to all the subsequent endeavours to extend, complete, and refine the theoretical framework, including its string interpretation and the addition of fermions.

As is well known, this first phase of early string theory was originally aimed at finding a viable theory of hadrons in the framework of the analytic S-matrix (or S-matrix theory) developed in the early 1960s.[8] The S-matrix programme, initially pursued in particular by Geoffrey Chew and his collaborators, was motivated by the difficulties arising in a field-theoretic description of strong interactions. Inspired by the earlier works of Wheeler and Heisenberg, its aim was to determine the relevant observable physical quantities – that is, the scattering amplitudes – only on the basis of some general principles such as *unitarity*, *analiticity*, and *crossing symmetry*, plus a minimal number of additional assumptions.

One of these assumptions, suggested by experimental data, was the *duality principle*, introduced in 1967 by Dolen, Horn, and Schmid. The meaning of this duality, also known as *DHS duality*, was that the contributions from resonance intermediate states and from particle exchange each formed a complete representation of the scattering process (so that they should not be added to each other to obtain the total amplitude).[9] In fact, the duality principle was seen as representing an effective implementation of two connected ideas defended, in particular, by Chew and his school: on the one hand, the idea of 'nuclear democracy', according to which no hadron is more fundamental than the others; on the other hand, the 'bootstrap idea', the idea of a self-consistent hadronic structure in which the entire ensemble of hadrons provided the forces, by hadron exchange, making their own existence possible.[10]

In 1968, Gabriele Veneziano achieved to obtain a first, brilliant solution to the problem of finding, in the framework of the S-matrix approach, a scattering amplitude that also obeyed the duality principle. This ground-breaking result is officially recognised as the starting point of early string theory. A period of intense theoretical activity immediately followed Veneziano's discovery, aimed at extending his dual amplitude so as to overcome its limits. First, his model considered only four particles, and of a specific type (mesons). Moreover, the model violated unitarity

---

[7] These two models were later understood as describing open and closed strings, respectively.

[8] On the S-matrix programme pursued by Chew and his collaborators, see in particular Cushing (1990), and Cappelli et al. (2012, Part II).

[9] See Cappelli et al. (2012, Part II, Section 5.4.3), for details.

[10] Whence the name of *dual bootstrap* for the DHS duality.

because of the narrow-resonance approximation.[11] It was also natural to search for different models that could include other neglected but important physical features (such as fermionic particles, beside bosons) and, more generally, to try to achieve a better understanding of the physical theory underlying the models that were being constructed.

This theory-building process was characterised by two particularly significant conjectures. First, the string conjecture in 1969: In independent attempts to gain a deeper understanding of the physics described by dual amplitudes, Nambu, Nielsen, and Susskind each arrived at the conjecture that the underlying dynamics of the dual resonance model entailed that of a quantum-relativistic oscillating string. Second, the conjecture or 'discovery' of extra spacetime dimensions: Consistency conditions in the developments of the dual theory led to the critical value $d = 26$ for the spacetime dimension (known as the *critical dimension*), which reduced to the value $d = 10$ when fermions were included.

Also in this second case, the conjecture was arrived at by following independent paths, although under the same general motivation – that is, to extend the original dual theory and overcome its initial limitations and problems. The result was a bold conjecture: 22 extra space dimensions. Undoubtedly, the fact that it was obtained in different, independent ways (and even from different starting points) was an influential reason for taking it seriously. In what follows, we briefly outline this significant discovery case, which is illustrative of both the rationale leading to apparently bold guesses and the kind of evidential support motivating a theory's progress.

10.3.1 Ways to the 'Critical Dimension'

In the first years of early string theory, we can distinguish three different approaches employed to arrive at the critical dimension. The first two ways have more to do with the building/discovery process; the third one also involves theory assessment.

Originally, the critical dimension emerged in the context of two independent programmes in the framework of the DRM: (1) the 'unitarisation programme' and (2) the 'ghost elimination programme'.

- In the first case, Claud Lovelace arrived at the conjecture $d = 26$ while addressing a problematic singularity case arising in the construction of the nonplanar one-loop amplitude.

---

[11] That is, the approximation corresponding to the fact that, in each channel, only single-particle exchanges were considered; for more on this issue, see Ademollo's contribution, in Cappelli et al. (2012, Chapter 6, Section 6.7).

• In the second case, the critical value $d = 26$ for the spacetime dimension issued from studying the DRM spectrum.

Let us have a closer look at these first two ways, keeping the details to a minimum.

Lovelace's Way (1971)

As mentioned earlier, the original dual amplitudes didn't respect the S-matrix unitarity condition. To go beyond the initial narrow-resonance approximation, the 'unitarization programme' was based on the analogy between this approximation in the dual theory and the so-called Born approximation (or 'tree approximation') in conventional quantum field theory. The programme was thus to generalise the initial amplitudes, considered as the lowest order or tree diagrams of a perturbative expansion, to include loops. As a first step for restoring unitarity, one-loop diagrams were constructed. In this building process, the calculation of a nonplanar loop diagram led to the conjecture of the value $d = 26$ for the spacetime dimension. This result was obtained by Claud Lovelace in a work published in 1971.

In some more detail: Lovelace arrived at the *critical dimension* $d = 26$ by addressing a singularity problem that emerged in the construction of the nonplanar one-loop amplitude.[12] He realised that the singularity could be turned into a *pole*, and thus interpreted as due to the propagation of a new intermediate particle, if the value of the spacetime dimension was $d = 26$. Lovelace conjectured that this pole was the Pomeron, the particle that was later understood as the graviton.[13] The decisive step, indeed, was to consider the possibility that the spacetime dimension d might be different from 4 and to treat it as a free parameter.[14]

A spacetime of 26 dimensions was not easy to accept, especially in the context of the phenomenology of strong interactions where it was proposed. In a recollection paper on his contribution to the dual theory, Lovelace describes the first reactions to his conjecture as follows: 'I gave a seminar ... which was attended by some powerful people as well as the Dual Model group. Treating the result as a joke, I said I had bootstrapped the dimension of spacetime but the result was slightly too big. Everyone laughed.'[15] As he himself acknowledges, one had to be 'very brave to suggest that spacetime has 26 dimensions'.

---

[12] In four spacetime dimensions, the amplitude had a singularity (a 'branch cut') in a certain channel, incompatible with unitarity. See Lovelace's own account of his 'discovery' in Cappelli et al. (2012, Chapter 15).

[13] For details, see Cappelli et al. (2012, Section 10.2.3).

[14] This was not the only condition, but there is no need to enter into such details here. For historical sake, see Olive's personal account concerning his own contribution to Lovelace's discovery, in Cappelli et al. (2012, Chapter 28, p. 398).

[15] Cappelli et al. (2012, Chapter 15, p. 228).

However, almost at same time, Lovelace's 'wild conjecture' (his words) was vindicated through another completely independent route: The very same number of spacetime dimensions made its appearance in the context of the ghost elimination programme.

The 'No Ghost' Way

Soon after Veneziano's discovery of his amplitude for the scattering of four scalar particles, endeavours started for its generalisation to the scattering of an arbitrary number N of scalar particles. In the ensuing studies of the properties of the resulting multi-particle Veneziano model, known as the Dual Resonance Model, a serious problem was represented by the presence of negative-norm states – or 'ghosts', as they were called at the time –[16] in the state spectrum of the model. Such kinds of states, leading to unphysical negative probabilities, had to be eliminated from the theory.

The strategy adopted for the ghost elimination programme was based on an analogy suggested by a similar situation encountered in the covariant quantisation of electrodynamics, where the unphysical negative-norm states were removed by using the gauge invariance of the theory and the 'Fermi condition' following from it (the equation characterising the positive-norm physical states). The DRM analogues of the conditions imposed by the gauge invariance were found to be given by the so-called Virasoro conditions – an infinite number of relations later understood as associated with the infinite-dimensional symmetry corresponding to the *conformal transformations* of the two-dimensional string world-sheet.[17]

At that point, a further step toward the elimination of the unwanted ghost states was the 1971 construction by Del Giudice, Di Vecchia, and Fubini of an infinite set of positive-norm states, known as the *DDF states*. Initially, these states did not seem to be sufficient to span the whole Hilbert space. But one year later, the result was obtained that the DDF states could indeed span the whole space of physical states if the spacetime dimension d was equal to 26 – the very same value as the one conjectured by Lovelace. Soon after, the proof of the No-Ghost Theorem, establishing that the DRM has no ghosts if $d \leq 26$, was achieved independently by Brower (1972) and by Goddard and Thorn (1972).[18]

---

[16] Note that this is a different meaning of the term 'ghost' with respect to how it is commonly used in quantum field theory (i.e., to indicate the unphysical fields associated with gauge invariance in functional approaches to field theory quantisation).

[17] See, in particular, Di Vecchia's contribution in Cappelli et al. (2012, Sections 11.3 and 11.4).

[18] By essentially same argument as in the case of the DRM, it was also proved that the Neveu–Schwarz dual model has no ghosts if $d \leq 10$, thus confirming the critical dimension as $d = 10$ in the case including fermions. A detailed description of the no-ghost result can be found, in particular, in Goddard's contribution to Cappelli et al. (2012, Chapter 20).

While initially almost no one had taken Lovelace's conjecture seriously, after the proof of the No-Ghost Theorem the attitude changed and the extra dimensions started to be gradually accepted in the dual model community. Further decisive support came from an immediately successive theoretical development, leading, among other things, to the same 'critical' value $d = 26$ for the spacetime dimension: namely, the 1973 work of Goddard, Goldstone, Rebbi, and Thorn (GGRT) on the quantisation of the string action.

The GGRT Way (1973)

As we have seen, the *S*-matrix approach was based on the construction of observable scattering amplitudes. Nonetheless, soon after Veneziano's result and its first generalisations, a physical interpretation of the dual amplitudes in terms of an underlying dynamics and an appropriate Lagrangian started to be investigated. In 1969, Nambu, Nielsen, and Susskind had each independently made the conjecture that the dynamics of the dual resonance model was that of an oscillating string. In the following year, Nambu (and then Goto) proposed the Lagrangian action for the string, formulated in terms of the area of the surface swept out by a one-dimensional extended object moving in spacetime, in analogy with the formulation of the action of a point particle in terms of the length of its trajectory.

It took some time, however, for the string interpretation of the DRM to be fully accepted. It was not clear, originally, whether the conjecture was a mere analogy,[19] useful for calculations, or whether it had deeper physical meaning. The work by Goddard, Goldstone, Rebbi, and Thorn on the quantisation of the string action had a decisive impact, in this respect: Thanks to their result, the string hypothesis became an essential aspect of the theory, revealing its underlying structure. The analogy thus turned out to provide an effective 'interpretation', playing an influential role in the transformation process from the original dual models to the string theoretical framework.

In the resulting description, all that had been previously obtained by proceeding according to a bottom-up approach and via various routes could now be derived in a more clear and unitary way. In particular, the critical dimension was obtained as a condition for the Lorentz invariance of the canonical quantisation of the string in the light-cone gauge: Only for $d = 26$ was the quantisation procedure Lorentz invariant.[20] Thus, in a certain sense, the GGRT way of arriving at the extra dimensions had to do with both discovery and theory assessment.

---

[19] The analogy between the structure of the DRM spectrum and that of a vibrating string was based on the harmonic oscillator.

[20] Details on this point, and in general on the quantisation of the hadronic string, are described by Di Vecchia and Goddard in their contributions to Cappelli et al. (2012, Chapter 11, Section 11.8, and Chapter 20, Section 20.7), respectively.

10.4 Conclusion

With respect to the acceptance of the extra dimension conjecture in the string theory community, the three independent ways of arriving at the one and same result illustrated in this chapter are not the whole story, of course. Further and more decisive support for this conjecture came from successive developments of string theory, especially after it was re-interpreted as a unified quantum theory of all fundamental interactions, including gravity.[21] In fact, today's way of understanding the critical dimension is in terms of what is technically called an 'anomaly': As shown in the seminal 1981 work by Polyakov, in the framework of his path-integral approach to the quantisation of the string action, the conformal symmetry of the classical string Lagrangian is 'anomalous' – that is, not conserved in the quantum theory – unless the value of the spacetime dimension is $d = 26$ ($d = 10$, in the case with fermions).[22]

Thus, with hindsight, the convergence of the different, independent ways of obtaining the critical dimension can be understood as one of the remarkable consequences of the strong constraints put on the theory by its conformal symmetry. At this point, one could object that to apply here the 'convergence argument' as a criterion for theory assessment appears a bit circular. Specifically, it looks like deriving (extra-empirical) support for the extra dimension conjecture on the grounds of another essential feature of the theory itself.

Let us say two things in this regard. First, what we are interested in, in this chapter, is the role of extra empirical criteria – like the convergence argument – in guiding the building of a theory and motivating its acceptance. The case of extra dimensions in early string theory surely illustrates this role. At the same time, to enter into the details of this case study allows us to show, from a methodological point of view, the concrete procedures followed in the development of the theoretical framework: originally a bottom-up activity, starting from a very phenomenological basis and successively progressing *via* generalisations, conjectures, and analogies, suggested and constrained by physical principles and data. This is a very typical example of scientific methodology.

Second, even if the convergence to the same surprising result of independent ways of investigation can be explained, afterward, as a natural consequence of the more mature theoretical framework – thus implying some circularity when using this fact as an assessment criterium for a full-fledged theory – this very fact speaks in favour of 'virtues' of the description obtained, such as its cohesiveness and consistency. In this case, we can say, the convergence criterion gives way to the

---

[21] See Cappelli et al., (2012, Part VI.)

[22] Polyakov, in his contribution to Cappelli et al. (2012, Chapter 44), gives his own personal recollection of these developments; for a more general overview, see also Cappelli et al., (2012, Part VII, Chapter 42).

other form of 'internal' criteria mentioned in Section 10.2, which are based on the theoretical virtues influential in theory assessment.

Acknowledgements

I am very grateful to the organisers and the audience of the 'Why Trust a Theory?' workshop in Munich. A special thanks to Erik Curiel, Radin Dardashti, and Richard Dawid for insightful and helpful discussions, and to the Munich Centre for Mathematical Philosophy for a visiting fellowship in December 2015, allowing me to enjoy its great and collaborative atmosphere.

References

Brower, R. C. (1972). Spectrum-generating algebra and no-ghost theorem in the dual model. *Phys. Rev.* D6:1655–62.

Cappelli, A., E. Castellani, F. Colomo, & P. Di Vecchia (eds.). (2012). *The Birth of String Theory*. Cambridge: Cambridge University Press.

Cushing, J. T. (1990). *Theory Construction and Selection in Modern Physics: The S-Matrix*. Cambridge: Cambridge University Press.

Dawid, R. (2013). *String Theory and the Scientific Method*. Cambridge: Cambridge University Press.

Goddard P., J. Goldstone, C. Rebbi, & C. B. Thorn. (1973). Quantum dynamics of a massless relativistic string. *Nucl. Phys.* B56:109–35.

Goddard, P., & C. B. Thorn. (1972). Compatibility of the Pomeron with unitarity and the absence of ghosts in the dual resonance model. *Phys. Lett.* B40:235–8.

Ivanova, M., & C. Paternotte. (2017). Virtues and vices in scientific practice. *Synthese* 194:1787–1807.

Lovelace, C. (1971). Pomeron form factors and dual Regge cuts. *Phys. Lett.* B34:500–6.

Schickore, J., & F. Steinle (eds.). (2006). *Revisiting Discovery and Justification: Historical and Philosophical Perspectives on the Context Distinction*. Dordrecht: Springer.

Veneziano, G. (1968). Construction of a crossing-symmetric, Reggeon behaved amplitude for linearly rising trajectories. *Nuovo Cimento* A57: 190–7.

11

What Can We Learn from Analogue Experiments?

KARIM THÉBAULT

11.1 Introduction

11.1.1 Types of Evidence

Different modes of inference can be understood as producing different *types of evidence* in various senses.[1] One sense is pragmatic: evidence produced by different modes of inference can be split into different *practical types* in terms of the different forms of action it rationally licences. Evidence based upon highly speculative inferences is of a different practical type than evidence based upon highly reliable inferences in that, for example, it cannot rationally licence risky actions. Even if one is a professional astrologer, one should not decide to cross a busy road while blindfolded based upon one's horoscope! As in everyday life, so in science. Evidence based upon principles of parsimony or judgments of mathematical beauty is of a different practical type than that based upon inductive generalisation of appropriately validated experimental results. While both have significant heuristic significance for science, only the latter can be defended as a rational basis for the design of a nuclear power plant. This much is difficult to contest.

Problems begin to emerge when we consider modes of inference that are neither highly speculative nor proven reliable. Take computer simulation. In what sense do computer simulations produce a different type of evidence than experiments? Are simulations any less a reliable guide to action than experiments? In some cases, it seems not. We can and do build things based upon the evidence of computer simulations. We might not, however, think about computer simulations as producing evidence of a form that can *confirm* a scientific theory. Instead, such an evidential role is often taken to be the province of experiment. That is, although, in terms of

---

[1] In our view, there is not presently available any fully philosophically satisfactory account of evidence in science. Similar sentiments are expressed by Cartwright et al. (2010). Here we offer our own proto-account based upon intuitions regarding the different types of evidence. More complete accounts, inadequate to our present purposes, are given by Achinstein (2001) and Roush (2005).

the practical actions it licences, evidence based upon computer simulations might be taken to be of the same type as that based upon experiments, prima facie, the two types of evidence should be understood as being of different *epistemic types*. Prima facie, only evidence drawn from experiment can be taken to be of the type that is potentially confirmatory.[2] This creates a particular problem when we are in a situation where experimental evidence is hard to come by. How do we test a theory when we are unable to perform experiments to probe the class of phenomena to which its predictions relate?

One idea, much discussed in the recent literature, is that we can obtain evidence that is of a (potentially) confirmatory epistemic type by *non-empirical* means. For example, arguments based on the observation of an absence of alternative theories are claimed to, in certain circumstances, confer confirmatory support upon theories, such as string theory, whose predictions we cannot currently directly empirically test (Dawid, 2013; Dawid et al., 2015). A different approach to the problem confirming theories beyond the reach of direct empirical testing is to consider analogue experiments. In this chapter, we consider the idea of using *analogue experiments* as a means of performing an *ersatz empirical test* of an important untested theoretical prediction of modern physics. We will consider whether evidence gained from analogue experiments can be of the same epistemic type as evidence gained from conventional experiment. We will do this by comparison with evidence produced by speculative inferences that is of a type that cannot be understood as confirmatory – namely, evidence based upon arguments by analogy. We will largely set aside questions concerning the characterisation of confirmation itself.[3]

11.1.2 Analogue Experiments

Our story starts with one of the most celebrated theoretical results of twentieth century physics: Hawking's (1975) semi-classical argument that associates a radiative flux to black hole event horizons. Although almost universally believed by contemporary theoretical physicists, testing of this prediction of gravitational Hawking radiation is verging on the physically impossible. The temperature of Hawking radiation associated with a solar mass black hole is of the order of 100 million times smaller than the temperature of the cosmic microwave background. Trying to detect astrophysical Hawking radiation in the night sky is thus like trying to see

---

[2] There are, in fact, good arguments that evidence from computer simulation and experiment *are* of the same epistemic type. See Parker (2009), Winsberg (2009), and Beisbart and Norton (2012) for discussion of this issue in the literature.

[3] Major approaches to confirmation theory (according to a relatively standard classification) are confirmation by instances, hypothetico-deductivism, and probabilistic or Bayesian approaches. See Crupi (2013) for more details. For related discussion of different concepts of evidence, see Achinstein (2001).

the heat from an ice cube against the background of an exploding nuclear bomb. Short of the construction of terrestrial micro-black holes (Scardigli, 1999; Dvali and Redi, 2008) any *direct* experimental test of Hawking's prediction is effectively physically impossible.

The genesis of analogue gravity comes from the problem of trying to *indirectly* test Hawking's argument. Inspired by the analogy with sound waves in a waterfall, Unruh (1981) showed that Hawking's semi-classical arguments can be applied to sonic horizons in fluids. This work spurred the creation of an entire sub-field of 'analogue gravity' that features theoretical, and in some cases experimental, condensed matter analogues to a variety of systems, including both Schwarzschild black holes and the entire universe (Barceló et al., 2011). The most crucial experimental results thus far have been achieved by Steinhauer (2014, 2016) working with Bose–Einstein condensates (BEC) at the Technion in Israel. The experiment reported in the 2016 paper is particularly significant and, as discussed in Section 11.5, can reasonably be interpreted as the first conclusive confirmation of the existence of quantum Hawking radiation in an analogue system.

It remains to seen what such analogue experiments actually tell us about gravitational systems. In particular, what kind of evidence regarding the 'target systems' (i.e., black holes) do we gain from experiments on the 'source systems' (i.e., BEC)? Can analogue experiments provide us with evidence of the same epistemic type as conventional experiments? Or should we think of them as speculative inferences, producing evidence of the same epistemic type as arguments by analogy? In this chapter, we attempt to answer such questions, partially drawing inspiration from the work of Dardashti et al. (2016, 2017).

Ultimately, we will conclude that there *is* a plausible theoretical basis to 'externally validate' analogue black hole experiments, such that a BEC analogue black hole can be taken to 'stand in' for an astrophysical black hole. This gives us one example where analogue experiments can provide us with evidence of the same confirmatory epistemic type as conventional experiments. It does not, however, speak to the question of the significance of confirmation. Is it reasonable to think, in quantitive terms, that analogue experiments can provide a comparable degree of confirmation as conventional experiments? Can they be *substantially* confirmatory rather than merely *incrementally* confirmatory? Such questions bring us beyond the scope of the present chapter. However, there are reasons to be optimistic. As shown by a recent analysis in terms of Bayesian confirmation theory (Dardashti et al., 2016), given experimental demonstration of an array of analogue Hawking effects across a variety of very different media, the degree of confirmation conferred can be amplified very quickly. It is thus very plausible to think of analogue experiments as prospective means for providing confirmatory support that is substantial, rather than merely incremental.

11.2 Analogy and Experiment

Consider the following two examples of successful scientific practice.[4] In 1934, a pharmacologist named Schaumann observed that the compound meperidine had the effect of inducing an S-shaped curved tail when given to mice. This effect was only previously observed when morphine was given to mice. Together with the similarity in their chemical structures, this analogy between the drugs led Schaumann to reason that meperidine might share morphine's narcotic effects when given to humans. This subsequently proved to be the case. A Bose–Einstein condensate BEC is an exotic form of matter that Bose (1924) and Einstein (1924), predicted to exist for a gas of atoms when cooled to a sufficiently low temperature. In 1995, the experimental demonstration of the existence of a BEC was provided using supercooled dilute gases of alkali atoms (Anderson et al., 1995). The crucial observation was a sharp increase in the density of the gas at a characteristic frequency of the lasers used for cooling.

The type of inference and the type of evidence involved in our two examples are very different. The inference that meperidine would have narcotic effects is based upon an *argument by analogy*. The inference of the existence of Bose–Einstein condensation is based upon an *experimental result*. Arguments by analogy are standardly understood as producing evidence of a type that is not capable of providing confirmatory support to scientific claims. Rather, they are speculative inferences. Contrastingly, experimental results are standardly understood as producing evidence of a type that is capable of providing confirmatory support to scientific claims. Philosophical analysis sheds light on the reasoning behind such intuitions. First, let us rationally reconstruct the argument by analogy made by Schaumann:

P1. Morphine is similar to meperidine in terms of having similar chemical structure and having the effect of inducing an S-shaped curved tail when given to mice.
P2. Morphine has the effect of being a narcotic when given to humans.
 C. Therefore, meperidine will also have the feature of being a narcotic when given to humans.

Clearly, such an argument is deductively invalid. Moreover, as an inference pattern, it does not met the epistemic standard usually expected of a reliable inductive inference. For this reason we should not think of arguments by analogy as producing evidence of the same epistemic type as reliable inductive inferences. Rather, it seems reasonable to use arguments by analogy to establish only the *plausibility* of a conclusion, and with it grounds for further investigation. This is in

---

[4] The first example is taken from Bartha (2013); the second example is taken from Franklin and Perovic (2015, Appendix 3).

line with the views of Salmon (1990) and Bartha (2010, 2013). From this perspective, the importance of analogical arguments is their heuristic role in scientific practice: They provide 'cognitive strategies for creative discovery' (Bailer-Jones, 2009).

Despite their significance and ubiquity within scientific practice, the philosophical consensus is that *epistemically speaking* the role of evidence produced by analogical arguments is null. In contrast, experimental results are usually taken as the key exemplar of epistemically valuable evidence in science. In particular, experimental evidence is the form of evidence invoked in the *confirmation* of theories. However, philosophical analysis reveals good reasons to think that experimental results *taken on their own* do not, in fact, provide epistemic evidence of such unalloyed quality.

Following the work of authors such as Franklin (1989), it is valuable to reconsider the epistemological foundations of experimental science.[5] In particular, we can ask questions such as the following: (1) How do we come to believe in an experimental result obtained with a complex experimental apparatus? and (2) How do we distinguish between a valid result and an artefact created by that apparatus? One of the key ideas in the epistemology of experiment is that to assess the evidence gained from experimentation, we must examine and evaluate the strategies that scientists use to *validate* observations within good experimental procedures. Following Winsberg (2010), we can make an important distinction between two types of validation in the context of experimental science: An experimental result is *internally valid* when the experimenter is genuinely learning about the actual system the experimenter is manipulating – when, that is, the system is not being unduly disturbed by outside interferences. By contrast, an experimental result is *externally valid* when the information learned about the system being manipulated is relevantly probative about the class of systems that is of interest to the experimenter.

Consider the case of the 1995 experiments that were taken by the scientific community to provide conclusive evidence for the existence of BECs on the basis of experiments on supercooled dilute gases of alkali atoms. The internal validity of the experiments relates to the question of whether the results obtained genuinely reflect the fact that the particular supercooled dilute gases of alkali atoms experimented upon were behaving as BECs. The external validity of the experiments relates to the question of whether the inferences regarding the particular *source systems* experimented upon (particular supercooled dilute gases of alkali atoms) can be reliably generalised to the wide class of *target systems* to which the theory of BECs refers. Experimental results can only be reasonably taken to constitute evidence that is of an epistemic type that makes it suitable for confirmation, based on the assumption that the results have been both internally and externally validated.

---

[5] See Franklin and Perovic (2015) for a full review.

While the idea that an experimental result must be internally validated is a familiar one, the notion of external validation is not as frequently discussed – particularly by actual experimental scientists. In the end, external validation is the crucial link from an experimental result to the use of this result as a token of the type of epistemic evidence relevant to the confirmation of general scientific statements. Clearly, an experiment that is not internally validated does not produce evidence of an epistemic type that renders it suitable for confirming any scientific hypothesis. Conversely, an experiment that is internally validated but not externally validated produces evidence of an epistemic type that can confirm specific statements regarding the particular system experimented upon, but that cannot confirm statements regarding the wider class of systems that the experiment is designed to probe. As such, theoretical arguments for external validation are almost always a necessary requirement for integrating experimental results into scientific knowledge.

This brief discussion of analogy and experiment is intended to provide a prospectus for a philosophical analysis of analogue experiments. The key question in the *epistemology of analogue experimentation* is whether there are arguments that can provide a suitable form of external validation. If no such arguments exist, then we *should not* consider analogue experiments as providing evidence regarding the relevant 'target systems' (black holes in the Hawking radiation case) that is of the same epistemic kind as (externally validated) experiments. In such circumstances, analogue experiments would have a more heuristic role, like arguments by analogy, and could not in principle be used to produce evidence used for the confirmation of general scientific statements regarding the target system. However, if arguments for external validation can be found and justified in cases of analogue experimentation, then we *can* legitimately think about analogue experiments as providing evidence of the same epistemic type as conventional (externally validated) experiments.

Significantly, this question is distinct from any question about the *strength of evidence*. It may be the case, in either a conventional experiment or an analogue experiment, that the validation procedure is not highly reliable, in which case the evidence will be of the same epistemic type, but of a different strength. While questions regarding strength of evidence suggest immediately an analysis in terms of Bayesian confirmation theory along the lines described by Dardashti et al. (2016), we hold that questions of type of evidence can be answered independently of such probabilistic modes of analysis.

11.3 The Hawking Effect

In this section, we present the formal details behind the Hawking effect as instantiated in gravitational systems, fluid dynamical systems, and Bose–Einstein condensates. For the most part, we will follow the account of Barceló et al. (2011), which is also a good source of supplementary details.

11.3.1 Gravitational Hawking Effect

In a semi-classical approach to gravity, we consider a quantum field within a fixed spacetime background. For this modelling framework to be valid, it is assumed that we are considering quanta of wavelengths much larger than the Planck length. In the simplest semi-classical model, we consider a massless scalar field operator $\hat{\phi}$ that obeys a wave equation of the form $g^{ab} \nabla_a \nabla_b \hat{\phi} = 0$. We can expand the scalar field in a basis of orthonormal plane wave solutions:

$$\hat{\phi} = \int d\omega (\hat{a}_\omega f_\omega + \hat{a}_\omega^\dagger f_\omega^*), \tag{11.1}$$

where $f_\omega = \frac{1}{\sqrt{2}} e^{-i(\omega t - kx)}$ and $\hat{a}_\omega$, $\hat{a}_\omega^\dagger$ are creation and annihilation operators for modes of some frequency ω. The creation and annihilation operators allow us to define both a vacuum state, $\hat{a}_\omega |0\rangle = 0$, and a number operator, $\hat{N}_\omega = \hat{a}_\omega^\dagger \hat{a}_\omega$, in this particular basis. We call a vacuum state defined for the scalar field at past null infinity, $\mathcal{J}^-$, the 'in' state, and a vacuum state defined for future null infinity, $\mathcal{J}^+$, the 'out' state. In general, the 'in' state need not appear as a vacuum state to observers at positive null infinity: It may contain a flux of 'out-particles' that one can calculate by determining the Bogoliubov coefficients between the solutions expressed in the 'in' and 'out' basis:

$$_{in}\langle 0|(\hat{N}_\omega^{out})|0\rangle_{in} = \int d\omega' |\beta_{\omega\omega'}|^2. \tag{11.2}$$

Hawking's 1975 calculation shows that, for a spacetime which features the establishment of an event horizon via gravitational collapse leading to a black hole, one can derive the asymptotic form of the Bogoliubov coefficients and demonstrate that it depends only upon the *surface gravity* of the black hole denoted by κ_G. Surface gravity is the magnitude of the acceleration with respect to Killing time of a stationary zero-angular-momentum particle just outside the horizon. Hawking's calculation implies that a black hole horizon[6] has intrinsic properties that are connected to a non-zero particle flux at late times. The spectrum of this flux obeys the following relation:

$$\langle \hat{N}_\omega^{\text{Black Hole}} \rangle = \frac{1}{e^{\frac{2\pi\omega}{\hbar\kappa_G}} - 1} \qquad T_{BH} = \hbar\kappa_G / 2\pi \tag{11.3}$$

---

[6] Giddings (2016) has argued that we should trace the origin of Hawking radiation to a 'quantum atmosphere' some distance away from the horizon. The implications of this idea for analogue black hole experiments are an interesting issue to consider.

Crucially, the functional form of this spectrum is *thermal* in the sense that it takes a characteristic Planckian blackbody energy form for the temperature T_{BH}. Black holes, it turns out, are hot!

11.3.2 Hydrodynamic Hawking Effect

Consider a classical fluid as a continuous, compressible, inviscid medium and sound as an alternate compression and rarefaction at each *point* in the fluid. The points are *volume elements* that are taken to be *very small* with respect to the overall fluid volume, but *very large* with respect to the inter-molecular distances. The modelling framework of continuum hydrodynamics is thus valid only when fluid density fluctuations of the order of molecular lengths can be ignored. The two fundamental equations of continuum hydrodynamics are the *continuity equation*, which expresses the conservation of matter:

$$\frac{\partial \rho}{\partial t} + \nabla \cdot (\rho \vec{v}) = 0, \tag{11.4}$$

and the *Euler equation*, which is essentially Newton's second law:

$$\rho\left(\frac{\partial \vec{v}}{\partial t} + (\vec{v} \cdot \nabla)\vec{v}\right) = -\nabla p, \tag{11.5}$$

where ρ is the mass density of the fluid at a particular point, $\vec{v}$ is the velocity of the fluid volume element, and p is pressure.

If the fluid is *barotropic* and *locally irrotational*, Euler's equation reduces to a form of the Bernoulli equation. We identify sound waves in the fluids with the fluctuations (ρ_1, p_1, ψ_1) about the background, which is interpreted as bulk fluid motion. The linearised version of the continuity equation then allows us to write the equation of motion for the fluctuations as follows:

$$\frac{1}{\sqrt{-g}} \frac{\partial}{\partial x_\mu} (\sqrt{-g} g^{\mu\nu} \frac{\partial}{\partial x^\nu} \psi_1) = 0, \tag{11.6}$$

where we have defined the acoustic metric

$$g_{\mu\nu}^{\text{acoustic}} = \frac{\rho_0}{c_{\text{sound}}} \begin{pmatrix} -(c_{\text{sound}}^2 - v_0^2) & \vdots & -(v_0)_j \\ \cdots & \cdot & \cdots \\ -(v_0)^i & \vdots & \delta_{ij} \end{pmatrix}. \tag{11.7}$$

Propagation of sound in a fluid can be understood as being governed by an acoustic metric of the form $g_{\mu\nu}$. The close similarity between the acoustic case and gravity can be seen immediately if we consider the Schwarzschild metric in Painleve–Gullstrand coordinates:

$$
g_{\mu\nu}^{\text{Schwarzschild}} = \begin{pmatrix} -(c_0^2 - \frac{2GM}{r}) & \vdots & -\sqrt{\frac{2GM}{r}}\,\vec{r}_j \\ \cdots & \cdot & \cdots \\ -\sqrt{\frac{2GM}{r}}\,\vec{r}_i & \vdots & \delta_{ij} \end{pmatrix}. \tag{11.8}
$$

This similarity can be transformed into an isomorphism (up to a factor) given certain conditions on the speed of sound in the fluid and the fluid density and velocity profiles. The role of the black hole event horizon is now played by the effective acoustic horizon, where the inward-flowing magnitude of the radial velocity of the fluid exceeds the speed of sound. The black hole is replaced by a *dumb hole*.

Unruh's crucial insight in his 1981 paper was that once the relevant fluid-spacetime geometric identification has been made, there is nothing stopping one from repeating Hawking's 1975 semi-classical argument, but replacing light with sound. The result is that while, in the gravitational Hawking effect a black hole event horizon is associated with a late time *thermal photonic flux*, in the hydrodynamic Hawking effect a dumb hole sonic horizon can be associated with a late time *thermal phononic flux*.

11.3.3 BEC Hawking Effect

Following the same line of reasoning as Unruh's original ideal fluid argument, Garay et al. (2000) derived a BEC Hawking effect using an appeal to the hydro-dynamic approximation of a BEC. Consider the Gross–Pitaevskii equation, which can be derived by applying a mean field approximation to the many-body QM description of a BEC:

$$
i\hbar \frac{\partial \psi(\mathbf{r}, t)}{\partial t} = -\frac{\hbar^2}{2m} \nabla^2 \psi(\mathbf{r}, t) + V(\mathbf{r})\psi(\mathbf{r}, t) + U_0 |\psi(\mathbf{r}, t)|^2 \psi(\mathbf{r}, t) \tag{11.9}
$$
$$
= H_{GP} \psi(\mathbf{r}, t), \tag{11.10}
$$

where $V(\mathbf{r})$ is an external potential. $U_0 = 4\pi\hbar^2 a/m$ is the effective two-particle interaction, with a and m being the scattering length and atomic mass, respectively. From this, one can obtain a Madelung-type expression of the following form:

$$
\frac{\partial \mathbf{v}}{\partial t} = -\frac{1}{nm}\nabla p - \nabla\left(\frac{v^2}{2}\right) + \frac{1}{m}\nabla\left(\frac{\hbar^2}{2m\sqrt{n}}\nabla^2\sqrt{n}\right) - \frac{1}{m}\nabla V, \tag{11.11}
$$

where $\psi = fe^{i\phi}$, $n = |\psi|^2$, $p = n^2 U_0/2$, and $\mathbf{v} = \frac{\hbar}{m}\nabla\phi$. Consider the quantum pressure term:

$$\frac{1}{m}\nabla\left(\frac{\hbar^2}{2m\sqrt{n}}\nabla^2\sqrt{n}\right).$$

When variations in the density of the BEC occur on length scales much greater than the *healing length*, $\xi = \hbar/(mnU_0)^{\frac{1}{2}}$, the quantum pressure term can be neglected and we recover the usual classical Euler equation for an irrotational fluid. Now consider a 'second quantised' field theoretic description of the (weakly interacting) BEC, where the Hamiltonian is expressed in terms of creation and annihilation operators for Bosons, $\hat{\psi}^\dagger(\mathbf{r})$ and $\hat{\psi}(\mathbf{r})$. We can decompose the quantum field into a classical 'bulk' field and a quantum fluctuation:

$$\hat{\psi}(\mathbf{r},t) = \psi(\mathbf{r},t) + \hat{\phi}(\mathbf{r},t). \tag{11.12}$$

The Gross–Pitaevskii equation can be recovered as the equation for the classical field $\psi(\mathbf{r},t)$, when the backreaction with the quantum fluctuation can be neglected. Garay et al. (2000) showed that in the limit where we have no backreaction and the quantum pressure term can be neglected, linearized fluctuations in the BEC will couple to an effective acoustic metric of the same form as that derived by Unruh for the hydrodynamic case. In this limit, the derivation of the BEC Hawking effect then follows the same pattern as that for hydrodynamics and gravity.

In the Bogoliubov description of a (weakly interacting) BEC, one assumes that the backreaction of the classical field with the quantum excitations is *small but non-negligible* and includes only terms that are at most quadratic in $\hat{\phi}(\mathbf{r},t)$.[7] In the Bogoliubov description, fluctuations are no longer governed by an effective acoustic metric; instead, elementary excitations correspond to eigenvalues of the Bogoliubov operator:

$$\mathcal{L} = \begin{pmatrix} H_{GP} - \mu + U_0|\psi|^2 & U_0\psi^2 \\ -U_0\psi^{*2} & -H_{GP} + \mu - U_0|\psi|^2 \end{pmatrix}, \tag{11.13}$$

where μ is the chemical potential. Impressively, Recati et al. (2009) have shown that one can still derive an analogue Hawing effect for an inhomogeneous BEC when treated in the Bogoliubov description. This result is particularly interesting since in this regime there is no longer technically an analogy with the gravitational case since the 'surface gravity' is formally infinite. The authors note:

It is remarkable to note that [the results] still give the typical thermal behaviour of Hawking radiation even though one is not allowed to use the gravitational analogy. (Recati et al., 2009, p. 6)

---

[7] See Pethick and Smith (2002) for a detailed textbook treatment of the Bogoliubov description of BECs.

Thus, there is an interesting sense in which the BEC Hawking effect is *robust* to the breakdown in the hydrodynamic description at length scales smaller than the healing length.

11.4 The Trans-Planckian Problem and Universality

In the standard calculation of the Hawking temperature, exponential gravitational red-shift means that the black hole radiation detected at late times (i.e., the outgoing particles) must be taken to correspond to extremely high-frequency radiation at the horizon. Such a 'trans-Planckian' regime is the dominion of theories of quantum gravity, and is thus well beyond the domain of applicability of the modelling framework we are using. This problem with 'trans-Planckian' modes has a direct analogue in the BEC case in terms of the failure of the hydrodynamic limit: We cannot assume that the perturbations have wavelengths much larger than the healing length. Of course, since the results of Recati et al. (2009) show that the BEC Hawking effect *is* suitably robust, we have good reason not to worry too much about the 'trans-Planckian' problem for the BEC analogue model. But whereas the Bogoliubov description gives us microphysical understanding of a BEC, we do not have an equivalently trustworthy theory for the microphysics of spacetime or, for that matter, fluids.

We can, however, attempt to model the effect of the underlying microphysics on linear fluctuations by considering a modified dispersion relation. This idea originally comes from Jacobson (1991, 1993) who suggested that one could use a modified dispersion relation to understand the breakdown of continuous fluid models due to atomic effects. The question of particular importance is whether an exponential relationship actually holds between the outgoing wave at some time after the formation of the horizon, and the wavenumber of the wave packet Unruh (2008). Using numerical simulations, Unruh (1995) showed that the altered dispersion relation in atomic fluids implies that the early time quantum fluctuations that cause the late-time radiation are not, in fact, exponentially large. A related analytical argument was later applied to the gravitational case by Corley (1998). What is more desirable, however, is a set of general conditions under which such effective decoupling between the sub- and trans-Planckian physics can be argued to take place. One interesting proposal in this vein comes from Unruh and Schützhold (2005).[8] The 'universality' results of Unruh and Schützhold show that possible trans-Planckian effects can be factored into the calculation of Hawking radiation via a non-trivial dispersion relation. In particular, they consider a generalised–Klein–Fock–Gordon

---

[8] For further work on these issues, using a range of different methodologies, see Himemoto and Tanaka (2000), Barceló et al. (2009), and Coutant et al. (2012).

equation for the scalar field of the following form:

$$\left(\partial_t + \partial_x v(x)\right)\left(\partial_t + v(x)\partial_x\right)\hat{\phi} = \left(\partial_x^2 + F(\partial_x^2)\right)\hat{\phi} \qquad (11.14)$$

with F being the non-trivial dispersion relation and v being *either* the local velocity of free falling frames measured with respect to a time defined by the Killing vector of the stationary metric *or* the position-dependent velocity of the fluid. They use complex analytic arguments to show that the Hawking flux is robust under the modified dispersion relation – essentially because the dominant terms in the relevant integral are entirely due to the discontinuity caused by the horizon of the black hole (or its analogue). What Unruh and Schützhold establish is that the Hawking effect does not, to lowest order, depend on the details of underlying physics, given certain modelling assumptions.[9]

11.5 The Technion Experiments

Based upon the most recent of a series of experiments[10] conducted at the Technion (Israel Institute of Technology), Steinhauer has claimed to have observed BEC Hawking radiation. This observation is based upon measurement of the density correlations in the BEC that are understood to 'bear witness' to entanglement between Hawking modes outside the sonic horizon and partner modes inside the sonic horizon. In this section, we provide some background for these results before concluding our analysis by considering the potential for the Technion experiments to function as evidence for gravitational Hawking radiation.

The first key idea we must introduce to understand the theoretical context of Steinhauer's work is the idea of entanglement as a 'witness' to Hawking radiation. It is plausible to think of the Hawking effect for black holes in terms of the creation of correlated pairs of outgoing quanta triggered by the event horizon formation (Balbinot et al., 2008; Parentani, 2010; Schützhold & Unruh, 2010). The particles that featured in our treatment of the Hawking effect in section 3 correspond to escaping modes that make up one half of each pair of quanta. These 'Hawking modes' are each correlated with (rather ill-fated) 'partner modes' that are trapped inside the event horizon and forever lost. It is thus understandable that in the gravitational context, the Hawking modes are the centre of attention. In contrast, when considering the Hawking effect in an analogue context, the regions separated by the horizon are not causally disconnected: Unlike in the case of an event horizon, in

---

[9] For example, the evolution of the modes is assumed to be adiabatic: The Planckian dynamics are supposed to be much faster than all external (sub-Planckian) variations.

[10] Three particular landmark experimental results achieved by Steinhauer (and in the first case collaborators) are the creation of a BEC analogue black hole (Lahav et al., 2010), the observation of self-amplifying Hawking radiation in an analogue black hole laser (Steinhauer, 2014), and the observation of quantum Hawking radiation and its entanglement (Steinhauer, 2016).

principle an observer can have access to both sides of a sonic horizon. The key idea is that by measuring the correlations between modes inside and outside the horizon, we can detect the signature of Hawking radiation in an analogue black hole. In particular, if we can establish the existence of entanglement between the modes, then we have strong evidence for the existence for *quantum* Hawking radiation within the relevant analogue system.

The next important idea is that in the case of a BEC analogue black hole, the entanglement between the Hawking and partner modes can be measured by considering the density–density correlation function between spatial points on either side of the horizon (Steinhauer, 2015). This work is highly important from a practical perspective since there exist high-precision techniques for measuring point-by-point densities of a BEC by considering phase shifts of a laser beam shone through it. Formally, the relationship between the entanglement of the modes and the density correlations can be established by rewriting a given measure of the non-separability of the quantum state, $\triangle$, in terms of density operators, ρ_k, for modes of wavenumber k. For example, in his treatment Steinhauer (2015, 2016) considers a simple measure:

$$\triangle \equiv \langle \hat{b}_{k_H}^{u\dagger} \hat{b}_{k_H}^{u} \rangle \langle \hat{b}_{k_p}^{d\dagger} \hat{b}_{k_p}^{d} \rangle - |\hat{b}_{k_H}^{u} \hat{b}_{k_p}^{d}|, \tag{11.15}$$

where $\hat{b}_{k_H}^{u}$ is the annihilation operator for a Bogoliubov excitation with wavenumber k_H, localised in the subsonic 'downstream' region outside the event horizon. The letters u, d, H, and p stand for upstream (supersonic), downstream (subsonic), Hawking modes, and partner modes, respectively. Using the Fourier transform of the density operator, we can express such a non-separability measure in terms of correlations between the u and d density operators for the k_H and k_p modes. We can thus measure $\triangle$ purely by measuring density correlations in the BEC on either side of the sonic horizon. If $\triangle$ is negative, then the correlations are strong enough to indicate entanglement, which is the quantum signature of Hawking radiation.

In his landmark experiment, Steinhauer used a BEC of [87]Rb atoms confined radially by a narrow laser beam. The horizon was created by a very sharp potential step, which is swept along the the BEC at a constant speed. Significantly, the length scales are such that the hydrodynamic description of a BEC is appropriate: The width of the horizon is of the order of a few times bigger than the healing length. The main experimental result consists of an aggregate correlation function computed based upon an ensemble of 4,600 repeated experiments that were conducted over six days. Given some reasonable assumptions (e.g., modes at different frequency are assumed to be independent of each other), the experiments can be interpreted as establishing an entanglement witness to Hawking radiation in BEC.

11.6 Dumb Hole Epistemology

Three notions of validation are relevant to the Technion experiments described in section 5. The first, and most straightforward, is internal validation. Was Steinhauer genuinely learning about the physics of the particular sonic horizon within the particular ^{87}Rb BEC that he was manipulating? Various sources of internal validation are apparent from the description of the experimental setup given, not least the repetition of the experimental procedure nearly 5,000 times. Given this, the evidence gained from the experiments conducted can be categorised as of the appropriate epistemic type to confirm specific statements regarding the particular BEC that was experimented upon.

The next question relates to external validation of the Technion experiments as experiments in the conventional sense. Can the particular sonic horizon that was constructed, within the particular ^{87}Rb BEC, stand in for a wider class of systems – for example, all BEC sonic horizons within the realm of validity of the hydrodynamic approximation to the Gross–Pitaevskii equation, regardless of whether the relevant systems have been (or even could be) constructed on Earth? Given that this set of systems obeys the 'reasonable assumptions' of the Steinhauer experiments, such as the assumption that modes at different frequency are independent of each other, then we can also externally validate the experiments. This means, with relevant qualifications, the evidence produced by the experiments conducted can be categorised as of the appropriate epistemic type to confirm general statements regarding the class of BECs with sonic horizons. Thus, so far as their status as conventional experiments goes, the Technion experiments can be considered internally and externally validated based upon purely upon Steinhauer's report.

The details provided in the report do not, however, function as external validation of the Technion experiments *as analogue experiments*. That is, without further argument we do not have a link from the class of source systems (BECs with sonic horizons) to the class of target systems (astrophysical black holes). Thus, when considered in isolation, the evidence from the Technion experiments cannot be categorised as of the appropriate epistemic type to confirm general statements regarding astrophysical black holes. Rather, when considered in isolation, such statements are of the same epistemic type as evidence derived from speculative inferences. Although the experiment results might reasonably be argued to be evidence that is, in some sense, more convincing or valuable than that produced by many such inferences, in confirmatory terms the findings are equally null.

The question of external validation of dumb hole experiments as analogue experiments can be addressed by the universality arguments of Unruh and Schützhold. If accepted, their theoretical universality arguments would function as external validation for the Steinhauer experiments. They provide a theoretical basis for accepting the source system of the Technion experiments (a BEC analogue black hole) as a stand-in for a wider class of target systems (black holes in general, including other analogue models and gravitational black holes). In this sense, they give us a possible basis for upgrading the epistemic type of the experimental evidence, such that it can be used to confirm general statements regarding astrophysical black holes.

This statement should not be as surprising as it sounds. After all, universality arguments are essentially theoretical statements about the multiple realisability of a given phenomenon, and so imply that certain features of the phenomenon in one exemplification will be present in all others. That said, it is difficult not to be wary of the speed and strength of this kind of conclusion. This mode of external validation is almost completely unlike those used in more conventional experiments and, as such, it should be treated rather sceptically for the time being. Arguably, the importance of the Technion experiments lies more in their future rather than in their immediate evidential value. In particular, given experimental demonstration of an array of analogue Hawking effects across a variety of very different media, one could become increasingly confident in the universality arguments and, in turn, in the external validation of the experiments as analogue experiments.[11]

We can thus conclude that there *is* a plausible theoretical basis to 'externally validate' analogue black hole experiments such that a BEC analogue black hole can be taken to 'stand in' for an astrophysical black hole. Thus, we have one example in which analogue experiments can provide evidence of the same confirmatory epistemic type as conventional experiments. This case does not, however, speak to the question of the significance of confirmation. Is it reasonable to think, in quantitive terms, that analogue experiments can provide a comparable degree of confirmation as conventional experiments? Can they be *substantially* confirmatory rather than merely *incrementally* confirmatory? Such questions bring us beyond the scope of the present chapter. However, there are reasons to be optimistic. As shown by a recent analysis in terms of Bayesian confirmation theory (Dardashti et al., 2016), given experimental demonstration of an array of analogue Hawking effects across a variety of very different media, the degree of confirmation conferred can be amplified very quickly. It is thus very plausible to think of analogue experiments as prospective means for providing confirmatory support that is substantial, rather than merely incremental.

---

[11] There are grounds for optimism in this regard on a number for fronts. See Philbin et al. (2008), Belgiorno et al. (2010), Unruh and Schützhold (2012), Liberati et al. (2012), and Nguyen et al. (2015).

Acknowledgements

I am deeply indebted to Radin Dardashti, Richard Dawid, Stephan Hartmann, and Eric Winsberg for discussions and collaborative engagement, without which this chapter could not have been written. I am also thankful to the audience at the "Why Trust a Theory?" workshop in Munich for insightful questions, to Bill Unruh for providing me with some useful details about the Technion experiments, and to Erik Curiel for various helpful discussions about black hole thermodynamics.

References

Achinstein, P. (2001). *The Book of Evidence*, Vol. 4. Cambridge: Cambridge University Press.

Anderson, M., J. Ensher, M. Matthews, C. Wieman, & E. Cornell. (1995). Observation of Bose–Einstein condensation in a dilute atomic vapor. *Science 269*, 14.

Bailer-Jones, D. M. (2009). *Scientific Models in Philosophy of Science*. Pittsburgh: University of Pittsburgh Press.

Balbinot, R., A. Fabbri, S. Fagnocchi, A. Recati, & I. Carusotto. (2008). Nonlocal density correlations as a signature of Hawking radiation from acoustic black holes. *Physical Review A 78*(2), 021603.

Barceló, C., L. J. Garay, & G. Jannes. (2009). Sensitivity of Hawking radiation to superluminal dispersion relations. *Physical Review D 79*(2), 024016.

Barceló, C., S. Liberati, M. Visser, et al. (2011). Analogue gravity. *Living Reviews in Relativity 14*(3).

Bartha, P. F. (2010). *By Parallel Reasoning: The Construction and Evaluation of Analogical Arguments*. Oxford: Oxford University Press.

Bartha, P. (2013, Fall). Analogy and analogical reasoning. In E. N. Zalta (ed.), *The Stanford Encyclopedia of Philosophy*.

Beisbart, C., & J. D. Norton. (2012). Why Monte Carlo simulations are inferences and not experiments. *International Studies in the Philosophy of Science 26*(4), 403–22.

Belgiorno, F., S. Cacciatori, M. Clerici, V. Gorini, G. Ortenzi, L. Rizzi, E. Rubino, V. Sala, & D. Faccio. (2010). Hawking radiation from ultrashort laser pulse filaments. *Physical Review Letters 105*(20), 203901.

Bose, S. N. (1924). Planck's gesetz und lichtquantenhypothese. *Zeitschrift für Physik 26*(3), 178.

Cartwright, N., A. Goldfinch, & J. Howick. (2010). Evidence-based policy: Where is our theory of evidence? *Journal of Chilaren's Services 4*(4), 6–14.

Corley, S. (1998). Computing the spectrum of black hole radiation in the presence of high frequency dispersion: An analytical approach. *Physical Review D 57*(10), 6280.

Coutant, A., R. Parentani, & S. Finazzi. (2012). Black hole radiation with short distance dispersion: An analytical *S*-matrix approach. *Physical Review D 85*(2), 024021.

Crupi, V. (2013, Winter). Confirmation. In E. N. Zalta (ed.), *The Stanford Encyclopedia of Philosophy*.

Dardashti, R., S. Hartmann, K. Thébault, & E. Winsberg. (2016). Confirmation via analogue simulation: A Bayesian account. http://philsci-archive.pitt.edu/12221/.

Dardashti, R., K. Thébault, & E. Winsberg. (2017). Confirmation via analogue simulation: What dumb holes could tell us about gravity. *British Journal for the Philosophy of Science 68*(1), 55–89.

Dawid, R. (2013). *String Theory and the Scientific Method*. Cambridge: Cambridge University Press.

Dawid, R., S. Hartmann, & J. Sprenger. (2015). The no alternatives argument. *British Journal for the Philosophy of Science*, 66(1):213–34.

Dvali, G., & M. Redi. (2008). Black hole bound on the number of species and quantum gravity at Cern LHC. *Physical Review D 77*(4), 045027.

Einstein, A. (1924). Quantentheorie des einatomigen idealen gases. *Sitzungsberichte der Preussische Akadmie der Wissenschaften*, 261–7.

Franklin, A. (1989). *The Neglect of Experiment*. Cambridge: Cambridge University Press.

Franklin, A., & S. Perovic. (2015, Summer). Experiment in physics. In E. N. Zalta (ed.), *The Stanford Encyclopedia of Philosophy*.

Garay, L., J. Anglin, J. Cirac, & P. Zoller. (2000). Sonic analog of gravitational black holes in Bose–Einstein condensates. *Physical Review Letters 85*(22), 4643.

Giddings, S. B. (2016). Hawking radiation, the Stefan–Boltzmann law, and unitarization. *Physics Letters B 754*, 39–42.

Hawking, S. W. (1975). Particle creation by black holes. *Communications in Mathematical Physics 43*(3), 199–220.

Himemoto, Y., & T. Tanaka. (2000). Generalization of the model of Hawking radiation with modified high frequency dispersion relation. *Physical Review D 61*(6), 064004.

Jacobson, T. (1991). Black-hole evaporation and ultrashort distances. *Physical Review D 44*(6), 1731.

Jacobson, T. (1993). Black hole radiation in the presence of a short distance cutoff. *Physical Review D 48*(2), 728–41.

Lahav, O., A. Itah, A. Blumkin, C. Gordon, S. Rinott, A. Zayats, & J. Steinhauer. (2010). Realization of a sonic black hole analog in a Bose–Einstein condensate. *Physical Review Letters 105*(24), 240401.

Liberati, S., A. Prain, & M. Visser. (2012). Quantum vacuum radiation in optical glass. *Physical Review D 85*(8), 084014.

Nguyen, H., D. Gerace, I. Carusotto, D. Sanvitto, E. Galopin, A. Lemaître, I. Sagnes, J. Bloch, & A. Amo. (2015). Acoustic black hole in a stationary hydrodynamic flow of microcavity polaritons. *Physical Review Letters 114*(3), 036402.

Parentani, R. (2010). From vacuum fluctuations across an event horizon to long distance correlations. *Physical Review D 82*(2), 025008.

Parker, W. S. (2009). Does matter really matter? Computer simulations, experiments, and materiality. *Synthese 169*(3), 483–96.

Pethick, C. J., & H. Smith. (2002). *Bose–Einstein Condensation in Dilute Gases*. Cambridge: Cambridge University Press.

Philbin, T. G., C. Kuklewicz, S. Robertson, S. Hill, F. König, & U. Leonhardt. (2008). Fiber-optical analog of the event horizon. *Science 319*(5868), 1367–70.

Recati, A., N. Pavloff, & I. Carusotto. (2009). Bogoliubov theory of acoustic Hawking radiation in Bose–Einstein condensates. *Physical Review A 80*(4), 043603.

Roush, S. (2005). *Tracking Truth: Knowledge, Evidence, and Science*. Oxford: Oxford University Press on Demand.

Salmon, W. (1990). Rationality and objectivity in science or Tom Kuhn meets Tom Bayes. In C. Wade Savage (ed.), *Scientific Theories*, Vol. 14, pp. 175–204. Minneapolis: University of Minnesota Press.

Scardigli, F. (1999). Generalized uncertainty principle in quantum gravity from micro-black hole gedanken experiment. *Physics Letters B 452*(1–2), 39–44.

Schützhold, R., & W. G. Unruh. (2010). Quantum correlations across the black hole horizon. *Physical Review D 81*(12), 124033.

Steinhauer, J. (2014). Observation of self-amplifying Hawking radiation in an analogue black-hole laser. *Nature Physics 10*, 364–9.

Steinhauer, J. (2015). Measuring the entanglement of analogue Hawking radiation by the density–density correlation function. *Physical Review D 92*(2), 024043.

Steinhauer, J. (2016, August). Observation of quantum Hawking radiation and its entanglement in an analogue black hole. *Nature Physics 12*, 959–65.

Unruh, W. (1981). Experimental black-hole evaporation? *Physical Review Letters 46*(21), 1351–3.

Unruh, W. (1995). Sonic analogue of black holes and the effects of high frequencies on black hole evaporation. *Physical Review D 51*(6), 2827.

Unruh, W. (2008). Dumb holes: Analogues for black holes. *Philosophical Transactions of the Royal Society A: Mathematical, Physical and Engineering Sciences 366*(1877), 2905–13.

Unruh, W. G., & R. Schützhold. (2005). Universality of the Hawking effect. *Physical Review D 71*(2), 024028.

Unruh, W., & R. Schützhold. (2012). Hawking radiation from 'phase horizons' in laser filaments? *Physical Review D 86*(6), 064006.

Winsberg, E. (2009). A tale of two methods. *Synthese 169*(3), 575–92.

Winsberg, E. (2010). *Science in the Age of Computer Simulation*. Chicago: University of Chicago Press.

12

Are Black Holes about Information?

CHRISTIAN WÜTHRICH

12.1 Introduction

Physicists working in quantum gravity diverge rather radically over the physical principles that they take as their starting point in articulating a quantum theory of gravity, over which direction to take from these points of departure, over what we should reasonably take as the goals of the enterprise and the criteria of success, and sometimes even over the legitimacy of different methods of evaluating and confirmating the resulting theories. Yet, there is something that most of them agree upon: that black holes are thermodynamical objects, have entropy, and radiate, and thus that the Bekenstein–Hawking formula for the entropy of a black hole must be recovered, more or less directly, from the microphysics of the fundamental degrees of freedom postulated and described by their theories.[1] Thus, regardless of which programme in quantum gravity they adhere to, physicists of all stripes take it as established that black holes are thermodynamic objects with temperature and entropy. Thus, they *trust* the Bekenstein–Hawking formula for the latter as a test that a theory of quantum gravity must survive.

Why do physicists accept the Bekenstein–Hawking result as a litmus test for quantum gravity? The confluence of a number of distinct aspects of known black hole physics all pointing toward thermodynamic behaviour is certainly remarkable. Hawking's calculation in particular uses rather common semi-classical methods of

---

[1] For instance, string theorists hail the result by Strominger and Vafa (1996) of recovering the Bekenstein–Hawking formula from the microphysics of a class of five-dimensional extremal (and hence unrealistic) black holes, advocates of loop quantum gravity celebrate its kinematic derivation (up to a constant) in Rovelli (1996), and proponents of causal set theory applaud the black hole model constructed in Dou and Sorkin (2003), in which the black hole's entropy can be seen as proportional to the area of the horizon. All of these results are partial and amount to proofs of concepts, rather than being independently motivated and completely general results. To date, Hawking radiation has not been derived in any of the approaches to full quantum gravity. Recently, however, it has been pointed out to me that the temperature of near-extremal holes can be calculated in string theory, as can the absorption and emission rates for stringy quanta, both in agreement with Hawking's calculation.

quantum field theory on curved spacetimes. To many physicists, it seems hard to imagine that those methods should not be valid at least in some relatively placid arena of mildly semi-classical physics sufficient to derive Hawking radiation. The robustness of these thermodynamic properties of black holes under various theoretical considerations thus inspires a great deal of confidence in them.

Yet none of these properties has ever been empirically confirmed. Given this lack of direct observations of Hawking radiation or of any other empirical signatures of the thermodynamics of black holes, the universality and the confidence of this consensus are prima facie rather curious. The aim of this chapter is to investigate the foundation of this seeming, and curiously universal, agreement; in particular, it analyzes Jacob Bekenstein's original argument that led him to propose his celebrated formula, as its recovery constitutes an important assumption of much research into quantum gravity. The unexamined assumption is not worth believing, after all.

Two caveats before we start. First, if non-empirical theory confirmation is legitimate, as has been suggested, for example, by Richard Dawid (2013), then the fact that thermodynamic behaviour of black holes has not been observed to date may not be a worry as long as we have convincing non-empirical reasons for believing the Bekenstein–Hawking formula. In revisiting Bekenstein's original argument, I propose to evaluate some of these reasons. Second, I appreciate that the case for taking the Bekenstein–Hawking formula seriously would be significantly strengthened by the observation of Hawking radiation in models of analogue gravity.[2] In fact, until Stephen Hawking (1975) offered a persuasive semi-classical argument that black holes radiate, and so exhibit thermodynamic behaviour like a body with a temperature, most physicists were not moved by Bekenstein's earlier case for black hole entropy. I will briefly return to this point in the conclusions, but leave the detailed analysis for another occasion.

Section 12.2 covers some of Bekenstein's original motivations and discusses the area theorem in general relativity (GR), which serves as the vantage point for Bekenstein. Section 12.3 offers a reconstruction of Bekenstein's main argument. Section 12.4 evaluates Bekenstein's attempt to endow the central analogy with physical salience and challenges his invocation of information theory to this end. Section 12.5 concludes the chapter.

12.2 The Generalized Second Law and the Area Theorem

Bekenstein (1972) is motivated by the recognition that black hole physics potentially violates the second law of thermodynamics, which states that the entropy of an

---

[2] As argued, for example, by Dardashti et al. (2017). Steinhauer (2016) claims to have observed Hawking radiation in analogue black holes in a Bose–Einstein condensate, but to my knowledge the result has not been replicated.

isolated physical system never decreases, but always remains constant or increases.[3]
He considers the case of a package of entropy lowered into a black hole from the
asymptotic region and asserts that once this has settled into equilibrium,

[T]here is no way for the [exterior] observer to determine its interior entropy. Therefore, he
cannot exclude the possibility that the total entropy of the universe may have decreased in
the process. ... The introduction of a black-hole entropy is necessitated by [this] process.
(pp. 737f)

Thus, Bekenstein contends that once the entropy package has vanished behind the
event horizon of the black hole and thus becomes in principle unobservable, it may
well be, for all we know, that this entropy has dissipated and thus that the total
entropy in the universe has decreased, *pace* the Second Law. One might choose
to resist his argument right here: In a classical account at least, an observer in the
asymptotic region will never see the entropy package actually disappear behind the
event horizon. So, for all they can tell, it will always remain there, and thus will in
no way threaten the Second Law.[4]

   Suppose, however, that the asymptotic observer has reason to believe that the
entropy package has been swallowed by the black hole's event horizon; perhaps
the observer knows general relativity and calculates that the package has in fact
passed the event horizon, even though this is not what the observer sees. So the
observer could have reason to believe that the Second Law is violated. This problem
would obviously be avoided if the black hole itself possessed entropy, and moreover
possessed an entropy that would increase by at least the amount of entropy that
disappears behind the event horizon. The ascription of entropy to a black hole[5] thus
permits a generalization of the traditional Second Law, as Bekenstein asserts:

The common entropy in the black-hole exterior plus the black-hole entropy never decreases.
(1973, p. 2339, emphasis in original)

At this point in the argument, the only justification for introducing black hole
entropy rests on saving (a generalized version of) the Second Law. Although this
is not the main focus of the present chapter, it is important to realize that this
amounts to a rather significant extension of the validity of the Second Law from
the firm ground of terrestrial physics to the precarious and speculative realm of
black holes. For all the evidence we have, black holes may well exhibit unusual and
unexpected thermodynamic behaviour – or none at all. It is certainly not automatic

---

[3] For an authoritative and philosophical introduction to black hole thermodynamics, see Curiel and Bokulich
(2018, §5).
[4] I wish to thank Niels Linnemann for pointing this out to me.
[5] Entropy is standardly predicated only of systems in equilibrium. The standard black hole solutions in classical
general relativity are stationary and in this sense in equilibrium; however, not every physical black hole is
thought to be in equilibrium, and so the ascription of entropy to *those* black holes would require nonstandard
thermodynamics.

that the Generalized Second Law holds. In fact, not only are many proofs of the Generalized Second Law lacking (Wall, 2009), but Aron Wall (2013) has shown that a singularity theorem results from it. Just as energy conditions constitute essential premises for the standard singularity theorems of classical GR, the Generalized Second Law seems to play this role for a new version of a singularity theorem, which no longer depends on energy conditions and, therefore, may hold well into the quantum realm. Thus, that something like the Generalized Second Law obtains is far from trivial, particularly if we think that the presence of singularities points to a failure of our theories.

Once we accept it, however, and with it the need for a black hole entropy, we face the question of what the black hole entropy is and how it relates to other physical properties of black holes. Bekenstein (1972) finds the answer in a theorem that Hawking (1971) proved just a year earlier: the area theorem. In the formulation of Robert Wald (1994, p. 138), the area theorem is the following proposition:

Theorem 1 (Area theorem [Hawking, 1971]). *For a predictable black hole satisfying $R_{ab}k^a k^b \geq 0$ for all null k^a, the surface area of the future event horizon, h^+, never decreases with time.*

Here, R_{ab} is the Ricci tensor, k^a is a null vector tangent to the geodesic generators of the event horizon, and h^+ is the future event horizon of the black hole, that is, the boundary of all events in the spacetime that can send light signals to infinity. The surface area of the horizon h^+ is understood as the area of the intersection of h^+ and some spacelike (or possibly null) hypersurface Σ suitable to evaluate the entropy. For a proof of the theorem, see Wald (1994, pp. 138f).

The area theorem seems to give us an obvious candidate for a physical quantity to act in lieu of standard entropy: the never-decreasing surface area of the black hole's event horizon. Bekenstein argues for the explicit analogy of the area theorem and the Second Law, and the roles played in them, respectively, by the area surface of the event horizon and the black hole's entropy. Although the propositions remain merely analogous, the intended connection at the level of the physical magnitudes is one of identity: The area of the event horizon really *is* the black hole's entropy – or so goes the thought. It is my main purpose in this chapter to analyze Bekenstein's argument for this connection. While Bekenstein's argument may not be the most decisive of arguments in favour of attributing an entropy to black holes, it gave birth to an industry and thus deserves scrutiny.

Before we analyze the premises of the theorem, it ought to be noted that the analogy breaks down right out of the gate, at least in some respects. If we accept its nomic reduction to statistical mechanics, then the Second Law is not a strict law, but is only statistically valid: It suffers from occasional, though very rare, violations. Not so for the area theorem, which is a strict theorem in the context of GR: The area of the event horizon *never* decreases. As a consequence, the time asymmetry grounded in the area theorem is also strict, whereas the time asymmetry based on the Second Law is not strict – again, at least if we accept that it is nomically reducible to statistical mechanics. Now while this is clearly a disanalogy, it may not be fatal to Bekenstein's endeavour: It may be that the physics of black holes is different from that of ordinary thermodynamic objects in that there just is no possibility of decreasing entropy; or it may be that the area law, just like the Second Law, will have to be replaced by a proposition in a more fundamental (quantum) theory of gravity that will permit rare exceptions to the increase of area. This second possibility is, of course, just what many physicists working in quantum gravity expect. In fact, Hawking radiation and the evaporation of black holes, if borne out, hand us a mechanism counteracting the exceptionless increase of entropy suggested by the area theorem. Hawking radiation thus renders the area theorem otiose, although this should not come as a surprise given that the area theorem is a theorem in classical GR and Hawking radiation a result in semi-classical gravity. This suggests that we should not expect the area theorem to hold strictly once we move beyond the domain of GR.

Either way, what really matters is the Generalized Second Law, not the area theorem, which is, to repeat, just a theorem in classical GR. Whether the Generalized Second Law holds – approximately or strictly or not at all – once quantum effects are included is a rather involved question.[6] So insisting that Bekenstein's argument breaks down already at this early stage may be taking the area theorem too seriously as a statement about the fundamental behaviour of black holes.

A further note of interest before we proceed. Insofar as we pick a highly improbable initial state, there is a sense in which the time asymmetry coming out of the Second Law is put in by hand. But the asymmetry of time grounded in the area theorem is no better off: It is equally stipulated when we demand that the horizon is a *future*, rather than a past, horizon – in other words, when we postulate that the singularity forms a *black*, rather than a *white*, hole.[7] In fact, the situation may be worse in the case of the area theorem: In the case of the Second Law, the existence of a dynamical process that *drives* the early universe toward such a state would overcome this unsatisfactory sleight of hand and give us a physical mechanism that renders stipulating a highly special initial state obsolete.

The area theorem, like any other theorem, rests on premises. The two central assumptions are that of 'predictability' and a positivity condition on the Ricci tensor. Following Wald (1994, p. 137), predictability is defined as follows:

---

[6] See Wald (2001, §4.2) for a discussion. Note that the loss of black hole entropy in the form of a shrinking area
     is thought to be compensated by the ordinary entropy of the escaping Hawking radiation.
[7] For the distinction between black and white holes, see Wald (1984, p. 155).

Definition 1 (Predictable black hole). *A black hole in a spacetime* $\langle \mathcal{M}, g_{ab} \rangle$ *is called* predictable *just in case it satisfies the following conditions, where* Σ *is an asymptotically flat time slice,* $I^{\pm}(X)$ *is the chronological future/past of the set* X *of events in* $\mathcal{M}$, $\mathscr{I}^{+}$ *is the future null boundary of* $\mathcal{M}$, *and* $D(X)$ *is the domain of dependence of the set* X:

   (i) $[I^{+}(\Sigma) \cap I^{-}(\mathscr{I}^{+})] \subset D(\Sigma)$,
   (ii) $h^{+} \subset D(\Sigma)$.

Equivalently, an asymptotically flat spacetime $\langle \mathcal{M}, g_{ab} \rangle$ comprises a predictable black hole if the region $\mathcal{O} \subset \mathcal{M}$ to the future of Σ containing the region exterior to the black hole as well as the event horizon h^{+} is globally hyperbolic. This explains the choice of terminology: In a spacetime containing a predictable black hole, all events outside the black hole and to the future of time slice Σ are 'visible' from (future null) infinity; generically, the physics in the region exterior to the black hole is 'predictable' in the sense that typical dynamical equations for sufficiently well-behaved matter such as the Klein–Gordon equation have a well-posed initial value problem (Wald, 1994, Theorem 4.1.2). This means that these spacetimes satisfy a version of the Cosmic Censorship Hypothesis (Wald, 1994, p. 135). Whether the relevant version of the Cosmic Censorship Hypothesis (or, for that matter, any version of it) is true is assumed, but not known, and a matter of ongoing research (Penrose, 1999). Without dwelling on the matter, it should be noted that this may well be a non-trivial limitation of the scope of the area theorem.

Matters are worse with respect to the second substantive premise of the area theorem. The positivity requirement on the Ricci tensor, $R_{ab}k^{a}k^{b} \geq 0$, is related to the 'null energy condition':

Definition 2 (Null energy condition). $T_{ab}k^{a}k^{b} \geq 0$, *with* k^{a} *any future-pointing null vector field, where* T_{ab} *is the stress-energy tensor.*

If the Einstein field equation holds, then the null energy condition obtains just in case $R_{ab}k^{a}k^{b} \geq 0$ obtains. So in GR, the null energy condition implies and is implied by the positivity condition on R_{ab}. Since there exists a long list of possible violations of the null energy condition (Curiel, 2017a, §3.2), the second premise of the theorem constitutes a more severe limitation. The theorem simply does not speak to those spacetimes that either contain a non-predictable black hole or where the null energy condition is violated; in those cases, even if the entropy of a possibly present black hole is proportional to the surface area of its event horizon, we have no guarantee that the area and hence the black hole entropy will never decrease.

One might object to my reservations here that the situations in which we face a non-predictable black hole or a violation of the null energy condition, while formally possible, are not in fact physically possible. I accept that we may ultimately come to this conclusion. However, we should be wary to rule out those cases based on not much more than our intuitions concerning what is physically possible. GR is a notoriously permissive theory – but also a notorious case for challenging our physical intuitions. For example, the existence of singularities in some of its models has undergone a transformation from being so unpalatable as to be ruled out *by fiat* to a necessary part of celebrated predictions of the theory. The challenge in GR is to distinguish cases when it is overly permissive in allowing unphysical solutions from those situations in which it presents genuinely novel, perhaps unexpected, and potentially important lessons that we should heed. In this spirit, it is important to keep the limitations of the area theorem in mind as we proceed.

12.3 The Structure of Bekenstein's Argument

Bekenstein's argument to the conclusion that black holes have entropy, and that this entropy is proportional to the area of the black hole's event horizon,[8] essentially contains two parts. First, it establishes a formal similarity between thermodynamic entropy and black hole entropy. As this formal analogy is, in itself, rather weak, a second step establishing the physical salience of the formal similarity is required.

In Bekenstein (1972), the formal similarity is noted, and it is argued that if we use it to endow black holes with entropies proportional to the surface area of their event horizon (in Planck units), then the appropriately reformulated Second Law will still be valid for processes involving black holes that would otherwise apparently violate it. Thus, the first step in the argument is essentially present in the 1972 paper. Although it also contains inklings of the information-theoretic justification of the second step, the full defence of that is only given in Bekenstein (1973). Let us analyze the two steps in turn, starting with the first.

As Bekenstein (1972) notes, it appears as if black holes permit interactions with their environment, such as lowering entropy packages into them that apparently result in a decrease of thermodynamic entropy, thus violating the Second Law. Thermodynamic entropy is an additive state variable and gives a measure of the irreversibility of the evolution of an isolated thermodynamic system. The entropy of such a system remains constant just in case the isolated system evolves reversibly; it increases if and only if it evolves irreversibly. This gives us the entropic form of the Second Law, according to which the entropy of an isolated system cannot decrease.[9] Quantitatively, a differential increment of thermodynamic entropy S_{TD} is given by a form of the Second Law,

---

[8] Bekenstein (1972) argues for only the proportionality of entropy and area. Although he tries to fix the proportionality factor in Bekenstein (1973), the generally accepted value of 1/4 is first proposed in Hawking (1975).

[9] For a useful introduction to entropy, the reader may consult Lemons (2013).

$$dS_{TD} = \frac{\delta Q}{T}, \tag{12.1}$$

where δQ is an indefinitely small quantity of heat that is transferred to or from the system, and T is its absolute temperature. To block the possibility that heat can be converted to work without attendant compensating changes in the environment of the system, Bekenstein generalizes the Second Law to include a term for the entropy of black holes, as outlined in Section 12.2. He then identifies the area of the horizon of a black hole as the only physical property of a black hole that, just like thermodynamic entropy, exhibits a tendency to increase, as shown by Theorem 1:

The area of a black hole appears to be the only one of its properties having this entropylike behavior which is so essential if the second law as we have stated it is to hold when entropy goes down a black hole. (Bekenstein, 1972, p. 738)

Thus, the sole basis on which Bekenstein introduces the claim that the area of its event horizon is (proportional to) a black hole's entropy is that it is the only property that is not obviously altogether *unlike* entropy. A weak similarity indeed.

Given the parallel tendency to increase, Bekenstein proposes that the entropy of a black hole S_{BH} be simply given as proportional to the area A of its horizon:

$$S_{BH} = \eta \frac{kA}{\ell_P^2}, \tag{12.2}$$

where the introduction of Boltzmann's constant k and the Planck length ℓ_P is "necessitated by dimensional considerations" (1972, p. 738).[10] η is a constant of order unity, and part of Bekenstein's ambition in his longer article of 1973 was to fix that proportionality factor; he ends up suggesting $\eta = \ln 2/8\pi$, or something "very close to this, probably within a factor of two" (1973, p. 2338). This number is, however, an order of magnitude smaller than Hawking's generally accepted value of 1/4, which then yields what has since become known as the 'Bekenstein–Hawking formula' for the entropy of a black hole:

$$S_{BH} = \frac{kA}{4\ell_P^2}. \tag{12.3}$$

At this point, then, there is a tenuous formal similarity between entropy and horizon area underwritten only by the fact that both tend to increase; from this formal similarity, Equation (12.2) is divined as a simple choice with the right dimensions if the analogy were physical. There are plenty of examples of formal analogies that we should not reasonably think of as amounting to more than just that – merely formal analogies. For instance, the mathematics of the Lotka–Volterra

---

[10] More precisely, they are necessitated by dimensional considerations together with an argument that these are "the only truly universal constant[s]" (Bekenstein, 1972, p. 738) of appropriate dimension.

equations can be used to model the 'predator–prey dynamics' of ecosystems of populations consisting of two species, just as they can describe the dynamics of two macroeconomic variables, the workers' share of the national income and the employment rate.[11] One of the most dramatic examples of a formal analogy relates the resonances in the scattering spectrum of neutrons off of atomic nuclei and the frequencies of large-scale oscillations in the distribution of prime numbers through the integers described by the zeroes of Riemann's zeta function.[12] That these analogies are purely formal should be clear; scattering neutrons and distributing prime numbers are hardly subject to the same laws of nature, and ecology and macroeconomics are arguably governed by different forces. In the light of this, why should one take seriously the formally even weaker analogy between horizon area and thermodynamic entropy?

I read Bekenstein (1973) as an attempt to make the case for a more robust analogy, a *physical* analogy. His argument to this effect – itself the second part of the overall argument as described at the beginning of this section – is best analyzed as consisting of two parts. First, the thermodynamic entropy of isolated systems is identified with Claude Shannon's information-theoretic entropy. Second, the black hole area is identified with the information-theoretic concept of entropy. Both steps combined then amount to an identification of the area of the black hole's event horizon with thermodynamic entropy. Since both steps do not just formally identify but in fact carry physical salience across the identifications, the area of the horizon *really is* an entropy akin to the usual thermodynamic entropy. The inferential path through the information-theoretic concept of entropy is thus what establishes the physical salience (Figure 12.1). Of course, this identification then underwrites the generalization of the Second Law: As black hole entropy is of the same physical kind as ordinary thermodynamic entropy, they can unproblematically be summated to deliver the total entropy of the universe.

In sum, Bekenstein's argument can be broken down as follows:

(i) The universe as a whole has an entropy.
(ii) The Second Law is a universal law: It applies to the universe as a whole; in particular, (a generalization of) it still obtains in the presence of black holes.
(iii) The Second Law can hold in the presence of black holes if the black holes present have an entropy contributing to the total entropy of the universe.
(iv) The area of the event horizon of a black hole is the only property of a black hole that is formally similar to ordinary thermodynamic entropy in that both tend to increase over time.

---

[11] Weisberg (2013, pp. 77f) considers Goodwin's appropriation of the Lotka–Volterra equations for this macro-economic model a strong example of "construal change" (ibid.) of a model. Note that Goodwin took the analogy to be more than merely formal; in this, he was arguably motivated by his Marxism, illustrating just how the interpretation of the salience of an analogy may depend on one's ideological stance.

[12] See Tao (2012); Tao gives other examples of such 'universal' behavior.

$$dS_{TD} = \frac{\delta Q}{T}$$

$$S_{BH} = \frac{kA}{4\ell_P^2}$$

$$S_S = -\Sigma_n \, p_n \ln p_n$$

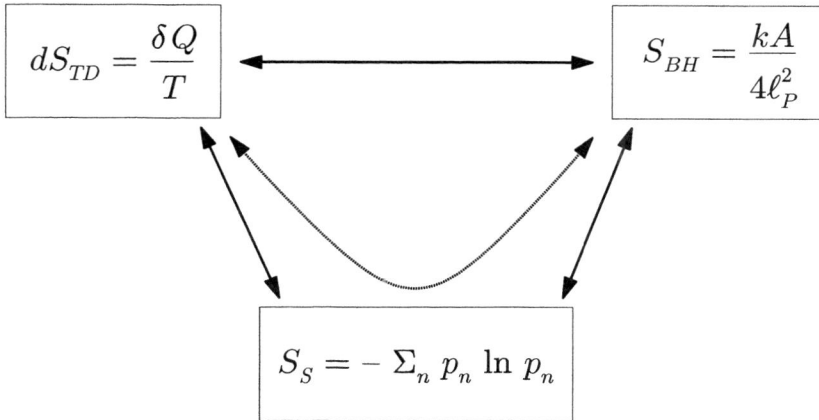

Figure 12.1 The path indicated by the dashed arrow through the information-theoretic concept of entropy (bottom) shows the inferential structure of Bekenstein's claim of physical salience of the analogy.

(v) The horizon area of a black hole is essentially the same kind of physical quantity as information-theoretic entropy.

(vi) Information-theoretic entropy is essentially the same kind of physical quantity as ordinary thermodynamic entropy.

(vii) Therefore, the horizon area of a black hole and ordinary thermodynamic entropy are essentially the same kind of physical quantity (which explains their formal similarity).

(viii) Therefore, black holes have an entropy contributing to the total entropy of the universe.

(ix) Therefore, (a generalization of) the Second Law still holds in the presence of black holes.

I take it that steps (vii) and (viii), assuming the transitivity of 'being essentially the same kind of physical quantity', are harmless. Steps (iii) and (ix) are perhaps not entirely automatic, but shall be granted for present purposes. One might argue that for claim (iii) to hold, an information-theoretic conception of entropy (and, moreover, an epistemic reading of that conception) is required for Bekenstein's motivating argument, that otherwise the Second Law may be violated, to have any force at all. Alternatively, one might rephrase (iii) as follows to be clear what that argument can establish at most:

(iii′) *We can only be certain that* the Second Law continues to hold in the presence of black holes if the black holes present have an entropy contributing to the total entropy of the universe.

While this complaint is perfectly justified, the point of my reconstruction is to isolate the move to an (epistemic reading of an) information-theoretic conception of entropy in a later step in the argument – in the hope of making the problematic nature of this step more evident. So let me accept (iii) as it stands for now.

Similarly to (iii) and (ix), (i) is a substantive assumption that may turn out to be false, or misguided – but we shall grant it for now, too. We have already seen in section 12.2 that (ii) is not unproblematic, as black hole physics may well violate the Second Law. In the same section, we also found that (iv) is not strictly true. Putting the concerns regarding (ii) and (iv) to the side, we are left with the most interesting claims of Bekenstein's argument: steps (v) and (vi), which are designed to secure the physical salience of the analogy. Before we address them in the next section, it should be noted that both (v) and (vi) could be interpreted in different, inequivalent ways. Both statements are of the form 'Thing *A* is essentially the same kind as thing *B*'. While Bekenstein and others often seem to think of these analogies as identities, strictly weaker statements of the form 'Thing *A* is essentially one of the things that make up the kind of things *B*', leaving open the possibility that the second placeholder designates a more general and encompassing category, would suffice for the argument to go through. I take it that these weaker statements would leave the transitivity necessary to infer to (vii) intact. One thing that is clear, though, is that the inverse direction would not work: It would not suffice to establish that thing *B* is a proper subset of the things that make up kind *A*.

12.4 The Quest for Physical Salience

In §2 of his often-cited paper of 1973, Bekenstein offers three formal analogies between black hole physics and thermodynamics. First, as already noted, both the surface area of black hole horizons and thermodynamic entropy tend to increase over time. Second, in thermodynamics, entropy can be considered a measure of the degradation of the energy of a system, in the sense that degraded energy cannot be transformed into work: If a thermodynamic system reaches maximum entropy, it can no longer be used to extract work. Surprisingly, if one considers interactions with a black hole by 'Penrose processes' on black holes with angular momentum, one can lower the mass (and the angular momentum) of the black hole, thereby extracting energy, up to a certain maximum.[13] Energy can be extracted from such black holes until their angular momentum is halted and the extractable energy thus is depleted. The 'irreducible mass' M_{irr} of a black hole represents the

---

[13] Penrose processes were originally articulated in Penrose (1969); cf. also Wald (1984, §12.4).

non-extractable mass–energy of the black hole. For a Kerr black hole, its square turns out to be proportional to the horizon area (Wald, 1984, eq. 12.4.12):

$$A = 16\pi M_{irr}^2. \tag{12.4}$$

Thus, the irreducible mass of a black hole acts similarly to the maximum entropy a thermodynamic system can attain. Throughout a Penrose process, where the mass of the black hole is held constant, its angular momentum decreases and its irreducible mass increases. Thus, the black hole's area as given by equation (12.4) increases, in concordance with the Area Theorem – until the Kerr black hole is 'spun down' to a Schwarzschild black hole of zero angular momentum, at which point no more energy can be extracted from it.[14]

The third formal analogy noted by Bekenstein arises from the possibility of finding a black hole analogue of the 'fundamental constraint' imposed by the first two laws in the thermodynamics of simple fluids:

$$dE = TdS - PdV, \tag{12.5}$$

where E is the internal energy, T is the temperature, S is the entropy, P is the scalar pressure, and V is the volume of the fluid (Lemons, 2013, §1.9). The analogy is tenuous, however – at least for black holes as described by general relativity – as it requires a black hole analogue of temperature T. Although such an analogue can formally be defined in classical general relativity as a function of the black hole's three parameters – its total mass M, its total angular momentum per mass a, and its total electric charge Q – the interpretation of this magnitude as a physical temperature is unavailable because classical black holes cannot radiate and as it thus makes little sense to attribute temperatures to them. Of course, this changes once quantum effects are taken into account. We will return to the question of the quantum nature of black hole thermodynamics in the concluding section.

Let us start with proposition (vi) in the reconstruction of Bekenstein's argument in the previous section, the association of information-theoretic entropy with ordinary thermodynamic entropy. First, assuming a nomic reducibility of thermodynamics to statistical physics, the macroscopic entropy for an ideal gas is replaced with the microscopic *Boltzmann entropy* defined as

$$S_B = k \log W, \tag{12.6}$$

where k is the Boltzmann constant and W is the 'size' of the given macrostate in the phase space of admissible microstates or its 'multiplicity' (or "Permutabilität" in Boltzmann's own words), the number of microstates corresponding to the macrostate at stake. Thus, the Boltzmann entropy in equation (12.6) presupposes

---

[14] I thank Erik Curiel for clarifications on this point.

an atomic hypothesis according to which the isolated macroscopic thermodynamic system really consists of microscopic parts. It furthermore presupposes that the microstates of the system, the fundamental states of its parts, are all equally probable. The microstates give rise to the system's 'macrostates' defined in terms of observable, macroscopic physical properties of the system. Importantly, the macrostates are multiply realized by the microstates.

This first identification of thermodynamic entropy with a statistical notion is already substantive, in that the two entropies are clearly conceptually, and sometimes (when out of equilibrium) numerically, distinct.[15] Consequently, there have been efforts, such as by Gour and Mayo (2001) and Prunkl and Timpson (2017), to establish the thermodynamic behaviour of black holes and to derive the Bekenstein–Hawking formula based on thermodynamic concepts and assumptions alone and thus cleanse the arguments from their usual detour through statistical notions. While such efforts establish that thermodynamic behaviour can consistently be attributed to black holes, they do not show this from independent assumptions about black hole physics: Instead, they *presuppose* that the laws of thermodynamics apply to black holes – and they assume that they emit Hawking radiation. Let's return to Bekenstein's original argument.

The fundamental postulate of statistical mechanics, a presupposition of equation (12.6), asserts that the microstates of an isolated thermodynamic system are all equally probable. For systems interacting with their environment, the equiprobability of their microstates cannot be assumed, and a generalization of equation (12.6) to thermodynamic systems with microscopic degrees of freedom assuming microstates of possibly unequal probability is required. The *Gibbs entropy* delivers just that:

$$S_G = -k\, \Sigma_i\, p_i \log p_i, \tag{12.7}$$

where p_i is the probability that the i-th microstate obtains.[16] Comparing the Gibbs entropy of a thermodynamic system with Shannon's dimensionless "measure of uncertainty," which quantifies the uncertainty lost by a message,

$$S_S = -\Sigma_n p_n \log_2 p_n, \tag{12.8}$$

we find another formal analogy: The right-hand side of equation (12.7) is proportional to that of equation (12.8).[17]

---

[15] See, for example, Callender (1999).

[16] Strictly speaking, the Gibbs entropy is defined on the basis of an *ensemble* of a large number – eventually an infinity – of independent copies of the given system. We will ignore this here. We will also ignore the more general case of continuously many microstates.

[17] Assuming, that is, binary states; this assumption is standard in information theory, where information is measured in bits, but also common in statistical mechanics.

In Shannon's information theory, S_S quantifies uncertainty reduced by the transmission of a message, or the attenuation in phone-line signals. Why should anyone think that it is essentially the same kind of physical quantity as thermodynamic entropy, originally introduced as a measure of the irreversibility of the dynamical evolution of isolated systems such as boxed-off ideal gases? Shannon himself, in a now famous recollection, testifies to the shaky origins of the association of his new formula with thermodynamic entropy:

My greatest concern was what to call it. I thought of calling it 'information', but the word was overly used, so I decided to call it 'uncertainty'. When I discussed it with John von Neumann, he had a better idea. Von Neumann told me, 'You should call it entropy, for two reasons. In the first place your uncertainty function has been used in statistical mechanics under that name, so it already has a name. In the second place, and more important, nobody knows what entropy really is, so in a debate you will always have the advantage.' (Tribus & McIrvine, 1971)

Thus, the first goal of a successful argument establishing the essential physical kinship of a black hole's horizon area and thermodynamic entropy must be to physically substantiate the association apparently so frivolously made by von Neumann. Attempts to do so predate Bekenstein's work by more than 15 years, having begun shortly after the publication of Shannon's field-defining classic of 1948. Based on Shannon's central insight that the new information received by the epistemic agent corresponds to the decrease of the agent's uncertainty about a system's internal state, Léon Brillouin (1956) identifies information with negative entropy. In the late 1950s and 1960s, Brillouin, Edwin T. Jaynes, Rolf Landauer, and others attempt to solidify the association between information theory, statistical physics, and thermodynamics. For instance, Jaynes (1957a, 1957b) first derives statistical mechanics from information-theoretic assumptions, and Landauer (1961) asserts that the erasure of information requires a minimum amount of energy of $kT \ln 2$ per bit of information erased, where T is the temperature of the circuit.[18]

The literature on this topic is vast and certainly not conclusive. I shall not attempt to do it justice here, but will merely give a few pointers. For instance, there have been claims of an experimental confirmation of 'Landauer's principle' (Bérut et al., 2012) – refuted by John Norton (2013, §3.7); advocacy of the principle (Bennett, 2003); and criticism of it (Norton, 2011). Some physicists have gone as far as claiming that the entire material universe is nothing but information.[19] In the opposite

---

[18] Maroney (2009) provides a critical introduction to the relationship between information processing or computation and thermodynamics.

[19] Vedral (2010), for example, has written a popular book advocating this position. One of the founding fathers of the physics-is-information school was, of course, John Archibald Wheeler – Bekenstein's PhD advisor! Wheeler's 1990 article is arguably his most pronounced articulation of the idea that physical objects depend in their very existence on information, as captured in the slogan 'It from bit'.

direction, Norton warns that the focus on information theory has produced an iden-
tifiable harm in the context of Maxwell's demon: Information-based arguments
against the demon have not only been unsuccessful, but have furthermore precluded
us from recognizing the straightforward violation of (the quantum analogue of)
Liouville's theorem as a definitive exorcism of the demon.[20]

One problem of the information-theoretic approach, also noted by Owen
Maroney (2009), is that radically different concepts and standards of proof are
operative in this literature, collectively obscuring the synthesizing lesson to be
drawn from it. Although it remains popular among physicists, there are excellent
reasons to resist the identification of thermodynamic entropy, which is grounded
in objective facts about the microphysics of physical systems, with information-
theoretic entropy, which concerns subjective facts about the epistemic state of
agents, as John Dougherty and Craig Callender (forthcoming, §5) insist. In fact, the
two entropies can have distinct numerical values when applied to the same system
at the same time. As an example, they offer an isolated ideal gas that has whatever
thermodynamic entropy it has regardless of whether or not we know its precise
microstate:

If a Laplacian demon told you the exact microstate of a gas, that would affect the value of
the Shannon entropy (driving it to zero) whereas it wouldn't affect the value of the Gibbs
entropy. (Dougherty & Callender, forthcoming, pp. 20f)

Thus, they conclude, it would be a big mistake to identify the two concepts.

I will not dwell on the profound issue of the relationship between physics and
information theory here. Nevertheless, for Bekenstein's argument to succeed, it
is necessary that all information-theoretic entropy essentially be thermodynamic
entropy, as asserted in (vi). As noted at the end of the previous section, while an
identity of the two concepts would certainly establish (vi), it would dialectically
suffice if information-theoretic entropy would turn out to be just one sort, or one
manifestation, of thermodynamic entropy among other possible types. It would not
suffice, however, if it turned out, as is generally assumed, that Shannon entropy is
more general than thermodynamic entropy in the sense of also assigning (Shannon)
entropy to systems without thermodynamic entropy. This would not be sufficient
precisely because it could then turn out that while black hole entropy is indeed
information-theoretic entropy, it is of the non-thermodynamic kind. If information-
theoretic entropy is a more general concept than thermodynamic entropy, then we
would have to establish not only that black-hole entropy is information-theoretic,
but also that it is of the thermodynamic kind. As a consequence, general proofs à
la Jaynes that statistical mechanics can be asymmetrically reduced to information

---

[20] See Norton (2013, §4) for the classical argument and Norton (2018) for the extension to the quantum case.

theory will not do all the work asked of proposition (vi) in my reconstruction of Bekenstein's case.

It should be noted that Bekenstein explicitly endorses what could be called the 'Jaynesian identification':

The entropy of a system measures one's uncertainty or lack of information about the actual internal configuration of the system. ... The second law of thermodynamics is easily understood in the context of information theory ... (1973, p. 2335)

Again, to subsume both the second law of thermodynamics and black hole physics into information theory will still require an argument as to why thermodynamic entropy and horizon area are relevantly similar – other than both being merely (and perhaps in relevantly distinct ways) information-theoretic.

The upshot is that it would be inadmissibly facile to think that proposition (vi) is firmly established. It isn't, as there remain open questions regarding the relationship between information theory and thermodynamics. But let us turn to proposition (v), the claim that the horizon area of a black hole is essentially the same kind of thing as information-theoretic entropy. The association between the horizon area and information-theoretic entropy articulated in proposition (v) is Bekenstein's original contribution and a main step in the overall argument.

Bekenstein sets the tone right away: in the abstract, he writes:

After a brief review of the elements of information theory, we discuss black-hole physics from the point of view of information theory. We show that it is natural to introduce the concept of black-hole entropy as the measure of information about a black-hole interior which is inaccessible to an external observer. ... The validity of [the generalized] second law is supported by an argument from information theory. ... (1973, p. 2333)

In the main body of the text, the theme is amply replayed:

In the context of information a black hole is very much like a thermodynamic system. ... Black holes in equilibrium having the same set of three parameters may still have different 'internal configurations'. ... It is then natural to introduce the concept of black-hole entropy as the measure of the *inaccessibility* of information (to the exterior observer) as to which particular internal configuration of the black hole is actually realized in a given case. At the outset it should be clear that the black-hole entropy we are speaking of is *not* the thermal entropy inside the black hole. (1973, pp. 2335f, emphases in original)

These quotes betray the same misconception: Fundamental physics is about the objective structure of our world, not about our beliefs or our information.

In fact, the problem may be considered to run deeper: Information, one might argue, is an inadmissible concept in fundamental physics. For information to exist in the first place, a communication system must be in place, a physical setup such that the concept of information is applicable. In Shannon's (1948) mathematical theory of communication, for there to be communication, there must be an *information source* of a message, a *transmitter* sending a signal, via a potentially *noisy channel*, to a *receiver*, which receives the signal and decodes it for the *destination*. This destination, according to Shannon, is "the person (or thing) for whom the message is intended" (p. 381). Even subtracting the intentionality, and abstracting from the personhood of the destination, we are still left with an ineliminable minimum level of complexity required for the signal to be interpreted as the transmission of *information*. As Christopher Timpson (2013, Ch. 2) shows, since 'information', in Shannon's theory, is "what is produced by a source that is required to be reproducible at the destination if the transmission is to be a success" (p. 42), we need to distinguish the abstract type of information from the tokens of its concrete physical instantiations. Hence, 'information', as understood in information theory, is an abstractum.[21] In addition to the category mistake, then, the complexity of the setup should not be required of a physical system described by a fundamental physical theory, which is, in general, too impoverished to incarnate all these roles at once at their level.

Moreover, unless one accepts an ultimately Platonist – indeed, Pythagorean – ontology of the material world, fundamental physics ought not be grounded in abstracta like information. The recently popular metaphor of the universe 'being' a quantum computer can survive as a claim of *physics* only if it is interpreted to mean that the universe is a physical system with a nature that is ultimately best described by a quantum theory, perhaps with a quantum dynamics that produces the 'steps' of the computing. A physical system is qualitatively distinct from an abstract object or structure. The more systematic assessment of the merits and limitations of an information-theoretic abstraction of fundamental physics shall be left for another day, but it should be clear that if we reject radical Pythagoreanism (i.e., the belief that the physical world is fundamentally mathematical in nature), then fundamental physics should not be cashed out in information-theoretic terms. Thus, the entropy of a black hole, if it is one, should not be identified with Shannon's information-theoretic entropy, and, a fortiori, neither should the surface area of a black hole's event horizon. Thus, proposition (v) is false.

Let us summarize the fate of Bekenstein's attempt to endow the analogy between the area of a black hole's event horizon and thermodynamic entropy as it appears in the Second Law with physical salience in the light of these considerations. If my reconstruction of Bekenstein's argument toward the end of Section 12.3 is correct, and if these considerations suffice to put in doubt propositions (ii), (iv), and particularly (vi), and to deny (v), then the propositions (vii) through (ix) cannot be maintained, at least not on the basis of this argument. Consequently, we can no

---

[21] For a discussion of different intepretative options concerning Shannon entropy and for the defence of an ulimately pluralist stance, see particularly the work of Lombardi et al. (2016). They also insist that Shannon information is not an absolute magnitude, but rather exists relative to an entire communication situation (§9).

longer conclude that the horizon area of a black hole and thermodynamic entropy are essentially the same kind of physical quantity and that, therefore, black holes have entropies that contribute to the universe's total entropy as captured by the Generalized Second Law. Bekenstein's attempt to give physical salience to the formal analogy between the horizon area of black holes and thermodynamic entropy via Shannon's information-theoretic entropy as depicted in Figure 12.1 does not succeed.

Before we arrive at the conclusion, let me emphasize that the failure of a particular argument to a particular conclusion does not in itself establish the falsity of the conclusion: Even though Bekenstein's original argument may fail to prove that black holes have entropies, this failure, importantly, does not imply that black holes are not thermodynamic objects. This global conclusion would be warranted only if *all* available arguments would fare equally badly.

12.5 Conclusion

As attempts to formulate a quantum theory of gravity remain bedevilled by a frustrating lack of empirical data to guide theoretical work, black hole thermodynamics in general, and the Bekenstein–Hawking formula for the black hole entropy in particular, are widely accepted as physical fact. This is so even though there is so far no empirical confirmation of any of the thermodynamic aspects of black holes.[22] This almost universal sentiment is, among many others, expressed by Jonathan Oppenheim is his obituary of Jacob Bekenstein:

Any potential theory of quantum gravity should correctly predict the value of the black hole entropy. Indeed, as quantum gravity is not driven by experiment, *black hole thermodynamics is the only really solid piece of information we have in our attempts to construct it.* (Oppenheim, 2015, p. 805, emphasis added)

Why, if there is no empirical confirmation of the thermodynamic behaviour of black holes available, is entropy so generally predicated of black holes and why do virtually all physicists – in quantum gravity and elsewhere – accept the Bekenstein–Hawking formula as correctly capturing its quantity? Since no direct empirical confirmation exists, the reasons physicists have for accepting black hole entropy and radiation must derive from theoretical considerations. If my argument presented in this chapter is sound, then the original argument by Bekenstein with its detour through information theory does not succeed in establishing the physical salience of the otherwise merely formal analogy between thermodynamic entropy and the

---

[22] But see Thébault (this volume, Chapter 11) for an analysis of whether analogue models can bestow confirmation on black hole radiation. Also see Unruh (2014) for an argument that Hawking radiation has, in fact, been measured (albeit indirectly).

black hole area, and so cannot offer the basis for accepting black hole thermodynamics as "the only really solid piece of information."

It may appear unfair, or even uninteresting, to attack an argument that could not, and did not, stand by itself. Bekenstein's papers making the case that black holes have entropy, and that furthermore, this entropy is measured by the event horizon's area, were not taken very seriously by physicists before Hawking's semi-classical calculation of his eponymous radiation (Curiel and Bokulich, 2018, §5.2). Rafael Sorkin (1998) expresses this sentiment when he writes that the "best known piece of evidence" for the association between the horizon area and entropy is Hawking radiation. As noted earlier, at the purely classical level, the thermodynamic analogy remains weak.[23] For classical black holes, the obvious disanalogy is that since they do not radiate, they should not be ascribed a temperature, and so they are not thermodynamic objects. Although Bekenstein's formulae were already quantum in the sense that the definition of S contained an $\hbar$, or, alternatively, the Planck length, this quantum trace originated simply from a classically completely free choice of a fiducial area for the event horizon, which might not have included any $\hbar$. While it is thus arguably the case that Bekenstein's argument did not stand by itself in the first place, I maintain that given the lack of direct empirical confirmation of any thermodynamic aspects of black holes, a careful analysis of all major arguments remains an important source of foundational insight.

As a final point, it should be remarked that even if black holes have entropy, they may lack microstates specified by degrees of freedom that are qualitatively distinct from the macroscopic degrees of freedom. Classically, black holes have 'no hair'; that is, the spacetime external to a black hole is uniquely determined by the black hole's mass, angular momentum, and charge. Thus, as far as classical GR in concerned, a black hole is fully characterized by its macrostate in terms of these three properties. In itself, it does not imply that a black hole also has microstates defined by additional properties. As stated earlier, the general expectation is that in a quantum theory of gravity, black holes will be shown to have microstates in some sense and the way they get their entropy is explained by counting the microstates in the appropriate way. On such a Boltzmannian view, one would expect the entropy to be huge, since presumably very many distinct microstates would give rise to one and the same simple macrostate.

However, as Chirco et al. (2014) have argued, black holes may well have entropy while lacking microstates characterized by distinct – 'hidden' – degrees of freedom. The entropy may be an entanglement entropy and arise from a coarse graining in the form of a split of the system into two components and the resulting restriction to

---

[23] Though Curiel (2015) demurs.

the observables of one of the two subsystems.[24] It is clear, however, that this line of reasoning requires the thermodynamic entropy of a black hole to be identified with the conceptually distinct von Neumann entropy, a claim that requires substantive defence.

To return to the main topic of this chapter and to conclude, let me emphasize once again that even though Bekenstein's detour through information does not succeed in endowing the formal analogy between thermodynamic entropy and black hole horizon area with physical salience, it does not thereby follow that black holes are *not* thermodynamic objects. For instance, gedanken experiments concerning the limits of the amount of thermodynamic work that can or cannot be extracted from black holes lend some support to the idea that black holes are thermodynamic in nature.[25] Moreover, the (direct or indirect) empirical confirmation of Hawking radiation would strongly support the idea, though in the case of an indirect detection questions regarding analogical reasoning would certainly arise. Ultimately, as with any substantive claim in empirical science, only the usual kind of experimental and observational work can establish that black holes are thermodynamic objects.

Acknowledgements

I wish to thank Karen Crowther, Erik Curiel, John Dougherty, Nick Huggett, Niels Linnemann, Keizo Matsubara, Tim Maudlin, John Norton, Daniele Oriti, Carlo Rovelli, and Karim Thébault for discussions and comments, as well as audiences in Munich and Düsseldorf. This work was performed under a collaborative agreement between the University of Illinois at Chicago and the University of Geneva and made possible by grant number 56314 from the John Templeton Foundation. Its contents are solely the responsibility of the author and do not necessarily represent the official views of the John Templeton Foundation.

References

Bekenstein, J. D. (1972). Black holes and the second law. *Lettere al Nuovo Cimento*, 4:737–40.
Bekenstein, J. D. (1973). Black holes and entropy. *Physical Review D*, 7:2333–46.
Bennett, C. H. (2003). Notes on Landauer's principle, reversible computation, and Maxwell's demon. *Studies in the History and Philosophy of Modern Physics*, 34:501–10.

---

[24] The idea that the black hole entropy is an entanglement entropy resulting from tracing out degrees of freedom hidden behind the event horizon goes back to Bombelli et al. (1986).

[25] I am thinking, in particular, of thought experiments unclouded by statistical and informational notions showing the consistent and beneficial application of thermodynamical concepts to black holes, such as those most recently developed by Prunkl and Timpson (2017).

Bérut, A., Arakelyan, A., Petrosyan, A., Ciliberto, S., Dillenschneider, R., & Lutz, E. (2012). Experimental verification of Landauer's principle linking information and thermodynamics. *Nature*, 483:187–90.

Bombelli, L., Koul, R. K., Lee, J., & Sorkin, R. D. (1986). Quantum sources of entropy for black holes. *Physical Review D*, 34:373–83.

Brillouin, L. (1956). *Science and Information Theory*. New York: Academic Press.

Callender, C. (1999). Reducing thermodynamics to statistical mechanics: The case of entropy. *Journal of Philosophy*, 96:348–73.

Chirco, G., Haggard, H. M., Riello, A., & Rovelli, C. (2014). Spacetime thermodynamics without hidden degrees of freedom. *Physical Review D*, 90:044044.

Curiel, E. (2015). Are classical black holes hot or cold? Manuscript.

Curiel, E. (2017a). A primer on energy conditions. In D. Lehmkuhl, G. Schiemann, & E. Scholz, eds., *Towards a Theory of Spacetime Theories*. Boston/Basel: Birkhäuser.

Curiel, E. and Bokulich, P. (2018). Singularities and Black Holes. In E. Curiel & P. Bokulich, eds., *The Stanford Encyclopedia of Philosophy*. https://plato.stanford.edu/archives/sum2018/entries/spacetime-singularities/

Dardashti, R., Thébault, K. P. Y., & Winsberg, E. (2017). Confirmation via analogue simulation: What dumb holes could tell us about gravity. *British Journal for the Philosophy of Science*, 68:55–89.

Dawid, R. (2013). *String Theory and the Scientific Method*. Cambridge: Cambridge University Press.

Dou, D., & Sorkin, R. D. (2003). Black hole entropy as causal links. *Foundations of Physics*, 33:279–96.

Dougherty, J., & Callender, C. (forthcoming). Black hole thermodynamics: More than an analogy? In A. Ijjas & B. Loewer, eds., *Guide to the Philosophy of Cosmology*. Oxford: Oxford University Press.

Gour, G., & Mayo, A. E. (2001). Why is the black hole entropy (almost) linear in the horizon area? *Physical Review D*, 63:064005.

Hawking, S. W. (1971). Gravitational radiation from colliding black holes. *Physical Review Letters*, 26:1344–6.

Hawking, S. W. (1975). Particle creation by black holes. *Communications in Mathematical Physics*, 43:199–220.

Jaynes, E. T. (1957a). Information theory and statistical mechanics. *Physical Review*, 106:620–30.

Jaynes, E. T. (1957b). Information theory and statistical mechanics II. *Physical Review*, 108:171–90.

Landauer, R. (1961). Irreversibility and heat generation in the computing process. *IBM Journal of Research and Development*, 5:183–91.

Lemons, D. S. (2013). *A Student's Guide to Entropy*. Cambridge: Cambridge University Press.

Lombardi, O., Holik, F., & Vanni, L. (2016). What is Shannon entropy? *Synthese*, 193:1983–2012.

Maroney, O. (2009). Information Processing and Thermodynamic Entropy, Edward N. Zalta, ed., *The Stanford Encyclopedia of Philosophy*. https://plato.stanford.edu/archives/fall2009/entries/information-entropy/

Norton, J. (2011). Waiting for Landauer. *Studies in the History and Philosophy of Modern Physics*, 42:184–98.

Norton, J. (2013). All shook up: Fluctuations, Maxwell's demon and the thermodynamics of computation. *Entropy*, 15:4432–83.

Norton, J. (2018). Maxwell's demon does not compute. In M. E. Cuffaro & S. C. Fletcher, eds., *Physical Perspectives on Computation, Computational Perspectives on Physics.* Cambridge: Cambridge University Press.

Oppenheim, J. (2015). Jacob Bekenstein: Quantum gravity pioneer. *Nature Physics*, 11:805.

Penrose, R. (1969). Gravitational collapse: The role of general relativity. *Rivista del Nuovo Cimento*, 1:252–76.

Penrose, R. (1999). The question of cosmic censorship. *Journal of Astrophysics and Astronomy*, 20:233–48.

Prunkl, C., & Timpson, C. G. (2017). Black hole entropy is entropy and not (necessarily) information. Manuscript.

Rovelli, C. (1996). Black hole entropy from loop quantum gravity. *Physical Review Letters*, 77:3288–91.

Shannon, C. E. (1948). A mathematical theory of communication. *Bell System Technical Journal*, 27:379–423 and 623–56.

Sorkin, R. D. (1998). The statistical mechanics of black hole thermodynamics. In R. M. Wald, ed., *Black Holes and Relativistic Stars*, pp. 177–194. Chicago: University of Chicago Press.

Steinhauer, J. (2016). Observation of quantum Hawking radiation and its entanglement in an analogue black hole. *Nature Physics*, 12:959–65.

Strominger, A., & Vafa, C. (1996). Microscopic origin of the Bekenstein–Hawking entropy. *Physics Letters B*, 379:99–104.

Tao, T. (2012). *E pluribus unum*: From complexity, universality. *Daedalus*, 141:23–34.

Timpson, C. G. (2013). *Quantum Information Theory and the Foundations of Quantum Mechanics.* Oxford: Oxford University Press.

Tribus, M., & McIrvine, E. C. (1971). Energy and information. *Scientific American*, 225(3):179–88.

Unruh, W. G. (2014). Has Hawking radiation been measured? *Foundations of Physics*, 44:532–45.

Vedral, V. (2010). *Decoding Reality: The Universe as Quantum Information.* Oxford: Oxford University Press.

Wald, R. M. (1984). *General Relativity.* Chicago: University of Chicago Press.

Wald, R. M. (1994). *Quantum Field Theory in Curved Spacetime and Black Hole Thermodynamics.* Chicago: University of Chicago Press.

Wald, R. M. (2001). The thermodynamics of black holes. *Living Reviews in Relativity*, 4:lrr–2001–6.

Wall, A. C. (2009). Ten proofs of the generalized second law. *Journal of High Energy Physics*, 06:021.

Wall, A. C. (2013). The generalized second law implies a quantum singularity theorem. *Classical and Quantum Gravity*, 30:165003.

Weisberg, M. (2013). *Simulation and Similarity: Using Models to Understand the World.* New York: Oxford University Press.

Wheeler, J. A. (1990). Information, physics, quantum: The search for links. In W. H. Zurek, ed., *Complexity, Entropy, and the Physics of Information*, pp. 309–36. Redwood City, CA: Addison-Wesley.

Part III

Cosmology and Testability

13

The Limits of Cosmology

JOSEPH SILK

13.1 Introduction

Cosmology has entered a new era in the twentieth century, having become a precision science. Nevertheless, big questions remain. I will attempt to address some key issues in this chapter that are central to our acceptance of the inflationary framework as a explanation of the beginning of the universe. I will focus on these topics:

- Understanding galaxy formation: Does brute force computing power suffice?
- Verifying inflation via B-mode polarization from gravity waves: Are there any guaranteed detections?
- Verifying inflation via non-Gaussianity: Conceptually appealing, but futuristically feasible?
- Dark energy and the multiverse: Is this physics or metaphysics?
- Dark matter: What next?

Modern cosmology began in 1922 with Alexander Friedmann, who developed the first models of the expanding universe. This was no more than a mathematical construction until 1928, when Georges Lemaître independently derived the current suite of cosmological models and realized that there was indeed tantalizing evidence for expansion of space from Vesto Slipher's measurements of galaxy redshifts and Hubble's published distance determinations. In 1929, Edwin Hubble, unaware of Lemaître's earlier work, announced his discovery of the law that relates redshift linearly to increasing distances of the galaxies. The expanding universe was born.

Lemaître himself was concerned about the problems of formally following his equations back to the singular state $t = 0$. He advocated the 'primeval atom' as a more physical starting point. The Lemaître–Eddington solution, which sought to avoid the initial singularity, starts from a static configuration but was later found to be unstable. Lemaître realised very early on that the initial state was the domain of quantum theory. Remarkably, as early as 1933, he argued that the universe was dominated by vacuum energy with equation of state $p = -\rho c^2$.

This set the scene, half a century later, for inflation along with other conceivable, but less inspiring, alternatives. After some 90 years of progress in observational cosmology, all that has really changed on the theory side is the addition of inflationary cosmology to our cosmic toolbox, allowing the extension of physical cosmology back to the Planck time, some 10^{-43} s after the Big Bang. There are vocal opponents of inflation, who point out that it is not a theory but rather a framework that is debatably unpredictive and lacks robust verifiability. Many feel strongly that the measured flatness of space and near scale-invariance of the density fluctuations are powerful and successfully verified predictions, but a vocal minority argues that these predictions arise from an infinitesimal subset of possible initial conditions. It is a lively and continuing debate that, as I will argue in this chapter, only highly challenging and undeniably futuristic observations can resolve. There may indeed be some as-yet-unknown unifying theory of matter and energy, of space and time, that lies outside our current perspective. Indeed, we should not forget that Lemaître himself argued forcefully for space and time to be emergent phenomena, and this still represents the holy grail of modern quantum cosmology.

13.2 History of Physical Cosmology: 1933–2000

13.2.1 The Standard Model

Little new happened in our understanding of cosmology for the first two decades following Hubble's announcement of the recession of the galaxies, and its rapid incorporation into evidence for the expansion of space. It required the entry into cosmology of nuclear physicists to facilitate the next major step forward. This was a consequence of the realization that the Big Bang was an ideal laboratory for studying nuclear physics, and in particular the awareness that the high density and temperature achievable in the first minutes of the universe might facilitate the nucleosynthesis of the chemical elements.

George Gamow dreamed that all of the chemical elements could be synthesized primordially, but he eventually realized that the lack of a stable element of atomic mass 8 made it impossible to go beyond lithium (mass 7). With Alpher and Herman, he did succeed in generating helium, the second most abundant element in the universe, in the Big Bang. This was a major accomplishment, as it was realized in the 1950s that ordinary stars could not generate enough helium as observed in the nearby universe [1]. Nor could stars account for what we now know to be helium's uniformity in abundance throughout the observable universe. It was soon shown, in a classic paper by Burbidge, Burbidge, Fowler, and Hoyle, that the ejecta of massive stars exploding as supernovae, as well as red giant mass loss, could account for the bulk of the heavy elements [2].

Modern calculations of primordial nucleosynthesis also account for the abundance of deuterium, an isotope of hydrogen that is only destroyed in stars. Lithium is also produced this synthesis, but the amount predicted is approximately 3 times more than observed in the oldest stars [3]. The cosmological origin of lithium is one of the very few serious questions that remain unresolved in cosmology. There is now such high precision in measurements of helium and especially deuterium that attempts to modify primordial lithium involving exotic particle decays to destroy Be^7, the source of lithium in the nuclear reaction chain, almost invariably create excessive deuterium.

Another important consequence of helium and deuterium synthesis is that it was demonstrated 40 years ago that the bulk of the universe is nonbaryonic [4]. If the bulk of the matter in the universe were baryonic, the nuclear reactions in the first few minutes would have been so efficient as to overproduce helium by a large amount. However, a quantitative baryon budget was to require Hubble Space Telescope observations of intergalactic and circumgalactic gas so as to measure the total density of matter with high accuracy (as described in [5]).

On the observational side, discovery of the cosmic microwave background radiation has provided a basis for undertaking the first precise measurements of the different components of the universe. In fact, precision cosmology emerged after 2000 as a consequence of several generations of cosmic microwave background experiments, following its detection by Penzias and Wilson in 1964 [6]; the measurement of its blackbody spectrum by COBE in 1990 [7]; the early ground-based and balloon experiments, notably BOOMERANG [8]; and culminating with the WMAP and Planck satellites.

The major theoretical development in cosmology, following the glory days of Lemaître, de Sitter, Einstein, and Eddington in the 1930s, was the invention of inflationary cosmology. This advance is attributed to Starobinsky, Guth, and Linde [9–11], although others played important roles (e.g., [12, 13]). Inflation accounted for the flat geometry of the universe, its size, and the presence of infinitesimal density fluctuations. These fluctuations were initially found to be too large in the early inflationary models, but a consistent theory was later developed by Mukhanov [14]. The inflationary density fluctuations were predicted to be adiabatic, Gaussian, and nearly scale-invariant. All of these properties were later confirmed to a high degree of approximation by the CMB satellite experiments, which provided the all-sky coverage essential for controlling foregrounds, especially on the largest angular scales.

13.2.2 Dark Matter

The case for dark matter in galaxies was effectively made with optical studies of M31 by Rubin and Ford in 1970 [15], but especially with the first deep 21-cm observations of M31 and other nearby galaxies by Roberts and others in 1972 [16]. The radio observations were essential for measuring the outermost parts of galaxies, where stars could not possibly account for the total enclosed mass. The presence of a dominant component of dark matter in galaxies did not come as a complete surprise. The arguments for the prevalence of dark matter on even larger scales, in clusters of galaxies, had been made four decades earlier by Fritz Zwicky, in 1933 [17]. Even earlier, Jan Oort [18] postulated that dark matter was present in the galactic disk, in the solar neighborhood, to account for nearby stellar motions.

The highest-precision case for nonbaryonic dark matter comes from primordial nucleosynthesis. The net effect is the following. Far more matter is needed than is allowed in the form of baryons, by a factor of about 6. A higher (lower) baryon density would have generated excessive (too little) helium and too little (excessive) deuterium. Modern observations of the abundances of these elements are so precise as to yield a baryon abundance as accurate as that obtainable independently, albeit more indirectly, from the cosmic microwave background.

Another mini-revolution in cosmology occurred following the discovery of inflation in 1981, when it became clear that the early universe was a fertile hunting ground for novel dark matter–related ideas in high-energy particle physics. Within two or three years, a number of particle physics candidates emerged for nonbaryonic dark matter. Now there are hundreds of candidates but still no detections, or even promising hints of detections, apart from one or two tantalizing anomalies such as the 3.5 keV x-ray line seen in diffuse emission in certain areas.

Two preferred candidates have attracted the most attention. One is the lightest neutralino, a left-over particle from the epoch when supersymmetry (SUSY) prevailed, proposed by Pagels and Primack in 1982 [19]. Neutralinos are an attractive candidate because if the natural scale for the breaking of SUSY is that of the W and Z bosons, around 100 GeV, one naturally obtains the observed relic density of dark matter.

Many experiments have been launched to search for neutralinos in the form of weakly interacting massive particles (WIMPs), inspired by the possibilities of direct detection via elastic or even inelastic scattering [20] and of indirect detection, via the annihilation products that include high-energy antiprotons, positrons, and gamma rays [21] and neutrinos [22]. Sadly, accelerator experiments have so far frustrated this dream by failing to find evidence for SUSY below a TeV.

A second candidate has been dubbed the invisible axion. Axions were invented by Wilczek and Weinberg in 1978 [23, 24] to solve an outstanding issue in QCD, the strong CP problem. Neutralinos and axions are intrinsically cold dark matter (CDM) – that is, dark matter that can respond to the pull of gravity on all mass scales as advocated by Blumenthal et al. in 1984 [25].

In 1985, Davis et al. [26] showed that CDM could account for large-scale structure in a way that hot dark matter (HDM) could not. HDM is epitomized by massive neutrinos. These particles are effectively nonbaryonic matter created late in the universe with a significant velocity dispersion. Neutrino dark matter is inferred to be hot initially, and cannot reproduce the intricate filamentary detail of large-scale structure. The dark matter is classified into cold, warm, and hot variants, depending on the epoch at which it first became nonrelativistic and underwent free streaming [27]. A modern variant that reproduces large-scale structure, more or less successfully, is warm dark matter in the form of keV mass sterile neutrinos

13.2.3 Acceleration

The case for acceleration of the universe from Type 1a supernovae was first presented in 1998 by Perlmutter, Riess, Schmidt, and colleagues [28, 29]. These results are interpreted as strong evidence for the cosmological constant, which is itself considered to be evidence for the presence of a dominant component of dark energy. The flatness of the universe, measured by CMB experiments such as BOOMERANG, was a notable factor in establishing the reality of the acceleration and the need for dark energy as well as for dark matter [8]. Indeed, it was not until 2000 that scientists decisively showed that the universe contained a near-critical density of mass–energy by measuring the curvature of space via the CMB. It soon became clear that the dominant energy–density component was not predominantly CDM. Had it been, galaxy-specific velocities would have been excessively large because of the clustering effects of dark matter. Dark energy dominates as a scalar field and remains nearly uniform, acting like antigravity on the very largest scales. It is not self-gravitating, so its effects on peculiar velocities are minimal.

13.2.4 Galaxy Formation

After robustly developing the framework for the Big Bang, the next challenge is to understand how galaxies, and more generally large-scale structures, formed. The seed fluctuations needed to form structure by gravitational instability in the expanding universe were first predicted in 1967 as temperature anisotropies in the cosmic microwave background [30] that damped out toward smaller scales [31, 32]. In fact, the first rigorous calculations of density irregularities in a coupled matter–

radiation fluid [33] identified what were later recognised as the baryon acoustic oscillations [34]. The early calculations were later improved by incorporating the growth-boosting effects of dark matter, leading to smaller fluctuations [35, 36]. It was to take more than another decade to detect these elusive temperature fluctuations, whose predicted amplitude came directly from the theory of galaxy-scale structure formation.

The first definitive indications of temperature fluctuations in the CMB came from the COBE satellite in 1992, but the measured temperature anisotropies were only on 7-degree angular scales. These are super-horizon fluctuations at their last scattering. Hence they demonstrate the imprint of quantum fluctuations arising at inflation and their strength can be predicted by extrapolation from galaxy formation scales.

Much finer angular scales must be studied to discern the direct traces of the seeds of the largest self-gravitating structures such as clusters of galaxies. These produce cosmic microwave background fluctuations on angular scales of a few arc-minutes. Such fluctuations were accurately measured a decade later, as the unique features of the predicted acoustic sound wave damping peaks detected by the WMAP satellite. Indeed, this telescope detected the first three peaks, and after another decade, the SPT and ACT telescopes measured a total of nine damping wiggles [37, 38]. These findings dispelled any doubt that we were seeing the seeds of structure formation imprinted on the microwave sky.

Observational surveys soon became the preferred tool for doing low redshift cosmology – that is. the cosmology of formation and evolution of observed structures over look-back times of up to 10 billion years. Million-galaxy surveys of the nearby universe, notably the 2DF (1997) and SDSS (2000) surveys, set out the observational basis for a CDM-dominated universe.

The next step toward connecting primordial fluctuations with galaxies required detection of the acoustic imprint of the primordial seed fluctuations on the matter distribution [39]. Improved, more accurate surveys were needed for detection, which eventually came with the next generation of galaxy redshift surveys.

Growth from seeds to structure formation involved the occurrence of gravitational instability. This operated effectively only once the radiation decoupled from the matter as the universe recombined 380,000 years after the Big Bang. In earlier epochs, the dominant inertia of the radiation effectively braked structure formation. Galaxy formation was greatly aided by the self-gravity of the dark matter. The case for CDM driving galaxy formation with the aid of baryonic dissipation was made by White and Rees in 1978 [40].

It also rapidly became apparent that complex baryonic physics was needed in addition to the effects of gravity to produce structures that resembled the observed distribution of galaxies. For one thing, the intergalactic gas had to condense and cool to make the dense cold clouds that can fragment and form the first stars. It soon became evident that the physics of supernova feedback was required to avoid excessive formation of large numbers of dwarf galaxies [41]. A similar problem for massive galaxies was avoided through the presence of supermassive black holes, which released prolific amounts of energy in a quasar phase that occurred contemporaneously with the formation of the oldest stars. These violent events left a fossil imprint on the relation between black hole mass and spheroid velocity dispersion [42].

13.3 Modern Cosmology

Theory has progressed enormously in the past decade, especially in the area of simulations of large volumes of the universe. At the same time, observational surveys of galaxies over ever-larger volumes are being implemented. While massive large-volume, large-scale structure simulations provide an impressive match to survey data, real issues remain with regard to small-scale cosmology, namely galaxy formation.

Of course, to explain the late universe requires a lot of complicated nonlinear physics, including calculations involving dust, chemistry, turbulence, and magnetohydrodynamics. In terms of complexity, it may be compared with predicting the weather over long periods. By comparison, the early universe is much simpler. Let me first summarize the current status of precision cosmology from the CMB.

13.3.1 Cosmic Microwave Background Radiation

All-sky mapping of the CMB has proved crucial for evaluating the parameters of cosmology with high precision. This has been the important contribution of the WMAP and especially the Planck experiments to cosmology [43]. Six numbers characterize the CMB fluctuations (seven, if one does not assume a spatially flat universe, but flatness is measured to percentage precision). Apart from the densities of dark energy Ω_Λ, dark matter Ω_m, and baryons Ω_b, one must consider the scalar spectral index n_s and normalization σ_8 of the density fluctuations, the (revised) optical depth τ [44], and the Hubble constant H_0 (Table 13.1). The inferred age of the universe is 13.799 ± 0.021 Gyr.

There is general consistency with other probes, notably baryon acoustic oscillations (BAO) and weak gravitational lensing, with some slight tension remaining in the latter case. A slight tension remains at the 2σ level with the slightly higher Planck 2015 determination of the fluctuation amplitude parameter σ_8 [45]. There is somewhat more of an issue with the Hubble constant measurements, which have

Table 13.1 *Cosmological parameters: 2016 [43, 44]*

Ω_b	0.0486 ± 0.0005
Ω_m	0.3089 ± 0.0062
Ω_Λ	0.6911 ± 0.0062
σ_8	0.8159 ± 0.0086
n_s	0.9667 ± 0.004
H_0	67.74 ± 0.46
τ	0.056 ± 0.009

been found locally to be slightly higher, 73.24 ± 1.74 km/s/Mpc [46], than the Planck 2015-preferred value (Table 13.1).

These differences may possibly be due to systematics between the very different distance scales probed, although more profound explanations cannot be excluded. It is already quite remarkable how well the very early universe, as mapped by Planck, matches the local universe, as charted via weak lensing, BAOs and supernovae, in the context of the Standard Model of cosmology, ΛCDM.

B-mode polarization is predicted to be a unique probe of the gravitational wave background imprinted by inflation [47, 48]. Many experiments are under way to address this measurement goal. The CMB temperature fluctuations have ushered in the modern age of precision cosmology. The low optical depth measured by Planck means that understanding reionization is much less of a challenge for theoretical models

13.3.2 Baryon Acoustic Oscillations and Weak Gravitational Lensing

The large-scale structure of the galaxy distribution is a fertile hunting ground for evaluating both the geometry of the universe and its rate of expansion [49]. The BAO scale of approximately 150 Mpc is a standard ruler for cosmology. It is the precise analogue of the photon oscillations at $z = 1,080$ on the last scattering surface but is observable in the large-scale distribution of the galaxies at the present epoch and out to $z \approx 1$ [50]. Its remarkable power is that measurement requires no knowledge of the type of tracer; for example, galaxies, quasars, and Lyman alpha absorption clouds are all powerful tracers. Comparison of the two length scales constrains the angular diameter distance and Hubble parameter at different redshifts. Hence deviations in dark energy due to possible variations in the cosmological constant assumption can be inferred.

The angular diameter distance measurements are especially sensitive to the geometry of the universe. Weak gravitational lensing, galaxy clustering, and redshift space distortions all measure the dark matter content of the universe to high precision, and at different redshifts constrain the growth rate of density fluctuations and its dependence on dark energy. Supernova measurements constrain the expansion rate of the universe. CMB measurements are powerful probes of the curvature of space. The combined analysis of multiple tracers is our most powerful probe of cosmology as degeneracies persist between cosmological parameters.

13.3.3 Dark Energy

No deviations with look-back time are detected from the cosmological constant Λ in the Standard Model to less than 5%. The small value measured for Λ represents what has been called the greatest problem in physics. There are two key questions: Why is Λ so small, by approximately 120 powers of 10 as compared to the vacuum density at the scale of grand unification? And why is it just becoming dominant today, at approximately 10^{17} s rather than at grand unification some 10^{-36} s after the Big Bang? Is the explanation due to particle physics or due to astrophysics? Or do we need to wait for a fundamental theory of quantum gravity to emerge?

There is a particle physics 'explanation', or rather the indication of a possible route to be further explored. Consider a topological classification of all Calabi–Yau manifolds. This idea seems promising, in reducing more than 10^{500} manifolds predicted by string theory to a much smaller number, those with small Hodge numbers, but needs further exploitation in terms of phenomenological string theory to be a useful guide [51]. The landscape may potentially be even more crowded: More recent studies of F-theory compactification result in even more candidate manifolds for populating the multiverse, some $10^{272,000}$ [52]. Nevertheless, this approach may provide a hint as to an eventual possible selection principle for a universe congenial to both the Standard Model of particle physics and a low value of the cosmological constant.

In addition, there is an astrophysics 'explanation', or more realistically a hint of a promising direction that merits in-depth study. Consider a universe with a few very large voids. Putting us near the center of a huge Gpc-scale void could explain our apparent acceleration, but at the price of producing excessive peculiar velocities and excessive kSZ signals [53]. Particularly disconcerting for this hypothesis is the remarkably tight limit on any gradient of residuals in the Hubble constant out to $z \approx 0.2$ [46]. Data do allow for a large but sufficiently shallow void, which could at least lead to biases when comparing local and distant measures of cosmological parameters [54].

A more controversial approach is to postulate a universe with many large voids, whose cumulative effect via back reaction may perhaps contribute to global acceleration, albeit probably not at a sufficient level to account for dark energy [55].

13.3.4 What Is Dark Matter?

Dark matter is not baryonic. This canonical result, emanating from nucleosynthesis considerations in the first minutes of the Big Bang, is confirmed by the observed abundances of helium and deuterium, in combination with the inferred effective number of neutrino species, $N_{eff} = 3.15 \pm 0.23$, versus the predicted value of 3.046. This differs from a pure integer in that neutrino decoupling precedes electron–positron pair annihilation but is close enough so that neutrino flavour oscillations generate neutrino spectral distortions that slightly enhance the number density of relic neutrinos [56].

Primordial nucleosynthesis is highly successful in accounting for light element abundances, albeit with a major question mark over Li^6 overproduction in the Standard Model. The dark matter fraction is $\Omega_m = 0.3089 \pm 0.0062$, whereas the baryon fraction is $\Omega_b = 0.02230 \pm 0.00014$. Hence 90% of the dark matter is not made of baryons.

13.3.4.1 Neutrinos As Dark Matter?

Nor is the dark matter made of standard model neutrinos. The sum of neutrino masses required to account for dark matter is $\Omega_\nu h^2 = \sum m_\nu / 93 \text{eV}$, with $h = H_0 / 100 \text{ km}^{-1}\text{s}^{-1}\text{Mpc}^{-1}$, whereas from the Planck (2016) data $\sum m_\nu < 0.23\text{eV}$. Including limits from the Lyman alpha forest of intergalactic hydrogen clouds extends the scales probed to smaller co-moving mass scales, albeit at the risk of introducing possible systematics associated with hydrodynamic modeling of the intergalactic medium, to set a limit of $\sum m_\nu < 0.12\text{eV}$ [57]. The measured contribution from neutrino oscillations is about 58 meV, so that $\Omega_\nu h^2 > 0.0006$. The maximum value is $\Omega_\nu h^2 < 0.0025$.

13.3.4.2 Scalar and Warm Dark Matter

Scalar fields have immense potential for devising dark matter candidates. The QCD axion in the μeV to meV mass range is one example that is motivating extensive experimental searches using magnetic cavities. Such axions act as CDM. Ultralight string theory–motivated axions with masses of order 10^{-22} eV are a fuzzy dark matter candidate, so called because the de Broglie wavelength is macroscopic and can potentially account for the dark matter cores of dwarf galaxies [58].

Warm dark matter (WDM), perhaps in the form of sterile neutrinos, remains a possible option that does not affect large-scale structure, unlike HDM, but improves predictions at dwarf galaxy scales via enhanced free-streaming relative to CDM that suppresses the numbers of dwarfs, consistent with the constraints from Lyman alpha clouds. The classical WDM candidate is a sterile neutrino of mass a few kilo electron-volts and is motivated in part by observations of the 3.5 keV diffuse x-ray line in diverse dark matter–dominated environments that motivates 7-keV sterile neutrinosas a candidate. This interpretation has been criticized, and there may be more mundane explanations [59, 60].

Moreover, it is difficult to reconcile the corresponding free-streaming mass, and associated abrupt cut-off of small-scale power, simultaneously with dwarf galaxy abundances and cores [61] as well as with the re-ionization of the universe [62].

The advantage of the ultralight axion is that while it has an equally abrupt cut-off in small-scale power, the characteristic de Broglie wave-like oscillations on horizon entry result in a much larger core, which can more easily be reconciled with dwarf galaxy rotation curves as well as their abundance. Lyman alpha forest data potentially constrain this hypothesis [63], although a rigorous constraint has yet to be imposed.

13.3.4.3 Primordial Black Holes As Dark Matter?

Primordial black holes (PBH) have long been postulated to be a form of dark matter, subject to ever-tightening observational constraints. This scenario was reinvigorated by the LIGO detection of black hole mergers, because the detected mass range was initially not expected to be as large as observed. Of course, theorists have since come up with plausible explanations and predictions for future event rates from astrophysically produced black hole mergers. At the same time, limits on primordial black holes as dark matter candidates have been refined.

The current consensus is that primordial black holes are viable dark matter candidates provided that their mass is spread over several decades of PBH mass. There are two windows of opportunity. One focuses on PBH that are irrelevant for LIGO and/or dwarf galaxies, and centered on approximately 10^{20} gm. I will not discuss these further here, other than to note that the relevant mass-scale and formation conditions are motivated in part by the possibility of a low reheating temperature after late inflation, preceded by a brief period of matter domination.

The mass range with implications for galaxy formation is of broader interest for cosmology and centers on $10–10^5 M_\odot$, with the lower end of this mass range being revived most recently by the LIGO results. The initial conditions for forming such PBH less well understood but focus, for example, on possible scale-dependent isocurvature fluctuations. Massive PBH play a dual role: not only as dark matter candidates but also as potential sites of feedback for outstanding problems in galaxy formation theory via the consequences of their inevitable accretion of baryonic matter once the universe is dominated by matter.

At the upper end of this mass range, PBH of $10^4–10^5 M_\odot$ are natural seeds for supermassive black hole formation at high redshift, where the mass density required in black hole seeds is as low as 10^{-9} of the critical density [64]. Gas accretion plays an important feedback role for dwarf galaxies, provided the PBH mass range is approximately 10^{-4} of the dwarf mass, and could account for the reduced abundance of dwarfs, the ubiquity of dark matter cores, resolution of the 'too big to fail' problem, and more generally the baryon deficiency of all galaxies formed via merging dwarfs [65]. All of this potential feedback is subject to the usual litany of increasingly refined constraints from CMB distortions, gravitational microlensing, and dynamical effects.

13.3.4.4 Indirect Searches

The mystery of dark matter is that despite intensive searches, both indirect [66] and direct [67], its nature has not been identified. There is little question as to its observational dominance on scales from tens of kiloparsecs up to horizon scales, assuming Einstein gravity. It most likely consists of massive weakly interacting particles. The most attractive ansatz for the WIMP particle has long been provided by supersymmetry, which motivates the so-called WIMP miracle. If the lightest stable SUSY particle was once in thermal equilibrium, $< \sigma v >_x \approx 3.10^{-26}$ cm^{3}s$^{-1} \approx 0.23/\Omega_x$, with the consequence that thermal WIMPs generically account for the dark matter if the cross-section is typical of weak interactions. A wide range of SUSY models give scatter in the cross-section at fixed WIMP mass, most notably via coannihilations of massive WIMPs, by several orders of magnitude while still giving the correct dark matter density. The 100 or more free parameters in SUSY have led theorists to explore minimal models [68], which are highly predictive and invaluable for guiding indirect detection experiments, but at the price of restrictions on the full range of dark matter candidate masses, mediators, and annihilation channels.

Two difficulties arise with the canonical WIMP hypothesis for dark matter. First, SUSY has not yet been detected, most notably by the LHC, despite intensive searches for missing energy events that are symptomatic of dark matter particle production if weakly coupled to the Standard Model of particle physics.

Second, no dark matter detections have occurred via direct or indirect searches. Concerning the latter, Fermi satellite observations of a possible excess (relative to standard templates) diffuse gamma ray flux in a few degrees around the galactic center region have motivated a mini-industry of dark matter interpretations, usually involving WIMP self-annihilations. The preferred WIMP mass is 40–50 GeV, although this depends on the adopted annihilation channels [69]. The dark matter interpretation requires a thermal cross-section, as favored by WIMP freeze-out in the very early universe.

However, an alternative interpretation involving gamma rays from an old population of millisecond pulsars has received much attention. It is even favored by recent studies of evidence via fluctuation analyses for a faint discrete source population [70, 71].

As the WIMP mass increases, the scaling $< \sigma v >_x \approx \alpha_W^2/m_x^2$ indicates that there is a maximum WIMP mass to avoid overclosing the universe, the so-called unitarity limit, of approximately 100 TeV [72]. The possibility of exploring such a large mass range, which is near the limit of most acceptable SUSY models, is a prime reason for encouraging the construction of a future 100-TeV collider [73].

13.3.5 Advances in Galaxy Formation

Some 380,000 years after the Big Bang, temperature fluctuations imprinted on the last scattering surface provide evidence for the spectrum of primordial adiabatic fluctuations that seeded galaxy formation about a billion years later. Gravitational instability of weakly interacting cold (or warm) dark matter enables the growth of structure in the form of galactic halos that culminates in the formation of the first dwarf galaxies. Gas accretion triggers star formation, and the subhalos merge together as gravitational instability relentlessly continues until the full range of galaxy masses has developed.

But this grand scenario hides the ad hoc subgrid physics. The evolution of massive gas clouds is controlled by the competition between self-gravity and atomic cooling. The building blocks of galaxies are clouds of approximately 10^6 solar masses, since these form at a redshift of order 20 and are the smallest self-gravitating clouds that are warm enough to excite trace amounts of H_2 cooling and allow the first stars to form.

The largest galaxies weigh in at approximately $10^{13} M_\odot$ in the CDM theory, and form as the dwarf galaxies merge together hierarchically in a bottom-up fashion, as dictated by the approximate scale-invariance (measured via the CMB) in Fourier space of the primordial spectrum of density fluctuations. One success is that simple scaling physics can explain the characteristic mass of a galaxy of approximately $10^{12} M_\odot$, as this is the largest halo that can effectively dissipate gas energy and form stars [74, 75].

Numerical simulations are used to explore the detailed process of galaxy formation, from that of the first stars to massive galaxies. However, we cannot resolve star formation or AGN feedback in cosmological zoom simulations. The sub-grid physics is modeled by local phenomenology. The current generation of simulations includes both supernova and AGN/quasar feedback, using jets and/or mechanical outflows. Many aspects of the observed galaxy population can be explained, but some cannot. No model yet accounts for all aspects of galaxy formation.

For example, an exploration of various supernova feedback recipes found that no current prescription was able to reconcile the observed star formation efficiency per unit of cold gas in star-forming Milky Way–type galaxies with the mass-loading observed for galactic outflows in a wide range of models [76]. Improvements might include, among others, more sophisticated turbulence prescriptions [77], AGN-loaded superbubbles [78], or cosmic ray pressure-induced cooling delays [79].

More globally, dwarf galaxy issues, including their predicted numbers, the question of cores versus cusps, and the 'too big to fail' problem, have been largely resolved by improved baryonic physics and high-resolution simulations [80]. Even here, though, not all studies agree on whether supernova feedback is sufficiently effective in a multi-phase interstellar medium in generating the low stellar content of the ultrafaint dwarfs [81] or in producing the recently discovered population of ultradiffuse galaxies in both cluster and group environments [82, 83]. The diversity of dwarf rotation curves is not easily reconciled with recent simulations that incorporate baryonic feedback [84].

Nor is it obvious how to reconcile the history of inefficient star formation and baryonic mass loss with the efficient chemical seeding observed of r-process elements [85] observed in at least one ultrafaint dwarf galaxy. Lastly, the AGN/high star formation rate connection at high redshift continues to be an enigma, if galaxy mergers are not a general panacea for this correlation [86]. More complex feedback physics, negative and/or positive, may be needed.

As observational facilities are built to probe ever higher redshifts, doubts persist regarding the physics of galaxy formation. The resulting numerical simulations are beautiful – but are they reliable, and more specifically, are they predictive? The simulations are becoming increasingly complex in terms of input physics. One can easily imagine that they will soon be as difficult to analyze as the observational data, in terms of extracting any fundamental understanding of how galaxies actually formed. Nor is it clear that even then we will have converged on the required resolution and accuracy needed for the missing baryonic physics. These uncertainties have led to proposed particle physics solutions to some of the enigmas surrounding dwarf galaxies. Such investigations include many discussions of modified dark matter as an alternative explanation [87–90]. However, ultimately one has to pose the question of whether the failure of the most naive baryonic recipes provides sufficient motivation for modifying fundamental physics, such as the nature of fundamental particles or of gravity,

Another alternative is to incorporate new modes of baryonic feedback into dwarf galaxy modeling. The most natural addition appears to be that of massive black hole feedback, increasingly motivated by discovery of intermediate mass black holes in dwarfs (as central AGN) and in globular star clusters (via millisecond pulsar dynamics).

How can one distinguish AGN feedback from SNe feedback, especially in dwarf galaxies – or in massive galaxies, for that matter? One new prediction for the next generation of spatially resolved spectroscopy is to search the high-surface-brightness regions of galaxies for kinematic stellar signatures of positive feedback.

Stellar kinematics has a long memory, and is imprinted in different ways by super-nova and central AGN feedback. One would expect signatures in the skewed positive velocity dispersion, visible near or beyond the escape velocitiy at kpc distances from a typical AGN (see the predictions [91] and observations [92, 93] of triggered star formation).

13.4 Ultimate Cosmology: To B or Not to B, and Beyond

One can expect the best CMB measurements to probe $N \approx \ell^2 \approx 10^6$ independent Fourier modes on the sky. These independent samples to $\ell \approx 1,000$ allow one to achieve $N^{-1/2} \approx 0.1\%$ precision on large scales, corresponding to the precursors of galaxy clusters, before damping of the primordial fluctuations sets in and eliminates any smaller scales from detectability on the last scattering surface.

The current goal that dominates most proposed experiments is to probe inflation via its gravitational wave imprint. Neutrino mass measurements are an important corollary, but will never replace direct experiments.

Polarization searches for the B-mode imprint of inflation are confused by the foreground B signals, most notably from galactic dust and from gravitational lensing of the CMB by dark matter. The immediate goal is to increase the sensitivity. The Planck satellite was good for microkelvin sensitivity. To achieve new frontiers in B-mode polarization detection, one needs to at least reach the nanokelvin level. Several ground-based experiments are under way using thousands or eventually tens of thousands of bolometers, or are being planned with hundreds of thousands of pixels, in contrast to the 32 bolometers on the Planck HFI instrument. These include terrestrial experiments, unified as CMB stage IV [94], balloon experiments such as EBEX and SPIDER, and space experiments including the proposed CoRE and LiteBIRD satellites.

Space has unique advantages with regard to overcoming systematic errors and foregrounds, especially at large angular scales, but there is a price to pay in terms of cosmic variance-limited cosmology. Indeed, CMB science is limited by the number of independent modes. One can never improve on $N \approx 10^6$. This may not be good enough, as there is no robust prediction of B modes. Indeed, multifield inflation, a corner of parameter space that is favored by current limits on the scale index and B-mode amplitude, generically predicts low B modes, because of the low reheating scale that many inflation models prefer. In fact, there is no lower bound on B-mode polarization. Most inflation models predict unmeasurably low B modes.

Another prediction arises from damping of fluctuations corresponding to scales one cannot probe directly in the CMB – namely, those of the precursors of dwarf galaxies. This is especially important because the hierarchical, bottom-up picture of structure formation via gravitational instability of CDM predicts that such small structures are the numerous building blocks of today's massive galaxies.

Most notably, spectral distortions in the blackbody spectrum arising from the standard model of structure formation due to dissipative processes are predicted take in the form of a negative chemical potential distortion at a level of a part in 10^8 [95]. Detection of μ-distortions at this level would probe fluctuation damping and primordial non-Gaussianity on dwarf galaxy scales [96]. Such values of μ are four orders of magnitude below the COBE FIRAS limits, however, and may be detected only with a dedicated future space experiment [97].

13.4.1 Future of Galaxy Survey Cosmology

The large-scale structure of the galaxy distribution is a fertile hunting ground for evaluating both the geometry of the universe and its rate of expansion [49]. The BAO scale of approximately 150 Mpc is a standard ruler for cosmology. It is the precise analogue of the photon oscillations at $z = 1,080$ on the last scattering surface but is observable in the large-scale distribution of the galaxies at the present epoch and out to $z \approx 1$ [50]. Its remarkable power is that measurement requires no knowledge of the type of tracer. Comparison of the two length scales constrains the angular diameter distance and Hubble parameter at different redshifts. Hence deviations in dark energy due to possible variations in the cosmological constant assumption can be inferred. The angular diameter distance measurements are especially sensitive to the geometry of the universe. Weak gravitational lensing, galaxy clustering, and redshift space distortions all measure the dark matter content of the universe to high precision, and at different redshifts they constrain the growth rate of density fluctuations and its dependence on dark energy. Supernova measurements constrain the expansion rate of the universe. Degeneracies persist between cosmological parameters, leaving scope for dynamical dark energy or modified gravity. Such theories will be probed by instruments such as DESI [98], telescopes such as LSST, and missions such as EUCLID [99].

Galaxy surveys allow one to approach smaller scales and increase the independent sample number and, therefore, precision. Moreover, one gains by working in three dimensions, with redshift information. Future surveys, such as with LSST, target as many as 10^{10} galaxies with photometric redshifts, and should allow at least $N \approx 10^8$ independent samples. This could provide an order of magnitude increase in precision over CMB measurements for cosmology. Nevertheless, galaxies are messy and biased, and larger numbers are surely needed per independent sample. Hence their advantage over the relatively clean CMB is limited by systematics.

13.4.2 Dark Ages Cosmology

There is only one option to truly enhance the accuracy of future cosmology experiments: 21-cm probes of the dark ages. These probes provide the ultimate precision by sampling Jeans mass HI gas clouds at $z \approx 30$ that are still in the linear regime [100]. These clouds are colder than the CMB, as long as we observe them before galaxies or quasars, or even the first stars, have formed – that is, at very high redshift.

Using 21-cm absorption against the CMB allows one to approach a very large number of independent modes, perhaps $N \approx 10^{10}$. This would enable 100 times better precision than is attainable with the CMB. With sufficiently high precision, one could attempt to measure the fluctuation power spectrum at hitherto unsampled scales.

One can now imagine trying to measure a guaranteed prediction of inflation, namely primordial non-Gaussianity. This outcome, which is generically expected in inflation, is measured by non-vanishing higher- order correlations. With sufficiently high precision, one could attempt to measure the fluctuation bispectrum. This would be an important probe of inflation; indeed, it is actually guaranteed, albeit at only a low level (the usual measure of second-order non-Gaussianity $f_{NL} \approx 1 - n_s \approx 0.02$) in single-field slow roll inflation [101, 102]. Failure to measure primordial non-Gaussianity offers the only prospect, in principle, of being able to falsify inflation.

Many inflationary models have been proposed, involving multiple inflation fields or inflationary features in the primordial power spectrum of density fluctuations [103]. There, the effects of scale-dependent non-Gaussianity could be substantial, especially if enhanced small-scale power is present that would not be constrained by the Planck limits $f_{NL} \approx \mathcal{O}(\pm 10)$ [104] on larger scales.

To attain such a large number of modes at high redshifts, ideally in the redshift range 30–60, one would need to use 21-cm measurements at low radio frequencies, below 100 MHz, combined with high angular resolution. These requirements necessitate a large array of dipoles in a telescope site in an area with low radio interference. For example, a dipole array would need to be of size $\ell\lambda/2\pi \approx 300$ km to resolve Fourier modes $\ell \approx 10^5$ or a few arc-seconds corresponding to a cloud size of order 1 Mpc and a corresponding bandwidth of 0.1 MHz [105]. Perhaps only the far side of the Moon would provide a suitable site. This would be a project to consider for the next century.

13.5 Summary

Remarkable progress in cosmology has occurred over the past two decades, and it is now a precision science. Even so, important questions remain to be addressed. Here is just a personal subset:

- Dark matter is here to stay – but what is it? Should we change our theory if we fail to detect dark matter within a decade?

- Does the low value of dark energy require fine-tuning? And does this justify the appeal to the multiverse with so many free parameters and little prospect of verifiability?
- Does dark energy vary with epoch? This might be the simplest way to avoid the appeal to anthropic arguments to understand its low value during the present epoch.
- Is galaxy formation more of the same at early epochs? Or is something radically new needed?
- How are supermassive black holes formed in the early universe? Are seeds needed? Will gravity waves eventually probe their formation directly?
- CMB: Should we prioritize, if funding constraints require this, B-mode polarization or spectral distortions as the next new frontier of the very early universe?
- Exploring the dark ages with 21-cm probes at $z > 20$ is the most promising but also the most challenging new frontier: Is this feasible, especially at the most radio-quiet site known, on the far side of the Moon?
- Is the cosmological principle valid? This certainly motivated Einstein, Friedmann, and Lemaître. We should soon be able to test its validity – for example, via all-sky surveys with the SKA.
- Our cosmological model, and in particular our assessment of its contents such as dark matter and dark energy, assumes general relativity. Is general relativity valid?
- Is inflation the correct description of the early universe? Can we ever falsify inflation, regarded as a family of theories, in the same spirit that earlier cosmologies, such as Einstein's static universe and the steady state theory, were falsified?

13.5.1 Observations and Experiments

Finally I turn to a complementary wish list of observational questions. Cosmology is a field where observations of the natural phenomena, so sparse for millennia, have now advanced far ahead of theory. Here what one might expect observations to provide, someday:

- Spatially resolved spectroscopic mapping of molecular gas- and star-forming regions in galaxies near the peak of cosmic star formation at $z \approx 2$. One could thereby attack the weakest link in galaxy formation theory, and understand the differences between early universe star formation, when galaxies are youthful and gas-rich, and the present epoch, which is witnessing the inevitable fading of the bright lights of the universe.
- Overcoming cosmic variance by targeting the scattering of cosmic microwave background photons by hot gas in millions of galaxy clusters as probes of the local

CMB quadrupole up to redshifts $z \approx 2$. One could then meaningfully address the large angular-scale anomalies in the cosmic microwave background, which are potential witnesses of a pre-inflationary past.

- With sufficiently high-precision measurement of CMB temperature fluctuations, determination of the spatial curvature of the universe on horizon scales. One could then address one of the most fundamental questions in cosmology: How big is the universe?

- Direct measurement of the expansion and acceleration of the universe by spectroscopy of distant objects at a resolving power of 10^6 or greater. The second-generation EELT CODEX spectrograph will be a first step, but one could do better with even larger telescopes in the future.

- Resolution of Jeans mass neutral hydrogen scales in the dark ages, before the first stars formed. These are the ultimate building blocks of galaxies. Detection in absorption against the CMB requires a low-frequency radio interferometer on the far side of the Moon with millikelvin sensitivity at 30 or 50 MHz.

- Extraction of a primordial non-Gaussianity signal from the seed fluctuations that generated large-scale structure. Even the simplest, single-field inflation models predict such a signal at a level about 100 times below current limits. Our best bet here again may be a low-frequency radio interferometer on the far side of the Moon. This may take a long time, but the burgeoning interest of the space agencies in establishing what is being called a Lunar Village over the next two decades will motivate the planning of futuristic lunar telescopes.

- Resolution of the central engine in quasars and active galactic nuclei by microarcsecond microwave imaging. This also may require development of a lunar observatory in combination with a lunar satellite.

- As for identification of dark matter, I see one promising strategy for exploring the final frontier of massive particle candidates. Let us assume the dark matter consists of a weakly interacting particle. The problem that currently affects all astrophysics searches via indirect detection signatures is the lamp-post strategy – that is, use of model predictions about electromagnetic signals for looking in the dark. Any Bayesian would admit that given any reasonable model priors, the probability of finding a signal would be low. History tells us that extending the energy frontier is a more fruitful approach for discovering new phenomena. No guarantees exist, of course, but one might opt for a future 100-TeV proton collider as the next step toward exploring a promising regime where heavy dark matter particles may be hiding. One may not even need to seek much higher energies, given that general unitarity arguments limit the mass range to about 50 TeV that one needs to search for hints of any new dark matter–related physics. A slightly less compelling strategy might be to focus on the hitherto poorly explored sub-GeV mass range for DM particle candidates.

- As our searches begin to eke out the last dregs of WIMP parameter space, as might be the case a decade or two from now, it would be useful to recall that only one dark matter candidate stays within the gloriously successful standard model of particle physics and invokes astrophysics that is confirmed beyond the shadow of any doubt: primordial black holes. Black holes, in their non-primordial mode, exist. To a limited extent under the stimulation of the LIGO black hole merger events, we even understand how they formed.

Primordial black holes require a more extreme approach to cosmology that many might regard as an act of desperation. I do not. Consider the infinitesimal part of parameter space over which the amplitude of primordial density fluctuations has been measured, essentially on galactic scales and beyond. We know very little about the scales between those of dwarf galaxies and those of the Planck mass in regard to their possible contributions to the dark matter density. Nor do we understand theories of density fluctuation generation via quantum fluctuations, such as inflation, well enough to make any assertions about the amplitude of primordial density fluctuations on small scales. All we have are limits on the power spectrum strength of around 10^{-4} on scales from around 10^{40} to 10^{80} gm. The mass range from 10^{-6} to 10^{40} gm is fertile territory for primordial black holes that contribute to the dark matter density. The beauty of this suggestion is that it can be tested astrophysically.

The 'Why Trust a Theory?' conference focused on our trust in theory, and on whether non-empirical evidence can provide a pathway toward ultimately developing a theory of the beginning of the universe. Only the dark ages offer the prospect of measuring enough independent modes to approach this latter goal.

We can never prove that inflation occurred, but we can provide evidence in its favor. But falsification is a proven pathway in physics, if not in philosophy. Primordial non-Gaussianity is a robust prediction of inflationary models. The most challenging, yet potentially the most robust pathway for refuting inflation is via probes of primordial non-Gaussianity: It remains our strongest hope for finding empirical evidence that strengthens the case for inflation.

Non-empirical tests cannot achieve this goal, no matter how large a cohort of physicists and philosophers assiduously explore the Bayesian paradise of the string theory landscape. They are content to meander in the multiverse, ruminating between subjective priors and free parameters, enlightened by the unreasonable effectiveness of string theory, and inspired by its dangerous irrelevance, while basking in the aura of a final theory of physics that beckons like a mirage from afar.

Acknowledgements

I thank my colleagues at the Institut d'Astrophysique de Paris, Johns Hopkins University, and Beecroft Institute for Particle Astrophysics and Cosmology for many discussions of these topics. This chapter is a substantially updated version of a talk presented at the 14th International Symposium on Nuclei in the Cosmos XIV, June 2016 (Niigata, Japan), and accepted for publication in the Japan Physical Society Conference Proceedings.

References

[1] Hoyle, F., & Tayler, R. J. (1964). The mystery of the cosmic helium abundance. *Nature, 203*(4950), 1108.

[2] Burbidge, E. M., Burbidge, G. R., Fowler, W. A., & Hoyle, F. (1957). Synthesis of the elements in stars. *Reviews of modern physics, 29*(4), 547.

[3] Coc, A. (2016). Primordial Nucleosynthesis Journal of Physics: Conference Series, 665, 2001.

[4] Steigman, G., Olive, K. A., & Schramm, D. N. (1979). Cosmological constraints on superweak particles. *Physical Review Letters, 43*(4), 239.

[5] Burles, S. (2000). Deuterium and big bang nucleosynthesis. *Nuclear Physics A, 663*(1), 861c–864c.

[6] Penzias, A. A., & Wilson, R. W. (1965). A measurement of excess antenna temperature at 4080 Mc/s. *The Astrophysical Journal, 142*, 419–421.

[7] Mather, J. C., Cheng, E. S., Eplee Jr, R. E., Isaacman, R. B., Meyer, S. S., Shafer, R. A., ... & Dwek, E. (1990). A preliminary measurement of the cosmic microwave background spectrum by the Cosmic Background Explorer (COBE) satellite. *The Astrophysical Journal, 354*, L37–L40.

[8] Lange, A. E., Ade, P. A., Bock, J. J., Bond, J. R., Borrill, J., Boscaleri, A., ... & Ferreira, P. (2001). Cosmological parameters from the first results of BOOMERANG. *Physical Review D, 63*(4), 042001.

[9] Starobinskii, A. A. (1979). Spectrum of relict gravitational radiation and the early state of the universe. *JETP Letters*, 30, 682–685.

[10] Guth, A. H. (1981). Inflationary universe: A possible solution to the horizon and flatness problems. *Physical Review D, 23*(2), 347.

[11] Linde, A. D. (1982). A new inflationary universe scenario: a possible solution of the horizon, flatness, homogeneity, isotropy and primordial monopole problems. *Physics Letters B, 108*(6), 389–393.

[12] Albrecht, A., & Steinhardt, P. J. (1982). Cosmology for grand unified theories with radiatively induced symmetry breaking. *Physical Review Letters, 48*(17), 1220.

[13] Sato, K. (1981). Cosmological baryon-number domain structure and the first order phase transition of a vacuum. *Physics Letters B, 99*(1), 66–70.

[14] Mukhanov, V. F., & Chibisov, G. V. (1982). Vacuum energy and large-scale structure of the universe. *Soviet Physics-JETP, 56*(2), 258–265.

[15] Rubin, V. C., & Ford Jr, W. K. (1970). Rotation of the Andromeda nebula from a spectroscopic survey of emission regions. *The Astrophysical Journal, 159*, 379.

[16] Roberts, M. S., & Rots, A. H. (1973). Comparison of rotation curves of different galaxy types. *Astronomy and Astrophysics, 26*, 483–485.

[17] Zwicky, F. (1933). Die rotverschiebung von extragalaktischen nebeln. *Helvetica Physica Acta, 6*, 110–127.

[18] Oort, J. H. (1932). The force exerted by the stellar system in the direction perpendicular to the galactic plane and some related problems. *Bulletin of the Astronomical Institutes of the Netherlands, 6*, 249.

[19] Pagels, H., & Primack, J. R. (1982). Supersymmetry, cosmology, and new physics at teraelectronvolt energies. *Physical Review Letters, 48*(4), 223.

[20] Goodman, M. W., & Witten, E. (1985). Detectability of certain dark-matter candidates. *Physical Review D, 31*(12), 3059.

[21] Silk, J., & Srednicki, M. (1984). Cosmic-ray antiprotons as a probe of a photino-dominated universe. *Physical Review Letters, 53*(6), 624.

[22] Silk, J., Olive, K., & Srednicki, M. (1985). The photino, the sun, and high-energy neutrinos. *Physical Review Letters, 55*(2), 257.

[23] Weinberg, S. (1978). A new light boson?. *Physical Review Letters, 40*(4), 223.

[24] Wilczek, F. (1978). Problem of Strong p and t Invariance in the Presence of Instantons. *Physical Review Letters, 40*(5), 279.

[25] Blumenthal, G. R., Faber, S. M., Primack, J. R., & Rees, M. J. (1984). Formation of galaxies and large-scale structure with cold dark matter. *Nature, 311*(5986), 517.

[26] Davis, M., Efstathiou, G., Frenk, C. S., & White, S. D. (1985). The evolution of large-scale structure in a universe dominated by cold dark matter. *The Astrophysical Journal, 292*, 371–394.

[27] Bond, J. R., & Szalay, A. S. (1983). The Collisionless Damping of Density Fluctuations in an Expanding Universe. *Astrophys. J., 274*, 443–468.

[28] Perlmutter, S., Aldering, G., Goldhaber, G., Knop, R. A., Nugent, P., Castro, P. G., ... & Hook, I. M. (1999). Measurements of Ω and Λ from 42 high-redshift supernovae. *The Astrophysical Journal, 517*(2), 565.

[29] Riess, A. G., Filippenko, A. V., Challis, P., Clocchiatti, A., Diercks, A., Garnavich, P. M., ... & Leibundgut, B. R. U. N. O. (1998). Observational evidence from supernovae for an accelerating universe and a cosmological constant. *The Astronomical Journal, 116*(3), 1009.

[30] Silk, J. (1967). Fluctuations in the primordial fireball. *Nature, 215*(5106), 1155.

[31] Silk, J. (1968). Cosmic black-body radiation and galaxy formation. *The Astrophysical Journal, 151*, 459.

[32] Wilson, M. L., & Silk, J. (1981). On the anisotropy of the cosmological background matter and radiation distribution. I-The radiation anisotropy in a spatially flat universe. *The Astrophysical Journal, 243*, 14–25.

[33] Peebles, P. J., & Yu, J. T. (1970). Primeval adiabatic perturbation in an expanding universe. *The Astrophysical Journal, 162*, 815.

[34] Eisenstein, D. J., Zehavi, I., Hogg, D. W., Scoccimarro, R., Blanton, M. R., Nichol, R. C., ... & Anderson, S. F. (2005). Detection of the baryon acoustic peak in the large-scale correlation function of SDSS luminous red galaxies. *The Astrophysical Journal, 633*(2), 560.

[35] Vittorio, N., & Silk, J. (1984). Fine-scale anisotropy of the cosmic microwave background in a universe dominated by cold dark matter. *The Astrophysical journal, 285*(2), L39–L43.

[36] Bond, J., & Efstathiou, G. (1984). Cosmic background radiation anisotropies in universes dominated by nonbaryonic dark matter. *The Astrophysical journal, 285*(2), L45–L48.

[37] Keisler, R., Reichardt, C. L., Aird, K. A., Benson, B. A., Bleem, L. E., Carlstrom, J. E., ... & De Haan, T. (2011). A measurement of the damping tail of the cosmic microwave background power spectrum with the South Pole Telescope. *The Astrophysical Journal, 743*(1), 28.

[38] Dunkley, J., Hlozek, R., Sievers, J., Acquaviva, V., Ade, P. A., Aguirre, P., ... & Bond, J. R. (2011). The atacama cosmology telescope: cosmological parameters from the 2008 power spectrum. *The Astrophysical Journal, 739*(1), 52.

[39] Sunyaev, R. A., & Zeldovich, Y. B. (1970). Small-scale fluctuations of relic radiation. *Astrophysics and Space Science, 7*(1), 3–19.

[40] White, S. D., & Rees, M. J. (1978). Core condensation in heavy halos: a two-stage theory for galaxy formation and clustering. *Monthly Notices of the Royal Astronomical Society, 183*(3), 341–358.

[41] Dekel, A., & Silk, J. (1986). The origin of dwarf galaxies, cold dark matter, and biased galaxy formation. *The Astrophysical Journal, 303*, 39–55.

[42] Silk, J., & Rees, M. J. (1998). Quasars and galaxy formation. *Astron. Astrophys, 331*, L1–L4.

[43] Ade, P. A., Aghanim, N., Arnaud, M., Ashdown, M., Aumont, J., Baccigalupi, C., ... & Battaner, E. (2016). Planck 2015 results-xiii: cosmological parameters. *Astronomy & Astrophysics, 594*, A13.

[44] Aghanim, N., Ashdown, M., Aumont, J., Baccigalupi, C., Ballardini, M., Banday, A. J., ... & Benabed, K. (2016). Planck intermediate results-XLVI: Reduction of large-scale systematic effects in HFI polarization maps and estimation of the reionization optical depth. *Astronomy & Astrophysics, 596*, A107.

[45] Hildebrandt, H., Viola, M., Heymans, C., Joudaki, S., Kuijken, K., Blake, C., ... & Morrison, C. B. (2016). KiDS-450: Cosmological parameter constraints from tomographic weak gravitational lensing. *Monthly notices of the Royal Astronomical Society, 465*(2), 1454–1498.

[46] Riess, A. G., Macri, L. M., Hoffmann, S. L., Scolnic, D., Casertano, S., Filippenko, A. V., ... & Chornock, R. (2016). A 2.4% determination of the local value of the Hubble constant. *The Astrophysical Journal, 826*(1), 56.

[47] Kamionkowski, M., Kosowsky, A., & Stebbins, A. (1997). A probe of primordial gravity waves and vorticity. *Physical Review Letters, 78*(11), 2058.

[48] Seljak, U., & Zaldarriaga, M. (1997). Signature of gravity waves in the polarization of the microwave background. *Physical Review Letters, 78*(11), 2054.

[49] Percival, W. J. (2015 Large-scale structure observations in Deffayet, C., Peter, P., Wandelt, B., Zaldarriaga, M., & Cugliandolo, L. F. (Eds.). *Post-Planck Cosmology: Lecture Notes of the Les Houches Summer School: Volume 100*, July 2013 (Vol. 100). Lecture Notes of the Les Houch. 317–352.

[50] Beutler, F., Seo, H. J., Ross, A. J., McDonald, P., Saito, S., Bolton, A. S., ... & Font-Ribera, A. (2016). The clustering of galaxies in the completed SDSS-III Baryon oscillation spectroscopic survey: Baryon acoustic oscillations in the Fourier space. *Monthly Notices of the Royal Astronomical Society, 464*(3), 3409–3430.

[51] Candelas, P., Constantin, A., & Mishra, C. (2018). Calabi-Yau Threefolds with Small Hodge Numbers. *Fortschritte der Physik, 66*(6), 1800029.

[52] Taylor, W., & Wang, Y. N. (2015). The F-theory geometry with most flux vacua. *Journal of High Energy Physics, 2015*(12), 1–21.

[53] Zumalacarregui, M., Garcia-Bellido, J., & Ruiz-Lapente, R., (2012). Tension in the void: cosmic rulers strain inhomogeneous cosmologies. *Journal of Cosmology and Astroparticle Physics, 10*, 009.

[54] Keenan, RC, Barger, AJ, & Cowie, LL (2013). Evidence for a 300 megaparsec scale under-density in the local galaxy distribution. *The Astrophysical Journal, 775*, 62–78.

[55] Buchert, T., Carfora, M., Ellis, G. F., Kolb, E. W., MacCallum, M. A., Ostrowski, J. J., ... & Wiltshire, D. L. (2015). Is there proof that backreaction of inhomogeneities is irrelevant in cosmology?. *Classical and Quantum Gravity, 32*(21), 215021.

[56] Mangano, G., Miele, G., Pastor, S., Pinto, T., Pisanti, O., & Serpico, P. D. (2005). Relic neutrino decoupling including flavour oscillations. *Nuclear Physics B, 729*(1–2), 221–234.

[57] Palanque-Delabrouille, N., Yèche, C., Baur, J., Magneville, C., Rossi, G., Lesgourgues, J., ... & Viel, M. (2015). Neutrino masses and cosmology with Lyman-alpha forest power spectrum. *Journal of Cosmology and Astroparticle Physics, 2015*(11), 011.

[58] Schive, H. Y., Chiueh, T., Broadhurst, T., & Huang, K. W. (2016). Contrasting galaxy formation from quantum wave dark matter, ΨDM, with ΛCDM, using Planck and Hubble data. *The Astrophysical Journal, 818*(1), 89.

[59] Shah, C., Dobrodey, S., Bernitt, S., Steinbrügge, R., López-Urrutia, J. R. C., Gu, L., & Kaastra, J. (2016). Laboratory measurements compellingly support a charhe-exchange mechanism for the "dark matter" ~ 3.5 keV x-ray line. *The Astrophysical Journal, 833*(1), 52.

[60] Jeltema, T., & Profumo, S. (2016). Deep XMM observations of Draco rule out at the 99 per cent confidence level a dark matter decay origin for the 3.5 keV line. *Monthly Notices of the Royal Astronomical Society, 458*(4), 3592–3596.

[61] Macciò, A. V., Paduroiu, S., Anderhalden, D., Schneider, A., & Moore, B. (2012). Erratum: Cores in warm dark matter haloes: A Catch 22 problem. *Monthly Notices of the Royal Astronomical Society, 428*(4), 3715–3716.

[62] Bozek, B., Marsh, D. J., Silk, J., & Wyse, R. F. (2015). Galaxy UV-luminosity function and reionization constraints on axion dark matter. *Monthly Notices of the Royal Astronomical Society, 450*(1), 209–222.

[63] Iršič, V., Viel, M., Haehnelt, M. G., Bolton, J. S., & Becker, G. D. (2017). First constraints on fuzzy dark matter from Lyman-α forest data and hydrodynamical simulations. *Physical review letters, 119*(3), 031302.

[64] Carr, B., Kühnel, F., & Sandstad, M. (2016). Primordial black holes as dark matter. *Physical Review D, 94*(8), 083504.

[65] Silk, J. (2017). Feedback by massive black holes in gas-rich dwarf galaxies. *The Astrophysical Journal Letters, 839*(1), L13.

[66] Gaskins, J. M. (2016). A review of indirect searches for particle dark matter. *Contemporary Physics, 57*(4), 496–525.

[67] Undagoitia, T. M., & Rauch, L. (2015). Dark matter direct-detection experiments. *Journal of Physics G: Nuclear and Particle Physics, 43*(1), 013001.

[68] Cirelli, M., Corcella, G., Hektor, A., Hütsi, G., Kadastik, M., Panci, P., ... & Strumia, A. (2011). PPPC 4 DM ID: a poor particle physicist cookbook for dark matter indirect detection. *Journal of Cosmology and Astroparticle Physics, 2011*(03), 051.

[69] Daylan, T., Finkbeiner, D. P., Hooper, D., Linden, T., Portillo, S. K. N., Rodd, N. L., & Slatyer, T. L. (2016). *Physics of the Dark Universe* 12 1. arXiv preprint arXiv:1402.6703.

[70] Lee, S. K., Lisanti, M., Safdi, B. R., Slatyer, T. R., & Xue, W. (2016). Evidence for unresolved γ-ray point sources in the inner galaxy. *Physical Review Letters, 116*(5), 051103.

[71] Di Mauro, M., Donato, F., Fornengo, N., & Vittino, A. (2016). Dark matter vs. astrophysics in the interpretation of AMS-02 electron and positron data. *Journal of Cosmology and Astroparticle Physics, 2016*(05), 031.

[72] Griest, K., & Kamionkowski, M. (1990). Unitarity limits on the mass and radius of dark-matter particles. *Physical Review Letters, 64*(6), 615.

[73] Arkani-Hamed, N., Han, T., Mangano, M., & Wang, L. T. (2016). Physics opportunities of a 100 TeV proton–proton collider. *Physics Reports, 652*, 1–49.

[74] Rees, M. J., & Ostriker, J. P. (1977). Cooling, dynamics and fragmentation of massive gas clouds: clues to the masses and radii of galaxies and clusters. *Monthly Notices of the Royal Astronomical Society, 179*(4), 541–559.

[75] Silk, J. (1977). On the fragmentation of cosmic gas clouds. I-The formation of galaxies and the first generation of stars. *The Astrophysical Journal, 211*, 638–648.

[76] Rosdahl, J., Schaye, J., Dubois, Y., Kimm, T., & Teyssier, R. (2016). Snap, crackle, pop: sub-grid supernova feedback in AMR simulations of disc galaxies. *Monthly Notices of the Royal Astronomical Society, 466*(1), 11–33.

[77] Semenov, V. A., Kravtsov, A. V., & Gnedin, N. Y. (2016). Nonuniversal star formation efficiency in turbulent ISM. *The Astrophysical Journal, 826*(2), 200.

[78] Keller, B. W., Wadsley, J., & Couchman, H. M. P. (2016). Cosmological galaxy evolution with superbubble feedback–II. The limits of supernovae. *Monthly Notices of the Royal Astronomical Society, 463*(2), 1431–1445.

[79] Salem, M., Bryan, G. L., & Corlies, L. (2015). Role of cosmic rays in the circumgalactic medium. *Monthly Notices of the Royal Astronomical Society, 456*(1), 582–601.

[80] Fattahi, A., Navarro, J. F., Sawala, T., Frenk, C. S., Sales, L. V., Oman, K., ... & Wang, J. (2016). The cold dark matter content of Galactic dwarf spheroidals: no cores, no failures, no problem. arXiv preprint arXiv:1607.06479.

[81] Bland-Hawthorn, J., Sutherland, R., & Webster, D. (2015). Ultrafaint dwarf galaxies – the lowest-mass relics from before reionization. *The Astrophysical Journal, 807*(2), 154.

[82] Van Dokkum, P. G., Abraham, R., Merritt, A., Zhang, J., Geha, M., & Conroy, C. (2015). Forty-seven milky way-sized, extremely diffuse galaxies in the coma cluster. *The Astrophysical Journal Letters, 798*(2), L45.

[83] Merritt, A., van Dokkum, P., Danieli, S., Abraham, R., Zhang, J., Karachentsev, I. D., & Makarova, L. N. (2016). The dragonfly nearby galaxies survey. II: ultra-diffuse galaxies near the elliptical galaxy NGC 5485. *The Astrophysical Journal, 833*(2), 168.

[84] Santos-Santos, I. M., Di Cintio, A., Brook, C. B., MacciO, A., Dutton, A., & Domínguez-Tenreiro, R. (2017). NIHAO–XIV. Reproducing the observed diversity of dwarf galaxy rotation curve shapes in ΛCDM. *Monthly Notices of the Royal Astronomical Society, 473*(4), 4392–4403.

[85] Ji, A. P., Frebel, A., Chiti, A., & Simon, J. D. (2016). R-process enrichment from a single event in an ancient dwarf galaxy. *Nature, 531*(7596), 610.

[86] Fensch, J., Renaud, F., Bournaud, F., Duc, P. A., Agertz, O., Amram, P., ... & Jog, C. J. (2016). High-redshift major mergers weakly enhance star formation. *Monthly Notices of the Royal Astronomical Society, 465*(2), 1934–1949.

[87] Bose, S., Hellwing, W. A., Frenk, C. S., Jenkins, A., Lovell, M. R., Helly, J. C., & Li, B. (2015). The Copernicus Complexio: statistical properties of warm dark matter haloes. *Monthly Notices of the Royal Astronomical Society, 455*(1), 318–333.

[88] Marsh, D. J., & Silk, J. (2013). A model for halo formation with axion mixed dark matter. *Monthly Notices of the Royal Astronomical Society, 437*(3), 2652–2663.

[89] Hui, L., Ostriker, J. P., Tremaine, S., & Witten, E. (2017). Ultralight scalars as cosmological dark matter. *Physical Review D, 95*(4), 043541.

[90] Schneider, A., Trujillo-Gomez, S., Papastergis, E., Reed, D. S., & Lake, G. (2017). Hints against the cold and collisionless nature of dark matter from the galaxy velocity function. *Monthly Notices of the Royal Astronomical Society, 470*(2), 1542–1558.

[91] Dugan, Z., Gaibler, V., & Silk, J. (2017). Feedback by AGN jets and wide-angle winds on a Galactic scale. *The Astrophysical Journal, 844*(1), 37.

[92] Cicone, C., Maiolino, R., & Marconi, A. (2016). Outflows and complex stellar kinematics in SDSS star-forming galaxies. *Astronomy & Astrophysics, 588*, A41.

[93] Maiolino, R., Russell, H. R., Fabian, A. C., Carniani, S., Gallagher, R., Cazzoli, S., ... & Cresci, G. (2017). Star formation inside a galactic outflow. *Nature, 544*(7649), 202.

[94] Abazajian, K. N., Adshead, P., Ahmed, Z., Allen, S. W., Alonso, D., Arnold, K. S., ... & Bischoff, C. A. (2016). CMB-S4 science book. arXiv preprint arXiv:1610.02743.

[95] Hu, W., Scott, D., & Silk, J. (1994). Power spectrum constraints from spectral distortions in the cosmic microwave background. arXiv preprint astro-ph/9402045.

[96] Pajer, E., & Zaldarriaga, M. (2012). New window on primordial non-gaussianity. *Physical Review Letters, 109*(2), 021302.

[97] Chluba, J., & Sunyaev, R. A. (2011). The evolution of CMB spectral distortions in the early Universe. *Monthly Notices of the Royal Astronomical Society, 419*(2), 1294–1314.

[98] Aghamousa, A., Aguilar, J., Ahlen, S., Alam, S., Allen, L. E., Prieto, C. A., ... & Baltay, C. (2016). The DESI experiment part I: science, targeting, and survey design. arXiv preprint arXiv:1611.00036.

[99] Amendola, L., Appleby, S., Avgoustidis, A., Bacon, D., Baker, T., Baldi, M., ... & Branchini, E. (2018). Cosmology and fundamental physics with the Euclid satellite. *Living reviews in relativity, 21*(1), 2.

[100] Loeb, A., & Zaldarriaga, M. (2005). Small-scale power spectrum of cold dark matter. *Physical Review D, 71*(10), 103520.

[101] Maldacena, J. (2003). Non-Gaussian features of primordial fluctuations in single field inflationary models. *Journal of High Energy Physics, 2003*(05), 013.

[102] Cabass, G., Pajer, E., & Schmidt, F. (2017) How Gaussian can our Universe be?. *Journal of Cosmology and Astroparticle Physics, 1*, 003.

[103] Renaux-Petel, S. (2015) Primordial non-Gaussianities after Planck 2015: an introductory review, *Comptes Rendus Physique, 16*, 969.

[104] Aghanim, N., Ashdown, M., Aumont, J., Baccigalupi, C., Ballardini, M., Banday, A. J., ... & Benabed, K. (2016). Planck intermediate results-XLVI. Reduction of large-scale systematic effects in HFI polarization maps and estimation of the reionization optical depth. *Astronomy & Astrophysics, 596*, A107.

[105] Muñoz, J. B., Ali-Haïmoud, Y., & Kamionkowski, M. (2015). Primordial non-Gaussianity from the bispectrum of 21-cm fluctuations in the dark ages. *Physical Review D, 92*(8), 083508.

14

The Role of Cosmology in Modern Physics

BJÖRN MALTE SCHÄFER

14.1 Introduction

Modern cosmology aims to describe the expansion dynamics of the Universe on large scales and the formation and evolution of cosmic structures with physical models. In this endeavour, cosmology links fundamental laws of physics, most importantly general relativity as the theory of gravity, with (fluid) mechanics for the formation of structures and statistics to provide a framework for interpreting cosmological observations.

Experimental results from the last decades have established the Lambda cold dark model (ΛCDM) as the standard model for cosmology and have measured the set of cosmological parameters at percent precision. Measurements of the distance redshift relation of cosmological objects from Cepheids on small distances (Freedman et al., 2001, Freedman & Madore, 2010) to supernovae on large distances (Perlmutter et al., 1998, Riess et al., 1998) have established a value of the Hubble constant and the presence of accelerated expansion on large scales (for a review, see Weinberg et al., 2013), while at the same time observations of the cosmic microwave background (Mather et al., 1990, 1991, Smoot et al., 1991), have constrained the spatial curvature to be very small and allowed a precision determination of cosmological parameters, in particular those related to inflation from the statistics of cosmic structures as a function of scale. Even more importantly, the adiabaticity of perturbations have ruled out structure generation models based on cosmic defects. Almost in passing, the Friedmann–Lemaître–Robertson–Walker (FLRW) class of cosmologies (Friedmann, 1924; Lemaître, 1931; Robertson, 1935) was able to provide a framework of thermal evolution with the epochs of element formation and atom formation in the early universe. Converging on the current value of the Hubble constant and adopting a non-zero value of Λ (Carroll et al., 1992, Carroll, 2001), it was possible to accommodate old astrophysical objects such as white dwarfs and slowly decaying atomic nuclei within the age of the Universe suggested by the cosmological model.

It is important to realise, however, that cosmology differs substantially from other branches of physics. This concerns in particular the passive role of observers, the process of inference from data, and the importance of assumptions for the cosmological model that are difficult to test or that cannot be tested at all. Furthermore, gravity as the fundamental interaction in cosmology is peculiar because its formulation is unique if certain physical concepts are accepted, so any observed deviation from the prediction of relativity will not shed doubt on a particular formulation of the gravitational model but rather goes directly to these fundamental physical concepts. Given these perspectives, cosmology brings together deductive reasoning (concerning the construction of gravitational theories) with statistical inference for selecting the true model, meaning the most economic model compatible with the data.

In this chapter, I discuss the relation between cosmology and fundamental physics on one side and the process of statistical inference on the other. Specifically, I discuss the concepts of general relativity in section 14.2, their application to cosmology in section 14.3, and the construction of ΛCDM as the standard model of cosmology in section 14.4. The peculiar role of statistics in cosmology and the process of statistical inference are treated in sections 14.5 and 14.6. Cosmic structure formation with its relations to relativity and statistics are the topic of section 14.7, and I conclude in section 14.8 with an overview of future developments, caveats, and a summary of the implications of cosmology for fundamental physics.

14.2 Gravity and General Relativity

Adopting the principle of extremised proper time as a generalisation of the principle of least action from classical mechanics to the relativistic motion in gravitational fields leads to the construction of metric theories of gravity (Einstein, 1915, 1916). The equivalence principle stipulates that all objects fall at the same rate in gravitational fields irrespective of their mass, which suggests that the proper time is affected through the metric as the expression of the gravitational field. In the absence of gravitational fields (or in an infinitesimally small, freely falling, and non-rotating laboratory without the possibility of observing distant charges), the metric assumes the Minkowskian shape, while the first derivative of the metric vanishes, reducing the covariant derivative to a partial one, so that the dynamics of all systems is determined locally by the laws of special relativity. The vanishing first derivatives of the metric is the reason, from a relativistic point of view, why pendulum clocks stop ticking in free fall.

The clearest physical setup in which one can experience gravitational fields is that of geodesic deviation – that is, the relative motion of two test particles that fall in a gravitational field. The second derivative of the distance between two test particles with respect to proper time is proportional to the Riemann curvature. In particular, if the curvature is zero, the test particles increase or decrease their distance at most linearly. By comparison, if the curvature is non-vanishing, one can see more complicated forms of relative motion – for instance, the acceleration or deceleration of distant galaxies in cosmology or the oscillatory motion of two objects in the field of a gravitational wave.

The metric as a generalisation to the gravitational potential is sourced by the energy-momentum density in a field equation, and the gravitational field has a relativistic dynamic on its own that is described by the (contracted) Bianchi identity. It is very impressive that the shape of the field equation is completely determined by fundamental principles and that "God did not have any choice in the creation of the world" as Einstein put it – although it was only rigorously proved by Cartan (1922), Vermeil (1917), and Lovelock (1969) that general relativity is the only possible metric theory of gravity based on Riemannian differential geometry in four dimensions that is based on a single dynamical field, which is locally energy-momentum conserving, with a local field equation, and which is based on derivatives up to second order (Lovelock, 1971; Navarro and Navarro, 2011).

It is worth discussing these five requirements in detail because they reflect fundamental principles of physical laws that make gravity so interesting. Because there is no alternative theory of gravity under these assumptions, any deviation from general relativity will directly call these fundamental principles into question. Cosmology plays an important role in this context because it is the simplest gravitational system, which, in contrast to the Schwarzschild solution or gravitational waves in vacuum, involves the field equation where the local Ricci curvature is directly coupled to the cosmological fluids.

A relation between the second derivatives of the metric and the source of the gravitational field is necessary for a number of reasons. Apart from analogies to other relativistic field theories such as electrodynamics, which feature identical constructions, second-order field equations are a very natural way to incorporate parity and time-reversal symmetry in a field theory. In fact, general relativity is perfectly $\mathcal{PT}$-symmetric, in particular in the weak field limit, where this symmetry is not already provided by general covariance. In addition, the definition of curvature is based (in the simplest constructions of parallel transport) on the non-commutativity of the second covariant derivatives, which ensures that the Bianchi identity applies to the description of the propagation dynamics of the gravitational field and enforces local energy momentum conservation. Lastly, the inclusion of higher-order derivatives would lead to the Ostrogradsky instability (Woodard, 2015) because systems are not necessarily bounded energetically from below, although some constructions might lead to stable theories despite the inclusion of higher derivatives. From a fundamental point of view, it is natural to base a gravitational theory on second derivatives of the metric because of their invariant nature, as tidal fields cannot be set to zero by transforming them into freely falling frames. Lovelock's original restriction to second derivatives can be understood as well through dimensional arguments (Aldersley, 1977) or through similarity transformations (Navarro and Sancho, 2008).

Arguing from fundamental physics adds a very interesting thought: The dynamics of fields is modelled on the dynamics of (relativistic) particles. For instance, one uses concepts like the action as a generalisation of the principle of extremised proper time in analogy to the causal transition from one field configuration to another. Lagrange functions for point particles are convex due to causality and yield again convex Hamilton functions that are bounded from below, resulting in stability. The construction of a convex Lagrange function involving squares of first derivatives is naturally suggested by special relativity. This construction principle applies equally to the gravitational field, where the formulation by Einstein and Palatini involving squares of Christoffel symbols is natural from this point of view (Palatini, 1919) and yields, in addition to the field equation, the metric connection. Assuming this relationship as an axiom allows the Einstein–Hilbert action as the second formulation in which general covariance and the locality of the field equation are apparent.

The four dimensions of spacetime are peculiar. While classical Newtonian gravity can be formulated in any number of dimensions, general relativity requires at least four dimensions, which can be seen from a number of arguments, the most elegant perhaps being the fact that the Weyl tensor, which contains the non-local contributions to the Riemann curvature and which describes the propagation of the gravitational field away from its sources, is only non-zero in four or more dimensions. If one assumes a larger number of dimensions, one needs immediately to provide an explanation why there are differences between the basic four dimensions and the additional ones. In particular, one must invoke the equivalence principle, which locally provides a reduction to flat Minkowski space, which has four dimensions. Conversely, empirical requirements limit the number of dimensions: For instance, there are no stable Kepler orbits in more than four dimensions and the Huygens principle is exactly valid only in spaces with an odd number of spatial dimensions (Balakrishnan, 2004).

Locality of the field equation is realised in general relativity because only the Ricci curvature is related to its source, the energy-momentum tensor, while the Weyl tensor is unconstrained. The fact that the Riemann curvature is only partially determined by the field equation and the Bianchi identity as a second ingredient for the field dynamics is the construction that allows the field to propagate away from its sources, realising Newton's definition of a field as "an action at a distance."

The origin of energy-momentum conservation is the symmetries of the Lagrange density, which determine the internal dynamics of the gravitating substance. If the dynamics of the substance is universal and does not depend on the coordinates, the Lagrange density defines together with the metric a conserved energy-momentum tensor. In this sense, general relativity is the gravitational theory for locally energy-momentum–conserving systems in the same way as electrodynamics is a field theory for charge-conserving systems, and one notices a remarkable consistency between the dynamics of the fields and the dynamics of the field-generating charges. This is realised in both gravity and electrodynamics in the antisymmetry of the curvature form.

Finally, general relativity uses the metric as the only dynamical degree of freedom of the gravitational field. While a larger number of gravitational degrees of freedom are conceivable (de Rham, 2014) and form the basis of the Horndeski class of gravitational theories (Horndeski, 1974), one must include an explanation of why the gravitational interaction of all particles is exclusively determined by the metric. Other gravitational degrees of freedom would lack the unique physical interpretation of the metric in terms of the principle of extremised proper time.

Apart from the uniqueness of the metric, this point seems the easiest assumption to overcome in comparison to the four others. Personally, I would be interested in knowing whether the introduction of new gravitational degrees of freedom has implications in relation to Mach's principle, which was the primary motivation for Brans and Dicke to introduce a scalar degree of freedom in their gravitational theory (Brans and Dicke, 1961).

It is remarkable that general relativity is the only consistent expression of a gravitational theory based on these five principles, and that the principles equally apply to other field theories. For instance, the five principles apply to electrodynamics, which is likewise unique as a $\mathcal{PT}$-symmetric, charge-conserving, local field theory with a single, vectorial field in four dimensions, for exactly the same reasons, with the only difference being the replacement of general coordinate covariance by Lorentz covariance. It is difficult to overstate the implications of this uniqueness, because any deviation from the predicted behaviour of the theories questions the fundamental assumptions for their construction. In addition, the cosmological constant Λ is, under these assumptions, a natural feature of the field equation, although the metric exhibits a different Weyl scaling than the Einstein tensor. From this point of view the cosmic coincidence is more remarkable, because the transition between the two scaling-regimes takes place in the current epoch (Navarro and Sancho, 2008, 2009).

The Universe is a unique testbed for strong, geometric gravity with the involvement of the field equation in a very symmetric setup with pure Ricci curvature, and a system that is large enough to display order-unity effects in relation to Λ. Currently, most of the research focuses on additional dynamical degrees of freedom of gravity as well as a possible non-locality of the interaction with the gravitational field source, where the influence of gravity on cosmic structure formation constitutes the physical system (Clifton et al., 2012, Joyce et al., 2016, Koyama, 2016).

14.3 Cosmology As an Application of Gravity

Cosmology provides a natural example of strong-field, geometric gravity in a very symmetric setup. Under the Copernican principle, which assumes spatial homogeneity and isotropy, the FLRW class of cosmologies exhibits the same degree of symmetry as the exterior Schwarzschild solution or the gravitational waves. These three cases are discussed in every textbook, because they allow exact solutions to the gravitational field equations due to their high degree of symmetry. In fact, if the degree of symmetry were any higher, there could not be curvature, and consequently no effects of gravity.

There is, however, one very important difference between these three solutions: Whereas the Schwarzschild solution and gravitational waves are vacuum solutions and exhibit curvature in the Weyl part of the Riemann tensor, the curvature in FLRW cosmologies is contained in the local Ricci part. As an immediate consequence, one needs the full field equation for linking the geometry and the expansion dynamics of the FLRW cosmologies to the properties of cosmological fluids, which is in contrast to the Schwarzschild solution or the gravitational waves, where the Bianchi identity is sufficient in its property of describing the dynamics of curvature. To put this in a very poignant way, cosmology investigates gravity in matter, and because the curvature is in the Ricci part of the Riemann tensor due to the symmetry assumptions of the metric, one does not have to deal with propagation effects of gravity, unlike in the case of the vacuum solutions.

Equating the Einstein tensor with the energy-momentum tensor of a homogeneous, ideal relativistic fluid leads to the Friedmann equations, which describe acceleration or deceleration of the cosmic expansion in its dependence on the density and the pressure–density relation (i.e., the equation of state) of the cosmic fluids. Under the symmetry assumptions of the Copernican principle, only the density and the equation of state can characterise a cosmic fluid as a source of the field equation. As a consequence of the field equation, which relates geometry to energy-momentum density, it is possible to assign these two properties as well to geometric quantities like spatial curvature or to the cosmological constant, although they do not correspond to actual physical substances. For instance, the density parameter associated with curvature can in principle be negative, and the density associated with Λ does not decrease with increasing scale factor. In these cases, the assigned density parameter and the equation of state only characterise the influence on the expansion dynamics, in contrast to an actual substance as a classical source of the gravitational field, where the equation of state is related to the relativistic dispersion relation of the particles that make up the fluid.

The cosmological constant is a natural term in the gravitational field equation under the conditions discussed in the last section, but it influences the expansion dynamics of the Universe only on scales in excess of 10^{25} meters – that is, on today's Hubble scale c/H_0 – due to its numerical value. The numerical similarity of the density parameters Ω_m and Ω_Λ constitutes the coincidence problem because spatial flatness only requires their sum to be close to 1 ($\Omega_m + \Omega_\Lambda = 1$), but there is no argument on the basis of which one can make statements on the individual values, in particular because they show a very different evolution with time.

Spatial curvature, referring to the curvature of spatial hypersurfaces at fixed time, is a degree of freedom that is allowed by the Copernican principle, but that is not realised in Nature at the current epoch. While there is no symmetry that fixes curvature to be zero, cosmological observations constrain the associated density parameter to be at most a percent, or put a lower limit of the curvature scale to be at least $\simeq 10c/H_0$ (Planck Collaboration et al., 2016a). The problem of small spatial curvature is exacerbated by the fact that curvature grows if the cosmic expansion decelerated, and it was in fact decelerating for most of the cosmic history under the dominating gravitational effect of radiation at early times and of matter at later times. A widely accepted mechanism that explains the smallness of spatial curvature is cosmic inflation, where spatial curvature is made to decrease by assuming an early phase of accelerated expansion.

Likewise, it is possible to assign density and pressure to the energy-momentum tensor of fields with self-interactions that form the basis of quintessence models of dark energy (Mortonson et al., 2013): In those models one obtains a conserved energy-momentum density from the coordinate independence of the Lagrange density. Identification of terms leads to relations between the fluid properties density and pressure and the kinetic and potential terms of the field, and constrains the equation of state to a physically sensible range, which in particular leads to finite ages of the cosmological models. If the dark energy density is chosen to be comparable to that of matter and if the equation of state is negative enough, one observes as the most important effect accelerated expansion at late times (Huterer and Turner, 2001, Frieman et al., 2008).

Geodesic deviation is again a very illustrative example of how gravity affects the motion of galaxies in cosmology. Each galaxy is considered to be a freely falling test particle whose relative acceleration depends on the spacetime curvature, which in turn is linked through the field equation to the density of all cosmological fluids and their equations of state. With distances in excess of c/H_0, one can see that galaxies increase their distance relative to the Milky Way in an accelerated way under the action of the cosmological constant, in the presence of dark energy, or through a modification of the gravitational field equation. Naturally, light from these distant galaxies reaches us on our past light cone as photons move along null geodesics. As a consequence, the measurement in this case consists of relating the distance measured on the past light cone to the redshift of the galaxies. It is essential that measurements in cosmology involve the relative motion of macroscopically separated objects: Any local, freely falling, and non-rotating experiment will experience the metric as being locally Minkowskian, and would conclude from the absence of curvature that there are no gravitational field generated.

One issue that defies a definitive formal solution and that is still a matter of current debate is the problem of backreaction (Wiltshire, 2007; Adamek et al., 2016; Green and Wald, 2013; Buchert et al., 2015; Bolejko and Korzyński, 2017). Due to the nonlinearity of the gravitational field equations, it matters how the gravitational field on large scales, where statistical homogeneity and isotropy should apply, arises from the structured matter distribution on small scales – in other words, how the Weyl curvature due to structures on small scales transitions to the Ricci curvature of the homogeneous FLRW universe. There are analytical arguments as well as simulations of the spacetime geometry, but a definitive answer is still not reached.

14.4 ΛCDM As the Standard Model

Modern cosmology has established ΛCDM as its Standard Model (Silk, 2016). In this model, the expansion dynamics of the Universe is described by general relativity with the assumption of the Copernican principle, leading to the FLRW class of homogeneous and isotropic cosmologies with very small spatial curvature and with radiation, (dark) matter, and the cosmological constant Λ as dominating, ideal fluids at early, intermediate, and late times, respectively. The confirmation of spatial flatness and accelerated expansion was achieved by combining supernova data with observations of the cosmic microwave background, which determine to leading order the sum and difference of the matter density parameter and the density parameter associated with Λ, while there are indications for isotropy from the Elhlers–Geren–Sachs theorem applied to the cosmic microwave background, even in the case of small perturbations (Ehlers et al., 1968; Stoeger et al., 1995; Clarkson and Maartens, 2010; Planck Collaboration et al., 2016b).

While the density parameters are well determined by observations and even the equation of state parameters have been constrained to values suggesting that the cosmological constant is the driving term behind accelerated expansion, the fundamental symmetries of the FLRW cosmologies are only partially tested. While tests of isotropy provide support for the Copernican principle, homogeneity is difficult to test and there are weak indications that it is in fact realised in Nature – for instance from the magnitude of the Sunyaev–Zel'dovich effect in Lemaître–Tolman–Bondi models. In a certain sense, deviations from homogeneity do not lead to a less complex cosmological model: After all, we as observers need to be positioned close to the centre of a large, horizon-sized void.

The bulk of gravitating matter is required to be non-interacting with photons (from the amplitude of structures seen in the CMB and their growth rate in comparison to the amplitude of large structures today), if it is to have a very small cross-section for elastic collisions (leading to the pressureless and inviscid "fluid" that is responsible for the typical core structure of dark matter haloes). The gravitating matter must also to have very little thermal motion, if it is to be dynamically cold (from the power-law behaviour of the halo mass function at small masses and lately from the abundance of substructure in dark matter haloes).

Fluctuations in the distribution of matter and in the velocity field at early times are thought to be generated by cosmic inflation (Tsujikawa, 2003; Baumann, 2009; Martin et al., 2013). At the same time, inflation remedies the flatness problems discussed earlier, and provides in addition an explanation of the horizon problem, referring to the uniformity of the CMB on large scales. Cosmic inflation generates fluctuations whose fluctuation spectrum follows a near scale-free law known as the Harrison–Zel'dovich spectrum, and generated mostly adiabatic perturbations (i.e., identical fractional perturbation in every cosmological fluid), with near-Gaussian statistical properties. This theory has been tested most importantly with observations of the cosmic microwave background.

In summary, ΛCDM as the Standard Model for cosmology is a model with low complexity but makes extreme assumptions at the same time, most importantly the Copernican principle and the properties of dark matter. It features cosmic inflation as a construction to explain the origin of fluctuations as well as the absence of spatial curvature.

Adopting the view that general relativity including the cosmological constant is the only possible metric theory of gravity under certain concepts allows one to estimate scales by which the Universe should be correctly characterised. Constructing a length scale $l_H = 1/\sqrt{\Lambda}$, a time scale $t_H = 1/(c\sqrt{\Lambda})$, and a density scale $\rho_H = c^3/(G\sqrt{\Lambda})$ from the speed of light c, the gravitational constant G, and the cosmological constant Λ leads to values that to a very good approximation correspond to the Hubble length c/H_0, the Hubble time $1/H_0$, and the critical density $\rho_{\text{crit}} = 3H_0^2/(8\pi G)$ if the curvature scale becomes infinitely large, or equivalently, the associated density parameter Ω_K approaches zero.

Comparing these scales to the Planck system of units with $l_P = \sqrt{G\hbar/c^3}$, $t_P = \sqrt{G\hbar/c^5}$, and $\rho_P = c^5/(G^2\hbar)$ constructed from the speed of light c, the gravitational constant G, and the Planck constant $\hbar$ shows large discrepancies, most notably the estimate $\rho_P = 10^{120}\rho_H$, which is usually stated as a motivation for dark energy models. If one associates the effects of the cosmological constant to the expectation value of the field theoretical vacuum, one obtains values that are much too large in comparison to ρ_H, but the problem seems to be constructed because it would apply to matter in exactly the same way: The matter density $\Omega_m\rho_{crit}$ is close to ρ_H instead of ρ_P. Personally, I am lost at the fact that with the inclusion of the cosmological constant Λ as a natural constant to general relativity (required by Lovelock, Cartan, and Vermeil), the definition of a natural system of units is not unique. Furthermore, the early universe is described by one choice of constants (namely, c, G, and $\hbar$) and the late, current universe by the other possible choice (c, G, and Λ). Adding the Boltzmann constant k_B to this argument, one can show that the number 10^{120} is the smallest non-trivial number that can be constructed from all fundamental constants of Nature.

In a certain sense, ΛCDM suffers from being a successful Standard Model, because it is difficult to find evidence for its failing, which is of course very motivating and fascinating. But in this respect, it is necessary to report all statistical tests irrespective of whether they provide evidence for or against ΛCDM, in the light of all available data. Moreover, perhaps one would need to be clearer about whether a certain model construction is part of the theoretical discussion or thought as a viable alternative. From what was discussed in Section 3, it might as well be that progress comes from conceptual work on gravity, in analogy to the construction of general relativity in the first place more than a century ago.

14.5 Statistics in Cosmology

As an astronomical discipline, observational cosmology restricts the observer to a purely passive role, because it is impossible to have an active influence on the dynamics of the cosmological expansion or the formation of cosmic structures. Because of this impossibility of carrying out experiments, one is forced to require that only questions can be answered for which experiments have been carried out by Nature. Concerning the expansion dynamics, this does not appear to be a large restriction because we can observe the behaviour of FLRW cosmologies over a wide range of values for the density parameters and for dominating fluids with different equations of state. Certain models – for instance, phantom dark energy models, where a very negative equation of state parameter leads to a infinite scale factor after a finite time – do not seem to be realised in Nature even though the gravitational dynamics of such a model would be very interesting to investigate. Similarly, the investigation of small values of curvature is difficult because the finite horizon size does not permit to access large enough scales.

The passive role of cosmological observers is a central issue in the observations of cosmic structures (Coles, 2003). There, one assumes that the random process of which the actual distribution of matter in our Universe is one realisation is ergodic, meaning that we can determine ensemble-averaged quantities that characterise the random process, by performing spatial averages over data taken from inside our horizon. Ergodicity cannot be experimentally verified, but there are strong theoretical indications that it is a good assumption. On theoretical grounds, a theorem by R. J. Adler stipulates that ergodicity is ensured in Gaussian random fields with continuous spectra. Ergodicity is an important assumption when investigating objects such as clusters of galaxies, because only sufficiently large volumes yield statistically unbiased samples because of a large enough set of initial conditions from which the objects have formed. On smaller scales, theoretical investigations indicate that relativistic effects in surveys related to the light propagation cause a violation of ergodicity on a small level.

While the statistical properties of the cosmic large-scale structure are very close to Gaussian in the early universe or today on large scales, nonlinear structure formation generates large deviations from Gaussianity, which are usually quantified in terms of n-point correlation functions or polyspectra of order n in the case of homogeneous random processes. It is unknown if there is an efficient description of non-Gaussianity introduced by nonlinear structure formation. Currently, one can predict all higher-order correlators in perturbation theory where they are sourced due to the nonlinearities in the continuity and Euler equations. Although many alternatives have been explored, such as extreme value statistics or Minkowski functionals, which might in principle provide an efficient quantification of non-Gaussianity, their relationship to the fundamental dynamical equations is less clear than in the case of correlation functions.

All estimates of statistical quantities suffer from cosmic variance; that is, they are limited in their precision by the law of large numbers. Any moment of a random variable can be estimated only with a finite variance that scales inversely proportional to the number of samples. In cosmology, this limits the detectability of correlation functions or spectra on large scales in a statistical way, even if the estimate itself is unbiased. In the measurement of angular spectra of the galaxy density, of the weak lensing signal, or of the temperature of the cosmic microwave background, there are only a finite number of modes available at a given angular scale, such that spectra can only be determined at an uncertainty inversely proportional to the angular scale. This limits cosmology in a fundamental way, in particular on the largest scales.

The cosmic large-scale structure is to a very good approximation described by classical fluid mechanics on an expanding background on scales smaller than the Hubble scale c/H_0 and in the linear regime of structure formation. The dynamical variables are the density and velocity field, if a fluid mechanical approach is adopted – which is conceptually not correct because of the collisionlessness of dark matter, which is required to have a very small, possibly vanishing cross-section for elastic collisions. The phase-space information contained in the density and velocity field are accessible at different degrees of precision into the directions parallel and perpendicular to the line of sight. An important result that should be mentioned in this context is the detection of baryon acoustic oscillations in the distribution of galaxies, which links the large-scale structure at a very early point in cosmic history to the present time and shows that the statistics of the cosmic matter distribution is conserved in linear structure formation.

Counting the cosmological parameters associated with the FLRW dynamics (the matter density Ω_m; the dark energy density Ω_w, which is replaced by Ω_Λ in the case of the equation of state being $w = -1$; the curvature Ω_K, which assumes very small values due to cosmic inflation; and the Hubble parameter h itself) and those describing structures – that is, fluctuations in the distribution of matter (the fluctuation amplitude σ_8 and the slope n_s of the spectrum) – results in six parameters, or seven if a time evolution of the dark energy equation of state is admitted. The spectral slope n_s and the amplitude f_{NL} of non-Gaussianities have to result consistently from the inflationary slow-roll parameters. As cosmological probes usually combine information from the homogeneous expansion dynamics with cosmic structures, measurements depend on a large set of at least seven physical parameters (Narlikar and Padmanabhan, 2001).

14.6 Inference in Cosmology

The process of inference – that is, the estimation of model parameters (or at least of bounds on model parameters) from cosmological data – differs substantially from other parts of physics: It is close to impossible to investigate certain aspects of the cosmological model independently from others. To the contrary, in elementary particle physics direct detection experiments measure, for example, mass, charge or spin of particles, and their couplings independently from other aspects of the Standard Model of particle physics. Only in indirect detection experiments for probing physics above the provided energy does one need to deal with simultaneous contributions from many physical processes and particles.

Any large-scale structure observable and its statistical properties depend in the most straightforward case on the density parameters of all cosmic fluids and their respective equations of state as well as on parameters that characterise the distribution of matter. Cosmological observations combine statistical properties of the fluctuations with their growth rates and the relation between observable redshift and distance. The investigation of models as complex as those in cosmology requires very strong signals; for instance, the spectrum of CMB temperature fluctuations or the weak lensing signals of galaxies constitute signals with a statistical significance of the order of $10^3\sigma$. There is a natural progression in cosmology from the observation of the homogeneous and isotropic expansion dynamics on large scales (e.g., with supernova measurements) to the statistics of linear fluctuations (in the cosmic microwave background) and finally to the evolved, nonlinear cosmic structure on small scales and at late times.

In addition, one needs to deal in most cases (with the possible exception of weak lensing) with a model that relates the observable to the underlying dark matter objects – for instance, biasing relation for the density of galaxies in relation to the dark matter densities or mass–temperature relations for the luminosity of clusters or their Sunyaev–Zel'dovich or x-ray signal. If the set of cosmological parameters along with effective models for the observables is to be constrained by data, large degeneracies are natural. Many algorithmic sampling techniques have been developed to treat likelihoods in high-dimensional parameter spaces. Sampling algorithms based on Monte Carlo Markov chains are optimised to deal with strong degeneracies and can sample at high efficiency from strongly non-Gaussian likelihoods with an implicit assumption about unimodality of the likelihood.

A further complication is hidden in the hierarchy in sensitivity toward cosmological parameters that is inherent to cosmology: Certain parameters are more difficult to measure than others, which means that a measurement needs to offer exquisite precision on easy parameters if it is to provide interesting statements about parameters that are difficult to measure. Examples might include weak lensing, which is strongly dependent on the matter density Ω_m and the fluctuation amplitude σ_8, and weakly dependent on the dark energy equation of state parameters w_0 or its time evolution w_a, such that one needs to achieve relative accuracies of 10^{-4} on Ω_m and σ_8 to answer questions in relation to dark energy – for instance, the equation of state parameters w_0 and w_a. A similar situation is found in the interpretation of data of the cosmic microwave background, where one central parameter is the optical depth τ, which needs to be constrained for fundamental parameters to be accessible.

It is breathtaking to think that the future generation of surveys will constrain parameters of a ΛCDM-type cosmology or of a basic dark energy cosmology to a level of 10^{-4}, in particular because Newton's gravitational constant is known at similar levels of precision from laboratory experiments. At this level, the difficulty will lie in achieving a detailed understanding of the relationship between the observable quantity and the fundamental cosmological model. On small scales, this involves an understanding of astrophysical processes governing the properties of observed objects and their relation to the ambient large-scale structure as well as nonlinear corrections to observables and a treatment of their covariances. For instance, gravitational lensing, recognised as a very clean probe with a straightforward functioning principle that is entirely determined by relativity, requires a precise prediction of second-order corrections to lensing, a modelling of intrinsic alignments of galaxies, and an accurate description of the statistical distribution of estimates of the spectrum. Quite generally one can expect that cosmology in the next generation of experiments will not be limited by statistical factors, but rather by systematical errors.

If one is in the situation of deciding between two viable theoretical models, one would like to set up a decisive experiment that is designed in a way to favour one of the models strongly over the other (Trotta, 2017). It is difficult to pursue this idea in cosmology, because one can at most (1) use an observational technique in which the difference between two models ist most pronounced, or (2) optimise an experiment in such a way that the two models in question show a large difference in their generated signal, or (3) use degeneracy-breaking combinations of probes.

These optimisations can only show quantitative differences due to the degeneracies involved, such that the likelihoods of models differ not very strongly in parameter space and typically show tensions at the level below 3σ, making a decision between models difficult. As a consequence, many methods for model comparison have been developed, which quantify the trade-off between models of different complexity in their ability to fit data and in their simplicity. In this sense, methods of Bayesian model comparison are a quantification of Occam's razor, preferring simpler models over more complex ones if both models fit data similarly well, but deciding in favour of more complex models only if the data require a higher complexity. Specifically, Bayesian evidence measures the probability that a model is able to explain the data integrated over the model's parameter space weighted with distribution, which quantifies the knowledge about the parameters prior to the experiment. Two competing models are compared by the ratio of their Bayesian evidences, which in turn can be quantified by, for example, the Jeffreys scale for a quantitative interpretation of the Bayesian evidence ratio, although the scale for the degree of confidence in a model is, like in all applications of statistical testing, arbitrary. Of course, one must realise that complexity of models is a human concept and we commonly assume that the laws of Nature are conceptually simple (but technically complicated at times!).

Although it might be a weird thought at first, the future generation of experiments will survey such large volumes that they harvest a significant part of the cosmologically available information. Moreover, there is a natural limit in the complexity of cosmological models that can at best be investigated. Therefore, it is imperative to develop new techniques and new probes to open up new windows, such as to structures in the high-redshift universe in 21-cm observations, or the observation of recombination lines from the epoch of atom formation. Alternatively, there is the fascinating possibility of real-time cosmology, where one directly observes changes in the recession velocity of distant quasars in cosmologies with an effective equation of state unequal to $w = -\frac{1}{3}$ or, at least in principle, small changes in the fluctuation pattern of the cosmic microwave background on the time scale of 100 years. An alternative route to accessing scales much smaller than typical galaxy separations or the scales responsible for the substructure in haloes is through the search for primordial black holes, whose abundance is determined by the inflationary model.

14.7 Structure Formation

One of the goals of cosmology is a theory for the formation and evolution of cosmic structures, with the ultimate aim of providing a quantitative theory of galaxy formation, including a physical picture of galaxy properties, their evolution, and their relation to the cosmic large-scale structure (Frenk and White, 2012). Due to nonlinearities in the underlying system of equations, structure formation is a difficult and not yet fully understood problem even for dark matter alone, which obeys comparatively straightforward physical laws: For predicting the scale dependence and time evolution of statistical quantifiers, one carries numerical simulations, where algorithmic progress has allowed to run simulations with $8,192^3$ particles (Alimi et al., 2012), or uses perturbative methods (Bernardeau et al., 2002), where recent developments have made fully analytic treatments possible (Bartelmann et al., 2014). The inclusion of baryonic matter into structure formation is realistically restricted to numerical simulations, where, depending on the physical processes considered, one deals with a multi-scale and multi-physics problem whose results are still influenced by the choice of algorithms for solving the coupled system of differential equations, the discretisation scheme, and the details of the implementation of physics on small scales (Springel, 2010; Genel et al., 2014).

In the process of increasing the data volume, it is obvious that much of the signal collected by future large-scale structure surveys is generated by nonlinearly evolving structures. For instance, the weak lensing signal measured by the future Euclid mission will amount to more than $10^3\sigma$ of statistical significance, of which about three quarters is generated by nonlinear structures. Clearly, this requires a detailed understanding of phenomena related to nonlinear structure formation, in particular the interdependence of statistical quantifiers. Structure formation in the linear regime (i.e., for small perturbations) is easily shown to preserve all statistical properties of the initial conditions, which is the reason why it is possible to investigate inflationary theories with observations of the cosmic microwave background or of the cosmic structure on large scales. By contrast, nonlinear structure formation generates an entire hierarchy of higher-order cumulants in a coupled system. Perturbative approaches are in principle possible, but technically very difficult. Indeed, it is only lately that progress in effective field theory and statistical field theory has provided a theory of statistical evolution of the large-scale structure that agrees with numerical simulations on small scales.

The internal properties of haloes, especially their rotation curves and their virial equilibria, provide support for dark matter to be gravitationally dominating. From the core structure of haloes and from the abundance of substructure, one places constraints on the collisionlessness and its thermal motion, respectively (Colafrancesco, 2010). Additionally, structure growth in the early universe clearly requires dark matter to be non-baryonic. Haloes have been found in numerical simulations to have a certain profile shape, which is universal, scales weakly with halo mass, and naturally accommodates the constant rotation velocity at large distances from the halo centre. In addition, observations from weak gravitational lensing show that the profile shapes of haloes correspond well to those found in numerical simulations, although there is not yet a complete physical understanding of their stability and of the origin of their universal shape from first principles. Although very strong arguments for the existence of dark matter can be cited from cosmology and from astrophysics, the relation to elementary particle physics is still unclear. Many particle candidates have been proposed, from very light particles like axions with a mass of $\simeq 10^{-22}$ eV to WIMPs close to the supersymmetric scale at a few TeV (Bergström, 2009; Fornengo, 2016; Queiroz, 2016).

Cosmic inflation is a construction that provides a natural framework for the generation of cosmic structures, which in the course of cosmic history grow by gravitational instability. Despite all the successes of inflation, it must be emphasised that this construction relies on non-gravitational physics to explain the symmetries of the FLRW cosmology as an entirely gravitational system, and that its predictions follow from the construction of the particular inflationary model and not generically from fundamental physics. Apart from the scale dependence of the fluctuation amplitude, cosmic inflation generates small non-Gaussianities that are dwarfed by nonlinear evolution on small scales and are measurable on very large, quasi-linearly evolving scales at high redshift. These small amounts of non-Gaussianity reflect derivatives of the inflationary potential and should be consistent with the slow-roll parameters that determine the scale dependence of the inflationary fluctuations (Planck Collaboration et al., 2016c). It seems to be difficult, but not fundamentally excluded, to construct models with large non-Gaussianity. An independent measurement of the two slow-roll parameters and the corresponding non-Gaussianity parameters has yet to be carried out.

It is difficult to overstate the importance of computer simulations of cosmic structure formation for modern cosmology, due to the nonlinearity of the structure formation equations that contain many different active physical processes and due to the difficulty of separating physical scales. The sophistication of numerical simulations has led to a framework for dark matter–driven structure formation from small to large scales and the thermal evolution of the baryonic component. This framework has been successfully applied to galaxy formation and star formation on small scales; it is able to predict many properties of galaxies with the dependence on time, scaling behaviour, morphology, and their statistical distribution.

The development of analytical methods was not quite commensurate with this evolution, because analytical methods are mostly constrained to the dark matter component and focus on predicting the evolution of non-Gaussianities in a perturbative way, and a detailed theoretical understanding of many numerical results is still missing – for instance, concerning the stability of dark matter haloes. The worry that in cosmology one is accepting the need to be strongly dependent on numerical simulations is justified: This concerns both high-precision predictions of linear perturbations, in particular with more involved particle models, as well as predictions for nonlinear structures. But the situation is not very different compared to elementary particle physics, where the search for particles outside the Standard Model or at high energies involves, field-theoretical computations with similar complexity. Finally, there is always the question of interpreting simulations, isolating individual structure formation processes, and quantifying degeneracies in model parameter choices.

14.8 Conclusion

Modern cosmology relies on three building blocks: general relativity as a theory of gravity, (fluid) mechanics as a theory of motions in the distribution of matter, and statistics for their description. It offers views on gravity as one of the fundamental forces, both for the dynamics of the Universe on large scales as well as the driving force for the formation of cosmic structures. The Universe is a system for testing the gravitational interaction on large scales in a fully relativistic, yet reasonably symmetric system, with new gravitational effects in relation to the cosmological constant Λ. The dark matter distribution in the Universe does not form a continuum, and for that reason the usage of fluid mechanics is conceptually inadequate, in contrast to the baryonic component.

General relativity as a theory of gravity follows from a number of fundamental physical principles and is unique as a metric theory under certain assumptions. It includes naturally the cosmological constant Λ as a term in the field equation; any measured deviation of the behaviour of gravity would question these fundamental principles. A consequence of assuming general relativity as the theory of gravity is the need for the existence of dark matter, which is substantiated by its role in structure formation and by its influence on the internal properties of dark matter structure, but whose relationship to particle physics is still unclear.

Cosmology is currently evolving in many interesting directions. Large-scale surveys cover significant fractions of the observable cosmic volume, and it is imperative to develop new probes, such as observations of the large-scale structure with the hyperfine transition of neutral hydrogen, which makes the high-redshift universe accessible (Bull et al., 2015; Kitching et al., 2015; Maartens et al., 2015) and shows structures in a close to linear stage of evolution. Another frontier comprises B-modes of the polarisation of the cosmic microwave background, which are directly related to the energy scale of inflation. As a completely different approach, changes in cosmological signals with time can be investigated. Many current cosmological probes are limited by astrophysical processes on small scales that are difficult to model and require simulations with a correct implementation of all relevant physical processes. If these limitations cannot be controlled, cosmology will be limited by systematic, rather than statistical, errors. By contrast, Nature provides very clean standard signals in the form of the gravitational wave signals of coalescing black hole binaries, which have the potential of surpassing supernovae in precision. Finally, one could dream of independent measurements of the slow-roll parameters and of finding consistency between the predictions of inflation in the shape of the spectrum, the amounts of non-Gaussianity, and the level of primordial gravitational waves.

At the stage of interpreting cosmological data, it should be kept in mind that according to Bayesian evidence, simpler models should be preferred over more complex ones, and one should accept more complex models only if they provide substantially better fits to data. While conceptually clear, it might be very difficult to apply this, as one has to deal in the next generation of experiments with very large data volumes and models that not only include physical parameters but also need to be sufficiently flexible to cope for astrophysical peculiarities of the probes employed.

On the theoretical side, it would be important to have an effective (and computationally efficient) theory of nonlinear structure formation with a prediction and an efficient description of non-Gaussianity, and possibly incorporating relativity into the prediction of structure growth. These analytical methods compete with very evolved and sophisticated numerical simulations for structure formation, which move in their aim away from fundamental cosmology to a quantitative model for galaxy formation and evolution.

This chapter has sought to summarise some of the fundamental questions currently being investigated in cosmology and illustrate their relationship to fundamental principles that are realised in the laws of Nature. Many aspects of cosmology rely on arguments from simplicity and can never be rigorously tested. Keeping in mind that physical cosmology is only 99 years old, counting from the discovery of redshifts of galaxies by Slipher soon after the formulation of general relativity, it has come very far in linking physical processes at the Hubble scale to those close to the Planck scale, and I am confident that many surprises wait for their discovery.

Acknowledgements

I would like to thank Matthias Bartelmann, Richard Dawid, Yves Gaspar, Frank Könnig, Karl-Heinz Lotze, Sven Meyer, Robert Reischke, and Norman Sieroka for very constructive and valuable comments on the draft.

References

Adamek, J., Daverio, D., Durrer, R., & Kunz, M. (2016). General relativity and cosmic structure formation. *Nature physics, 12*(4), 346.

Aldersley, S. J. (1977). Dimensional analysis in relativistic gravitational theories. *Physical Review D, 15*(2), 370.

Alimi, J. M., Bouillot, V., Rasera, Y., Reverdy, V., Corasaniti, P. S., Balmes, I., ... & Richet, J. N. (2012). DEUS Full Observable Λ CDM Universe Simulation: the numerical challenge. *arXiv preprint arXiv:1206.2838.*

Balakrishnan, V. (2004). Wave propagation: Odd is better, but three is best. *Resonance, 9*(6), 30–38.

Bartelmann, M., Fabis, F., Berg, D., Kozlikin, E., Lilow, R., & Viermann, C. (2014). Non-equilibrium statistical field theory for classical particles: Non-linear structure evolution with first-order interaction. *arXiv preprint arXiv:1411.1502.*

Baumann, D. (2009). Cosmological inflation: Theory and observations. *Advanced Science Letters, 2*(2), 105–120.

Bergström, L. (2009). Dark matter candidates. *New Journal of Physics, 11*(10), 105006.

Bernardeau, F., Colombi, S., Gaztanaga, E., & Scoccimarro, R. (2002). Large-scale structure of the Universe and cosmological perturbation theory. *Physics reports, 367*(1–3), 1–248.

Bolejko, K., & Korzyński, M. (2017). Inhomogeneous cosmology and backreaction: Current status and future prospects. *International Journal of Modern Physics D, 26*(06), 1730011.

Brans, C., & Dicke, R. H. (1961). Mach's principle and a relativistic theory of gravitation. *Physical review, 124*(3), 925.

Buchert, T., Carfora, M., Ellis, G. F., Kolb, E. W., MacCallum, M. A., Ostrowski, J. J., ... & Wiltshire, D. L. (2015). Is there proof that backreaction of inhomogeneities is irrelevant in cosmology?. *Classical and quantum gravity, 32*(21), 215021.

Bull, P., Camera, S., Raccanelli, A., Blake, C., Ferreira, P. G., Santos, M. G., & Schwarz, D. J. (2015). Measuring baryon acoustic oscillations with future SKA surveys. *arXiv preprint arXiv:1501.04088.*

Carroll, S. M. (2001). The cosmological constant. *Living reviews in relativity, 4*(1), 1.

Carroll, S. M., Press, W. H., & Turner, E. L. (1992). The cosmological constant. *Annual review of astronomy and astrophysics, 30*(1), 499–542.

Cartan, E. (1922). Sur les équations de la gravitation d'Einstein. *Journal de Mathématiques pures et appliquées, 1*, 141–204.

Clarkson, C., & Maartens, R. (2010). Inhomogeneity and the foundations of concordance cosmology. *Classical and Quantum Gravity, 27*(12), 124008.

Clifton, T., Ferreira, P. G., Padilla, A., & Skordis, C. (2012). Modified gravity and cosmology. *Physics reports, 513*(1–3), 1–189.

Colafrancesco, S. (2010, January). Dark matter in modern cosmology. In *AIP Conference Proceedings* (Vol. 1206, No. 1, pp. 5–26). AIP.

Coles, P. (2003). Statistical cosmology in retrospect. *Astronomy & Geophysics, 44*(3), 3–16.

de Rham, C. (2014). Massive gravity. *Living reviews in relativity, 17*(1), 7.

Ehlers, J., Geren, P., & Sachs, R. K. (1968). Isotropic Solutions of the Einstein–Liouville Equations. *Journal of Mathematical Physics, 9*(9), 1344–1349.

Einstein, A. (1915). *Die Feldgleichungun der Gravitation.* Sitzungsberichte der Preussischen Akademie der Wissenschaften zu Berlin, p844–847.

Einstein, A. (1916). Die grundlage der allgemeinen relativitätstheorie. *Annalen der Physik, 354*(7), 769–822.

Fornengo, N. (2016). Dark matter overview. *arXiv preprint arXiv:1701.00119.*

Freedman, W. L., & Madore, B. F. (2010). The hubble constant. *Annual Review of Astronomy and Astrophysics, 48*, 673–710.

Freedman, W. L., Madore, B. F., Gibson, B. K., Ferrarese, L., Kelson, D. D., Sakai, S., ... & Huchra, J. P. (2001). Final results from the Hubble Space Telescope key project to measure the Hubble constant. *The Astrophysical Journal, 553*(1), 47.

Frenk, C. S., & White, S. D. (2012). Dark matter and cosmic structure. *Annalen der Physik, 524*(9–10), 507–534.

Friedman, A. (1924). Uber die Krummung des Raumes. *Zeitschrift fur Physik, 21*, 326.

Frieman, J. A., Turner, M. S., and Huterer, D., 2008, Dark Energy and the accelerating universe, *Annual Review of Astronomy and Astrophysics, 46*, 385–432.

Genel, S., et al., 2014, Introducing the Illustris Project: the evolution of galaxy populations across cosmic time, *Monthly Notices of the Royal Astronomical Society, 445*, 175–200.

Green, S. R., and Wald, R. M., 2013, Examples of backreaction of small scale inhomogeneities in cosmology, *Physical Review D, 87*(12) 124037.

Horndeski, G. W., 1974, Second-order scalar-tensor field equations in a four-dimensional space, *International Journal of Theoretical Physics, 10*, 363–384.

Huterer, D., and Turner, M. S., 2001, Probing the dark energy: methods and strategies, *Physical Review D, 64*, 123527.

Joyce, A., Lombriser, L., and Schmidt, F., 2016, Dark energy versus modified gravity, *Annual Review of Nuclear and Particle Science, 66*, 95.

Kitching, T. D. et al. 2015, Euclid and SKA Synergies, arXiv:1501.03978.

Koyama, K., 2016, Cosmological tests of modified gravity, *Reports on Progress in Physics, 79*, 046902.

Lemaître, G., 1931, A homogeneous universe of constant mass and increasing radius accounting for the radial velocity of extra-galactic nebulae. *Monthly Notices of the Royal Astronomical Society, 91*, 483.

Lovelock, D., 1969, The uniqueness of the Einstein field equations in a four-dimensional space, *Archive for Rational Mechanics and Analysis, 33*, 54.

Lovelock, D., 1971, The Einstein tensor and its generalizations, *Journal of Mathematical Physics, 12*, 498.

Maartens, R., Abdalla, F. B., Jarvis, M., and Santos M. G., 2015, Overview of cosmology with the SKA PoS AASKA14 (2015) 016.

Martin, J., Ringeval, C., and Vennin V., 2013, Encyclopaedia Inflationaris, Phys.Dark Univ. 5–6.(2014) 75–235.

Mather, J. C., et al., 1990, A Preliminary measurement of the cosmic microwave background spectrum by the Cosmic Background Explorer (COBE) satellite, *Astrophys.J. 354*, L37–L40.

Mather, J. C., et al., 1991, *Advances in Space Research*, http://adsabs.harvard.edu/abs/1991AdSpR..11..181M *11, 181*.

Mortonson, M. J., Weinberg, D. H., and White, M., 2013, Dark energy: A short review, arXiv/1401.0046.

Narlikar, J., and Padmanabhan, T., 2001, Standard cosmology and alternatives: a critical appraisal, ARAA, 39, 211.

Navarro, A., and Navarro, J., 2011, Lovelock's theorem revisited, *Journal of Geometry and Physics, 61*, 1950.

Navarro, J., and Sancho, J. B., 2008, On the naturalness of Einstein's equation, *Journal of Geometry and Physics, 58*, 1007.

Navarro, J., Sancho, J. B., 2009, A characterization of the electromagnetic stress-energy tensor, *Proceedings of the Spanish Relativity Meeting, 1122*, 360–363.

Palatini, A., 1919, Deduzione invariantiva delle equazioni gravitazionali dal principio di Hamilton, *Rend. Circ. Matem. Palermo, 43*, 203.

Perlmutter, S., et al., 1998, Measurements of Omega and Lambda from 42 high redshift supernovae, *Astrophysical Journal, 517*, 565–586.

Planck Collaboration, 2016a, Planck 2015 results. XIII. Cosmological parameters, *Astronomy & Astrophysics, 594*, A13.

Planck Collaboration, 2016b, Planck intermediate results. XXXIV. The magnetic field structure in the Rosette Nebula, *Astronomy & Astrophysics, 594*, A16.

Planck Collaboration, 2016c, Planck intermediate results. XLVII. Planck constraints on reionization history, *Astronomy & Astrophysics, 594*, A17.

Queiroz, F. S., 2016, Dark matter overview: collider, direct and indirect detection searches, arXiv/1605.08788.

Riess, A. G., et al., 1998, Observational evidence from supernovae for an accelerating universe and a cosmological constant, *Astronomical Journal, 116*, 1009–1038.

Robertson, H. P., 1935, Kinematics and world-structure, *Astrophysical Journal, 82*, 284–301.

Silk, J., 2016, Challenges in cosmology from the big bang to dark energy, dark matter and galaxy formation, arXiv/1611.09846.

Smoot, G. F., et al., 1991, First results of the COBE satellite measurement of the anisotropy of the cosmic microwave background radiation, *Advances in Space Research*, http://adsabs.harvard.edu/abs/1991AdSpR..11..193S 11, 193.

Springel. V., 2010, E pur si muove: Galilean-invariant cosmological hydrodynamical simulations on a moving mesh *Monthly Notices of the Royal Astronomical Society*, 401, 791.Stoeger, W. R., Maartens, R., and Ellis, G. F. R., 1995, Proving almost homogeneity of the universe: An Almost Ehlers-Geren-Sachs theorem *Astrophysical Journal 443*, 1.

Stoeger, W. R., Maartens, R., and Ellis, G. F. R., 1995, *ApJ*, 443, 1.

Trotta, R., 2017, Bayesian Methods in Cosmology, arXiv/1701.01467.

Tsujikawa, S., 2003, Introductory review of cosmic inflation, arXiv, hep-ph/0304257.

Vermeil, H., 1917, Notiz über das mittlere Krümmungsmaß einer n-fach ausgedehnten Riemann'schen Mannigfaltigkeit, *Nachrichten von der Gesellschaft der Wissenschaften zu Göttingen, 1917*, 334.

Weinberg, D. H., Mortonson, M. J., Eisenstein, D. J., Hirata, C., Riess, A. G., and Rozo, E., 2013, Observational Probes of Cosmic Acceleration, *Physics Reports, 530*, 87.

Wiltshire, D. L., 2007, Cosmic clocks, cosmic variance and cosmic averages, *New Journal of Physics, 9*, 377.

Woodard, R. P., 2015, Ostrogradsky's theorem on Hamiltonian instability, arXiv/1506.02210.

15

Theory Confirmation and Multiverses

GEORGE ELLIS

The core of the success of science is its base, since the times of Galileo and Newton, in the observational or experimental testing of proposed scientific models or theories. Those that work are accepted. and those that do not, in the sense that they fail to successfully characterise a physical phenomenon as probed by observation or experimental testing, are rejected. This is what gives science its rock-solid basis in terms of describing the way matter actually behaves. Thus Galileo measured the rates of descent of rolling balls down inclined planes and observationally determined their laws of motion; on this experimental basis, he asserted, 'The book of Nature is written in the language of mathematics.' Newton devoted sections of *The Principia* to a discussion of the experimental method as the basis of science (Levy, 2009, pp. 104–5).

A succession of proposals as to how to adequately formalise this method have been made by Popper, Kuhn, Lakatos, and others, taking into account the fact that a test of any specific model or theory necessarily involves assuming the validity of a range of subsidiary hypotheses and concepts. This means it may be difficult to determine whether any contradiction with evidence is a result of a problem with the proposed model, or with the surrounding belt of auxiliary hypotheses.[1] In each case, however, it still involves testing the theory against observational data, at the core.

When observational and experimental tests of a theory are not possible, as is the case for most multiverse proposals (Carr, 2005), a question arises as to whether the theory is a genuine scientific theory or instead a scientifically based philosophical proposal that will never attain the status of established science.[2] This chapter discusses this issue, considering the major arguments put forward for existence of

---

[1] See section 15.2.2 of https://plato.stanford.edu/entr es/lakatos/.

[2] All scientific theories are, in principle, provisional – but some (e.g., Newtonian gravitational theory and special relativity) are so well established within their domain of validity that they can be taken to be correct descriptions of the world as it is, within the confines of their domain of applicability.

multiverses and their relation to possible observational or experimental tests. While some tests are, in principle, possible for some specific multiverse proposals, many other multiverse models are not testable in any realistic experimental context in the foreseeable future (say, the next 50,000 years). This casts a shadow on their proposed scientific status, particularly because some multiverse proposals predict anything whatever can occur, so they cannot be disproved by any observation of any kind.

Attempts to show specific observations are probable in this context are necessarily based in the validity of specific probability measures for universe models in some envisioned ensemble. The measures themselves are untestable, so they can simply be chosen to give whatever outcome one desires, hence justifying them post hoc.

The chapter concludes by considering if this situation is changed by introducing Richard Dawid's proposals for non-empirical theory assessment (Dawid, 2013), and in particular through the use of Bayesian arguments that are a core feature of recent approaches to scientific confirmation.

15.1 Testability in Science and Cosmology

The rise of the physical sciences was due to the application of the experimental method to physical phenomena:

1. On the basis of current knowledge, generalised as seems appropriate to include the current application, propose a physical or mathematical theory or model for a natural phenomenon,
2. Test that theory or model by experiment or observation to see how well it corresponds to physical reality. In the case of the geographic or historical sciences, only the latter will be possible.
3. Modify or abandon theories or models that fail the test.
4. For those that succeed, try alternative options, to check how unique the proposed theory or model is.

The difference between a theory and a model is that a theory applies to a wide range of phenomena (its domain of applicability), whereas a model describes some specific instance. A key feature is that[3] 'A theory is a system of explanations that ties together a whole bunch of facts. It not only explains those facts, but predicts what you ought to find from other observations and experiments'.

This method has been enormously successful. Nevertheless, in the face of problems in testing some models and theories at the boundaries of present-

---

[3] Kenneth R. Miller, www.nytimes.com/2016/04/09/science/in-science-its-never-just-a-theory.html?_ r=0.

day physics, some proponents of string theory on the one hand, and of the cosmological multiverses on the other hand, have proposed that the experimental criteria for what constitutes a scientific theory should be weakened (Susskind, as quoted by Kragh, 2013; Carroll, 2014), leading to the idea of 'non-empirical theory confirmation' (Dawid, 2013). This approach should be treated with great caution.

15.1.1 The Rise of the Scientific Method

Galileo developed the idea of a systematic quantitative experimental study leading to a general theory through his observations of a ball rolling down an inclined plane. He was able to show that this occurred with uniform acceleration. Thus, when starting from rest, the distance travelled s is given by

$$s = (g/2)t^2 (h/l), \tag{15.1}$$

where t is the time of travel, h is the height of the inclined plane, l is the length of its base, and g is a constant that is the same for all inclined planes. This becomes a model for motion on a specific inclined plane when the relevant values of h and l for that plane are entered into formula (15.1). When the model was tested by measurements on a variety of different inclined planes, it passed that test. Thus, this model has been experimentally demonstrated to be the way things work.

This model became a special case of Newton's laws of motion combined with his theory of gravity, which could also be used to explain and predict the motion of the bodies in the solar system to high accuracy. Thus, these laws were confirmed by thousands of astronomical observations. Furthermore, Newton carried out many experiments on light, in particular his *experimentum crucis*, using a prism to demonstrate that white light is composed of many different colours of light combined together; different colours of light are bent differently by a prism. Once they have been separated, they can be recombined to again give white light.

15.1.2 The Success of the Scientific Method

Once established, this method led to the development of all of the well-tested branches of present-day physics, chemistry, and fundamental aspects of biology:

- Dynamics and thermodynamics
- Optics
- Electromagnetism
- Gravitation

- Quantum theory and particle physics
- Astronomy and astrophysics
- Physical chemistry
- Evolutionary theory
- Molecular biology, genetics, and epigenetics
- Neuroscience

Each is supported by a vast amount of data. Of course, there is always variability in physical systems, so a key part of the development of practical experimental tests has been the development of statistical tests characterising how well established a result is.

15.1.3 Cosmology

Cosmology has changed from a data-poor subject in the last century to an extremely data-rich subject at present, with a large variety of telescopes probing distant parts of the universe at all frequencies from radio to gamma rays. This has led to a well-established standard model of cosmology (Dodelson, 2003; Mukhanov, 2005; Durrer, 2008; Ellis et al., 2012; Peter & Uzan, 2013), in which models of physical conditions in the early universe. together with current theories of nuclear and atomic physics, thermodynamics, astrophysics, and gravitation. lead to cosmological models in which just a few free parameters can be chosen to reproduce with great accuracy a whole range of observations – particularly those of cosmic background radiation anisotropies and discrete source surveys on the one hand, and of primordial element abundances on the other hand.

There are fundamental limits to testing these models (Ellis, 1975, 1980) because of the existence of visual horizons on the one hand (we can't see further than the matter we observe on the last scattering surface, because the universe was opaque before then) and physics horizons on the other hand (we can't test in laboratories or colliders the physics relevant to the very early universe).

Nevertheless, these models can be regarded as well tested in the observable domain as well as close to our past world line back to the time of nucleosynthesis, but less so at early times before nucleosynthesis, and even less so beyond the visual horizon. Notably, we can only really test *models* of cosmology, rather than *theories* of cosmology (as categorised earlier). The key problem here is that the difference between a theory and a model is no longer clear in the case of cosmology because of the uniqueness of the universe. If we propose theories for the universe, we cannot test them, because we have no observational or experimental access to any other universes that might exist; all we can test is how our model fits the one universe domain to which we have observational access. However, we can examine that specific domain in a large variety of ways.

15.2 Multiverses

15.2.1 The Idea of a Multiverse

Proposals for an ensemble of universes or of universe domains – a multiverse – have received increasing attention in cosmology in recent decades (Carr, 2005). They can be suggested to exist in separate places (Andrei Linde, Alex Vilenkin, and Alan Guth), at separate times (Lee Smolin's [1997] Darwinian proposal, and a variety of cyclic universes including Penrose's cyclic conformal cosmology), as the Everett quantum multi-universe: other branches of the quantum wavefunction (David Deutsch [1997]), through the cosmic landscape of string theory, embedded in a chaotic cosmology (Leonard Susskind), or in totally disjoint form (David Lewis, Dennis Sciama, and Max Tegmark). Typical quotes are,

From *Our Cosmic Habitat,* by Martin Rees:

'Rees explores the notion that our universe is just a part of a vast "multiverse," or ensemble of universes, in which most of the other universes are lifeless.

What we call the laws of nature would then be no more than local bylaws, imposed in the aftermath of our own Big Bang. In this scenario, our cosmic habitat would be a special, possibly unique universe where the prevailing laws of physics allowed life to emerge'.

From *The Cosmic Landscape: String Theory and the Illusion of Intelligent Design*, by Leonard Susskind (2005): 'Susskind concludes that questions such as "why is a certain constant of nature one number rather than another?" may well be answered by "somewhere in the megaverse the constant equals this number: somewhere else it is that number. We live in one tiny pocket where the value of the constant is consistent with our kind of life. That's it! That's all. There is no other answer to the question'.

The *Scientific American* May 2003 issue on Cosmology states on the cover regarding an article by Max Tegmark, 'Parallel Universes: Not just a staple of science fiction, other universes are a direct implication of cosmological observations'.

At the present time, such proposals are frequently the topic of popular science articles and various blogs.

15.2.2 Varieties of Multiverses

Clearly, a large variety of multiverse proposals have been introduced. Tegmark (2014) has four; Brian Greene in his book *The Hidden Reality: Parallel Universes and the Deep Laws of the Cosmos* (2011) discusses nine different types of multiverse proposals:

1. Invisible parts of our universe
2. Chaotic inflation
3. Brane worlds
4. Cyclic universes
5. The landscape of string theory
6. Branches of the quantum mechanics wave function
7. Holographic projections
8. Computer simulations
9. All that can exist must exist – 'the grandest of all multiverses'

Probably the most popular version is chatic inflation, in conjunction with the string theory landscape. It is claimed that an inflating universe domain can have bubbles form by various mechanisms – for example, Coleman–de Lucia tunneling to form a new domain, where the new bubble is a 'pocket universe' like ours. The number of pocket universes continually increases, and the resulting configuration depends on competition between the expansion rate and the nucleation rate (neither of which can be determined from tested fundamental physics). It is claimed this process repeats an infinite number of times, giving an overall fractal-like structure. If the string theory landscape is correct, one can in principle have some mechanism resulting in different physical properties, such as different values for fundamental constants, being realised in the different expanding universe domains. Such mechanisms have been hypothesized, but they are far from natural.

The introduction of so many multiverse proposals might at first seem to make their existence almost a certainty: One or other of them must be true. In reality, the opposite is the case because of the simple fact that these varied proposals can't all be true: Many of them conflict with others, so how do we choose between them? If they conflict with each other, why should any of them be true?

No one will deny case 1, because it is not remotely plausible that the universe ends at the visual horizon. Nevertheless, calling this a multiverse is somewhat stretching the concept (we don't have many Earths just because we can't see beyond the horizon; we have one Earth and a limited view of it). The further possibility is:

10. Perhaps none of the possibilities 2–9 above is true – there may be just one universe.

Indeed, the evidence available to us together with Occam's razor supports this view. Postulating innumerable invisible macroscopic entities (domains as large as the entire visible universe) in addition to what we can see seems to clearly violate his dictum 'do not multiply entities unnecessarily'.

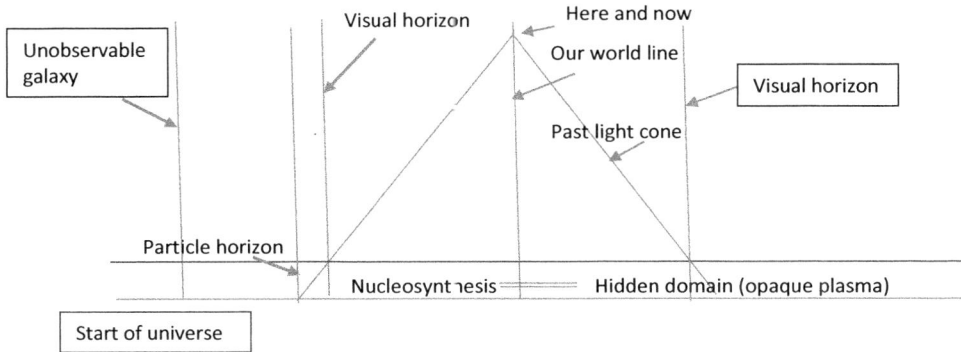

Figure 15.1 Conformal diagram of our past light cone and the visual horizon.

15.3 The Problem of Confirmation

The basic problem with the multiverse idea is that it is, with a few possible exceptions (see Sections 15.3.2–15.3.4), an observationally and experimentally untestable idea. In that case, is it actually a scientific proposal? Or is it a form of scientifically based philosophy? Its proponents claim the former – but they seem to be stretching the definition of science to do so.

15.3.1 Observational Horizons

The key observational point is that the domains considered in multiverse proposals are beyond the visual horizon and are therefore unobservable. To see this clearly, one can use a conformal diagram for a standard cosmological model (Figure 15.1), where use of comoving coordinates for the matter together with conformal time enables one to clearly see the causal structure (light cones are at $\pm 45°$).

Matter world lines are vertical lines with our past world line being at the centre. Our past light cone stretches back from the spacetime event "here and now" on that world line to the last scattering surface (LSS), where matter and radiation decoupled, and then to the start of the universe. All events between the LSS and the start of the universe are hidden to us because the mean free path for photons in this plasma was very small; thus, the LSS is the locus of the farthest matter we can see by any electromagnetic radiation, regardless of its wavelength. The cosmic microwave background (CMB) radiation was emitted at the LSS; the two-dimensional images of that radiation obtained from the COBE, WMAP, and Planck satellites[4] are actually images of the 2-sphere where our past light cone intersects the LSS, which is the farthest matter we can see. The world lines of that

---

[4] www.cosmos.esa.int/web/planck.

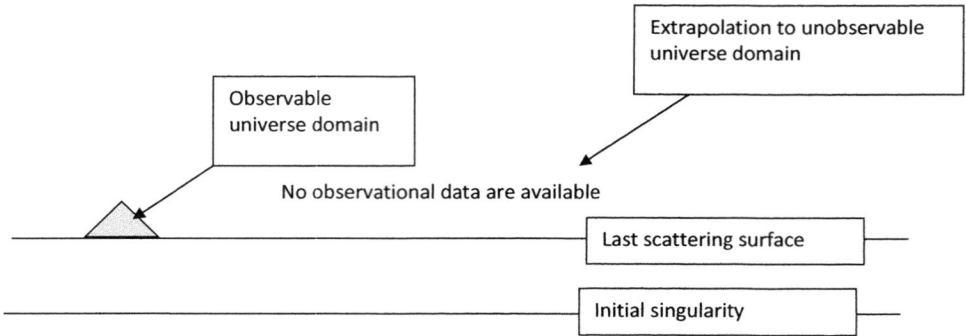

Figure 15.2 The observable region of the universe. Multiverse theories claim to tell us about the unobservable domains. No data whatever are available about these regions.

matter (at present approximately 42 billion light-years from us) form our visual horizon; we cannot obtain any information about any (co-moving) matter beyond that point, because light from that matter has not had time to reach us since the universe became transparent to radiation. Thus the galaxy indicated on the left of Figure 15.1 is invisible to us, though it can be influenced by data coming from distances up to the particle horizon. Similar horizons exist for neutrinos and gravitational radiation, but we are unlikely to be able to get useful cosmological information out to those distances beyond the visual horizon because of the weakness of the corresponding cosmologically relevant signals.[5] Finally, we do get information about conditions near our past world line in the very distant past through relics from that era; for example, element abundances give us data about nucleosynthesis processes in the very early universe (after inflation but long before the last scattering).

Now it is clear what the observational and causal limits in cosmology are (Figure 15.2). We have, in principle, observational access to everything within our past light cone but nothing outside it.

Multiverse theories claim to tell us what is true in the regions for which no observational data whatever are available! These theories are not modest in their aims. The assumption is that we can extrapolate from the observable domain to 100 Hubble radii, $10^{1,000}$ Hubble radii, or much much more (the word 'infinity' is liberally used). This is like an ant standing on the top of a sand dune in the Sahara and declaiming a sand dune–based theory of the entire world (Ellis, 1975). There might be some planets (e.g., Mars) where this would be essentially correct. For others (e.g., Earth), it is not.

---

[5] Note that event horizons have nothing to do with observational limits in cosmology (Ellis & Uzan, 2015).

Thus multiverse claims are essentially untestable. Perhaps the universe carries on the same way forever (as proposed by the cosmological principle; Bondi, 1960), or has similar domains with different properties (as in chaotic inflation: the multiverse case; Linde, 1982), or is an isolated domain of matter in an asymptotically flat spacetime (as is implied by the large literature in general relativity theory on asymptotically flat spacetime – for example, in black hole studies), or is surrounded by an eternal timelike singularity at a finite distance, or closes up on itself at a finite distance to give an S^3 or more complex spatial topology. Take your pick, and you cannot be proved wrong. Consequently, as far as observations are concerned, the multiverse may or may not exist

Notably, there are three cases where this gloomy prognosis as regards testing does not hold.

15.3.2 Caveat 1: A Proof Possibility?: Circles in the CMB Sky

In chaotic inflation models, different expanding universe bubbles might collide if the rate of nucleation is large relative to the rate of expansion at that time. This would be observable, in principle, by seeing specific kinds of circles in the CMB sky, resulting from another bubble intruding on ours – that is, from bubble collisions (Aguirre & Johnson, 2011; Kleban, 2011). It has even been suggested these circles might have already been seen, but this is very disputed; the WMAP data argue against such bubbles (Feeney et al., 2011), and the Planck 2015 data release has not changed this situation. Nevertheless, it would be quite convincing evidence for the multiverse proposal if we detected anomalous filled circles in CBR anisotropy observations where either the fine structure constant or the cosmological constant could be shown to be different within these circles. This would confirm the basic idea is correct, even if it was shown for only one other bubble.

A similar thing occurs in Penrose's conformal cyclic cosmology (Penrose, 2011), where a multiverse occurs in time, with different aeons being followed by different aeons, leaving traces of the previous aeon due to black hole explosions in circles in the CMB sky. Again, it has been claimed these phenomena have been seen, and again this claim has been treated with scepticism by CMB experts.

From the overall point being argued here, detecting such circles would be solid evidence toward one or other form of multiverse. But they are not required for a generic multiverse to exist, so not detecting them leaves the possibilities completely open either way. Some fraction of multiverse models could be testable in this way, but many could not.

15.3.3 Caveat 2: A Disproof Possibility?: Small Universes

Chaotic inflation version can be disproved if we observe that we are in a small universe. In these spatially closed universes, the scale of spatial closure is smaller than that of the visual horizon, so we have already seen a round the universe since the time of decoupling of matter and radiation (Ellis & Schreiber, 1986). There are a great many such possibilities for the spatial topologies (Lachieze-Rey & Luminet, 1995; Luminet et al., 2003). To check whether this is the case, we can search for identical circles in the CBR sky resulting from such identification (we will see the same event on the LSS in different directions in the sky; Cornish et al., 1998), as well as for CMB low anisotropy power at large angular scales, which is indeed what is observed. This is a very important test, as a positive result would observationally disprove the chaotic inflation variety of multiverse. However, not seeing them would not prove a multiverse exists. Their non-existence is a necessary but not sufficient condition. Such small universes have not been identified so far, but not all possibilities have been tested (analyzing the data to search for them is a very time-consuming procedure).

15.3.4 Caveat 3: A Disproof Possibility?: Positive Spatial Curvature

Some multiverse supporters (e.g., Freivogel et al., 2006) claim that the spatial curvature of each universe in a multiverse must be negative as a result of the bubble universe–generating mechanism envisaged (Coleman–de Luccia tunneling); hence, measuring positive spatial curvature would disprove the multiverse hypothesis. With other multiverses, this criterion does not hold. Thus, even if we do eventually determine the sign of the spatial curvature of the universe and find it to be positive (it is at present unknown), this evidence will not be decisive.

15.3.5 Conclusion in Regard to Direct Testing

Because of the existence of visual horizons, the universes or universe domains claimed to exist in a multiverse cannot be observed by any observational technique whatever, unless suitable bubble collisions take place, which happens in only a small fraction of multiverses. The proposal can be disproved if we demonstrate that we live in a small universe – but if that is not the case, the issue remains open. Some multiverse proposals demand space sections of negative spatial curvature in the universe bubbles, whereas others do not. The usual process of observational testing, as in the rest of cosmology, is therefore not possible for multiverse proposals: We cannot set out an observational program making definite predictions that will uniquely support the theory, or others that will contradict it. It is therefore not a scientific proposal in the usual sense.

15.4 Why Believe in Multiverses?

Given that we cannot test the proposals about multiverses in the usual experimental or observational way, what are the scientific arguments and evidence for existence of a multiverse? Why do some major scientists support the idea so strongly? The main motivation is twofold:

1. *Physical processes*: Multiverses are claimed to be the inevitable outcome of the physical originating process that generated the universe (e.g., as an outcome of the chaotic inflationary scenario)
2. *Anthropic arguments*: Multiverses are proposed as an explanation for why our universe appears to be fine- tuned for life and consciousness, particularly through setting the value of the cosmological constant in our observed universe domain.

These two bases are linked by the assumption that the laws of physics, or at least the constants of nature and/or cosmological parameters, are different in the different universe domains. Otherwise, they are essentially just part of the same universe rather than something deserving the name 'multiverse'.

15.4.1 Implied by Known Physics That Leads to Chaotic Inflation

Two kinds of physics are said to lead to multiverses: eternal new inflation, based on tunneling between false vacua, and eternal chaotic inflation, based on quantum fluctuations in a quadratic potential (Guth, 2007). The key physics in the eternal new inflation version of the multiverse (Coleman–de Luccia tunneling, the string theory landscape, plus the link between them; Susskind, 2005) is extrapolated from known and tested physics to new contexts. This extrapolation is unverified and indeed unverifiable; it may or may not be true.

The physics is hypothetical rather than tested! The situation is not

$$\textbf{Known physics} \rightarrow \textbf{Multiverse} \tag{15.2}$$

but rather

$$\textbf{Known physics} \rightarrow \textbf{Hypothetical physics} \rightarrow \textbf{Multiverse} \tag{15.3}$$
$$\textit{Major extrapolation}$$

This scenario is illustrated in Figure 15.3. Well-established theory is indicated by the large circle. One specific part of this theory (feature 1) is extrapolated to give extrapolation 1, while a second part (feature 2) is extrapolated to give extrapolation 2. Neither of these extrapolations is supported by experiment, even though they are developed from aspects of physics that are. Thus, the conclusion that the multiverse exists is the result of an untested set of extrapolations from known physics. These extrapolations are untested, and probably untestable by experiment:

Extrapolation 1   **Extrapolation 2**

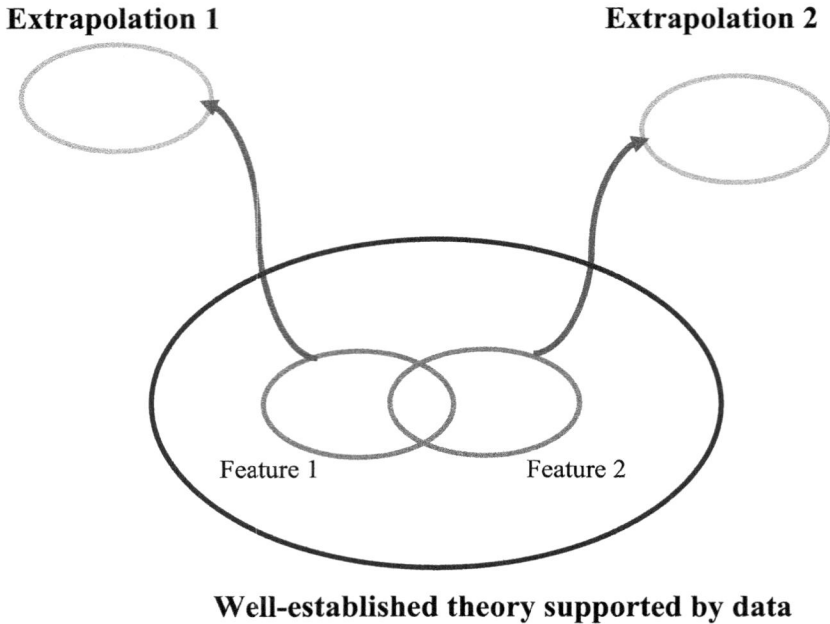

Well-established theory supported by data

Figure 15.3 Extrapolation from the known to the unknown. The large circle indicates theory well supported by data. Features 1 and 2 are subsets of this experimentally supported theory. Feature 1 is extrapolated to get extrapolation 1. Feature 2 is extrapolated to get extrapolation 2. Both have left direct connection to data far behind.

They may or may not be correct. Indeed, Banks (2012) suggests the string landscape concept should be treated with great caution. Furthermore, he states, 'If they exist as models of quantum gravity, they are defined by conformal field theories, and each is an independent quantum system, which makes no transitions to any of the others. This landscape has nothing to do with CDL tunneling or eternal inflation'. Whether one agrees or not, this analysis shows that the supposed underlying physics is certainly not well established, particularly the KKLT mechanism for constructing stabilised vacua with a positive cosmological constant (Kahcru et al., 2003). The part of the relevant physics that can be tested to some degree is the inflationary dynamics, which is supposed to lead to the existence of many bubble universes, as only some forms of inflation lead to this result. The Planck data on CMB anisotropies test the nature of structure formation in such scenarios: They disfavour the φ^2 potential that gives eternal chaotic inflation, instead supporting a Starobinski flat potential (Martin et al., 2014; Ade et al., 2015). However, neither potential corresponds to any field that has been shown to exist through laboratory or accelerator experiments. In that sense there is no solid link to established particle physics of even this part of the proposal. The potential proposed is a phenomenological fit chosen to 'save the phenomenon'. In any case, the multiverse proposal is not part of mainstream inflationary cosmology, as represented for example by the Planck team observations and data analysis. The possibility is not even mentioned by Martin et al. (2014) and Ade et al. (2015).

15.4.2 Fine-Tuning: The Anthropic Issue

A major philosophical question that concerns some cosmologists is, 'Why is the universe fine-tuned for life?' (Barrow & Tipler, 1988), both as regards the laws of physics and associated fundamental constants (Tegmark, 2003a) and as regards the boundary conditions of the universe (Rees, 2001). A relatively small change in many of these features will make any form of life whatever impossible, so why is our universe of such a nature as to support life? A multiverse with varied local physical properties is one possible scientific explanation: An infinite set of universe domains allows all possibilities to occur, so somewhere things will work out OK (the constants of nature (Uxan, 2010) will all simultaneously lie in the subspace that allows living systems to come into being and function). This solution to the problem has been supported by a number of prominent scientists, including Weinberg, Rees, Susskind, and Carroll (see Carr, 2005).

Note that an *actually existing* multiverse is necessary for this solution to work: This is essential for any such anthropic argument, and a theoretically existing ensemble will not suffice. Furthermore, a considerable number of constants must be fine-tuned for life to be possible (Rees, 2001; Tegmark, 2014; Lewis & Barnes, 2016). Those constants can be chosen and represented in many ways, and even their number is open to debate, but they include limits on parameters related to fundamental physics such as the fine structure constant in relation to the strong force coupling constant, the down minus up quark mass in relation to electron mass, the gravitational constant in relation to the fine structure constant, the number of spatial and temporal dimensions, and parameters related to the expanding universe such as a parameter Q describing the inhomogeneity of the universe at early times, the matter density in the universe, and the value of the cosmological constant.

15.4.3 The Cosmological Constant

A particularly significant application of a limit that has been made is explaining the small value of the cosmological constant by anthropic argument (Weinberg, 1987;

Rees, 2001; Susskind, 2005). The point here is that too large a value for Λ results in no structure and hence no life, while a negative value can lead to collapse of inhomogeneities to form black holes rather than stable stars and planets. Thus, anthropic considerations mean that the value of Λ we observe will be small (in fundamental units), justifying our adoption of an actual value extremely different from the 'natural' one predicted by physics, which is larger by between 70 and 120 orders of magnitude (Weinberg, 1989; Carroll, 2001). Hence the observed value of the cosmological constant is claimed to be confirmation of the existence of the multiverse. However, it is not a very tight limit, as noted by Weinberg (1987):

In recent cosmological models, there is an 'anthropic' upper bound on the cosmological constant Λ. It is argued here that in universes that do not recollapse, the only such bound on Λ is that it should not be so large as to prevent the formation of gravitationally bound states. It turns out that the bound is quite large. A cosmological constant that is within 1 or 2 orders of magnitude of its upper bound would help with the missing-mass and age problems, but may be ruled out by galaxy number counts. If so, we may conclude that anthropic considerations do not explain the smallness of the cosmological constant.

Efstathiou (1995) tightened the argument for an anthropic explanation of Λ by adding further assumptions, and Weinberg (2001) tightened the argument later to the form *the universe is no more special than need be to create life*. Davies (2007) gives a nice summary of the argument. At its heart, this is a deduction varying only Λ, and not other parameters at the same time; if you vary other parameters as well, as is required for life to exist (see the previous subsection), the argument is no longer clear. Aguirre (2005) gives a very careful analysis of how complex matters become if one looks at the full range of variations.

The justification for multiverses via the value of the cosmological constant makes clear the true multiverse project: making the extremely improbable appear probable, using a probabilistic argument based on assumed existence of an ensemble of universes and probability measure. It is a legitimate form of explanation. But a series of issues arise.

1. *Ensembles and statistics.* The statistical argument implied by saying the outcome is very probable applies only if an ensemble of actual universes (a multiverse) exists. It is simply inapplicable if there is no such ensemble; in that case, we have only one object we can observe, or do statistical calculations on. We can make many observations of that one object, and consider statistics of those many observations; even so, it is still only one object (one universe) that is being observed, and you can't do statistical tests as regards universes per se if only one such entity exists. Thus, this argument assumes the outcome it wishes to prove. It is not a proof of existence of a multiverse, but a consistency check on the idea.

2. *The logic of the argument.* In fact, no value L of the cosmological constant Λ can *prove* a multiverse M either exists or does not exist. This is elementary logic:

   - As regards proof:

   $$\text{If } \{M \rightarrow L\}, \text{it does not follow that } \{L \rightarrow M\} \qquad (15.4)$$

   - As regards disproof:

   $$\text{If } \{M \rightarrow L\} \text{probabilistically, it does not follow that } \{\text{not } L\} \rightarrow \{\text{not } M\}, \qquad (15.5)$$

   although it may shorten the odds if there is a valid context in which probability applies, which is not the case if there is no actual physically existing multiverse (see item 1 in this list).

   Hence there is no value of Λ that proves a multiverse exists (nor any that proves it does not exist). The multiverse can explain any value of the cosmological constant, so no specific value is uniquely predicted by the hypothesis. The argument made is actually a *consistency test* on multiverses, which is indicative but not conclusive (a probability argument cannot be falsified). Consistency tests must be satisfied, but they are not considered confirmation unless no other explanation is possible. That leads to Dawid's no-alternatives argument (see Section 15.5.3).

3. *Theoretical, not empirical, argument.* The argument from the value of Λ is a valid supportive argument for the existence of a multiverse, but it is a theoretical explanation rather than a physical test, based on untestable assumptions about the assumed probability measure. Any value of the cosmological constant that (in conjunction with values of other cosmological parameters) permits structure formation, and hence our existence, will do. This makes it essentially tautological, because we can only measure specific values of Λ if we exist.

4. *The measure problem.* Putting aside these issues, and assuming there is a really existing ensemble, there is no unique probability measure to use – yet the result depends critically on this measure. Numerous measures have been proposed (see, for example, Linde & Noorbala, 2010), but problems arise with them because of the infinities in the theory, such that 'natural' measures diverge and do not give unique answers. Consequently, reverse engineering is used in practic: Measures are chosen to give the outcome that is wanted, and then the claim is made that that they are 'natural' measures.

However the key point underlying is that whatever measure you may propose, it can't be observationally or empirically tested, hence it's a philosophically justified element of the theory. Also, virtually all the literature on measures looks only at measures in Friedmann–Lemaître–Robertson–Walker (FLRW) models, which are infinitely improbable within the space of all universe models. Consequently, they do not tackle the real problem at hand. Starkman (2006) claims that because of the measure problem, the argument does not work: He suggests that different ways of assigning probabilities to candidate universes lead to totally different anthropic predictions. Thus, anthropic reasoning within the framework of probability as frequency is ill defined. In the absence of a fundamental motivation for selecting one weighting scheme over another, the anthropic principle cannot be used to explain the value of any physical parameter, including Λ. This is, in effect, a tightening of the argument made by Aguirre (2005).

15.4.4 The Conclusion

The various arguments for the multiverse are sound motivations for exploring multiverse it as a scientific theory, which is a useful thing to do (and has led to many interesting results in terms of unearthing anthropic connections). However, only a small subset of multiverse models offer the possibility of observational support or disproof of the hypothesis. In all other cases, we have useful theoretical arguments that support the existence of multiverses, but are not testable. In each case, the argument is based on untestable assumptions, and hence has a philosophical flavour.

15.5 The Big Issue

In the face of this situation, some multiverse enthusiasts want to weaken the requirement of testability of a scientific theory: They propose weakening the nature of scientific proof in an effort to claim that multiverses are full-blooded scientific theories. This is a dangerous tactic. The very nature of the scientific enterprise is at stake. Multiverses are valid hypotheses to be developed and tested as far as possible, which is indeed a worthwhile enterprise – but they are not tested theories, or even in general testable. It is the latter point that makes their status open to query.

15.5.1 Weakening the Requirement for a Scientific Theory

In defence of string theory and multiverses, Susskind explicitly states the criteria for scientific theories should be weakened (see Kragh, 2013), as do Richard Dawid (2013) and Sean Carroll (2014). These attempts to justify 'non-empirical theory confirmation' (Dawid, 2013) amount to a proposal for weakening the nature of scientific proof, and hence of science, to a serious degree. It is a dangerous move because of current scepticism about science in many parts of society – in particular as regards vaccines, global warming, genetically modified organisms (GMOs), and nuclear power. In each case, a solid scientific argument can be made. For that argument to be persuasive, apart from the emotional and political issues to be faced, it is crucial that science itself is seen to be on solid ground – and particularly so physics, which is the model for the rest.

Two central scientific virtues are testability and explanatory power; in the cosmological context, these are sometimes in conflict with each other (Ellis, 2006). The extreme case is multiverse proposals, where no direct observational tests of the hypothesis are possible, as the supposed other universes cannot be seen by any observations, and the assumed underlying physics is also untested and probably untestable, as discussed. In this context, one must re-evaluate what the core of science is: Can one maintain one has a genuine scientific theory when direct and indeed indirect tests of that theory are impossible? By making such a claim, one alters what is meant by science. One should be very careful before so doing (Ellis & Silk, 2014). Many other theories waiting at the door also want to be called science (astrology, intelligent design, and so on). Their claims need careful consideration.

15.5.2 Falsifiability, Popper, and Lakatos

Carroll in his article "What Scientific Ideas Are Ready for Retirement?: Falsifiability" (2011) states:

The falsifiability criterion gestures toward something true and important about science, but it is a blunt instrument in a situation that calls for subtlety and precision. It is better to emphasize two more central features of good scientific theories: they are *definite*, and they are *empirical*. By 'definite' we simply mean that they say something clear and unambiguous about how reality functions. ... The relevant parameter space might be inaccessible to us, but it is part of the theory that cannot be avoided.

This claim that multiverses cannot be avoided is essentially Dawid's no-alternatives claim, which is discussed in the next section, applied in this context. As regards the empirical criterion, Carroll comments that even though we can't observe other parts of the multiverse directly, their existence has a dramatic effect on how we account for the data in the part of the multiverse we do observe, because those data inform the value of the cosmological constant. This point was considered earlier and the data found to not be adequate to draw that far-reaching conclusion. The existence of components of the multiverse may be interpreted as giving such support, but that is a philosophical decision taken in the face of a lack of any direct data.

Susskind (2005) gives rather similar arguments for the multiverse, although not so formalised. Again it is useful supporting argument, but many would not regard it as sufficient to be decisive. Nevertheless, one aspect of the debate deserves special notice: Some of the comments by multiverse proponents use phrases such as 'Popperazi'[6] and 'Falsifiability Police'.[7] This use of emotional rhetoric and name-calling instead of rational argument tends to confirm the sociology of science claims that social and emotional factors significantly affect outcomes of scientific discussions. This is a regrettable outcome.

In fact, Popper's approach was not simplistic, as implied by these attacks. Thornton (2016) explicates it as saying that a theory is scientific if and only if it divides the class of observation-reports into two non-empty sub-classes: (a) those with which it is inconsistent, which is the class of its potential falsifiers, and (b) those with which it is consistent, or which it permits. In these terms, the problem with the multiverse is that because it can predict anything (class [a] is empty), it does not satisfy this criterion.

Lakatos improved on Popper's analysis by using an approach with the core theme of scientific research programmes (Musgrave & Pigden, 2016). According to this view, the unit of scientific evaluation is no longer the individual theory, but rather the sequence of theories that constitute a scientific research programme. The issue is whether the research programme is both theoretically progressive, predicting novel and hitherto unexpected facts, and empirically progressive, predicting at least some new facts that turn out to be true. In this case it is good science. Otherwise, it is degenerating – successive theories do not deliver novel predictions, or the novel predictions that they deliver turn out to be false – and hence is bad science.

Thus the emphasis on experimental predictions as well as theoretical developments remains a component of this approach. In recent years, the multiverse project has produced novel theories about multiverse measures, and about possible bubble collisions or rebirths leading to circles in the CMB sky. The latter form the most progressive element of the project.

15.5.3 Non-Empirical Theory Confirmation

A more developed proposal for justifying as scientific new theories such as string theory and multiverse theories is given by Richard Dawid in his book *String Theory and the Scientific Method* (2013). He gives three non-empirical criteria for a scientific theory: NAA, MIA, and UEA.

1. The *no-alternatives argument* (NAA: Dawid, 2013; Dawid et al., 2015) proceeds as follows:

P1: A hypothesis H satisfies data D subject to constraints C.

P2: The scientific community has not yet found an alternative to H for explaining D.

P3: Hence we have confirmation of H.

---

[6] www.math.columbia.edu/~woit/wordpress/?p=312.
[7] www.math.columbia.edu/~woit/wordpress/p=7413.

Here in the application by Carroll discussed earlier, H is multiverse hypothesis (*many universes*), and the data D adduced are the value of the cosmological constant Λ (Section 15.5.2). However, there are at least two alternatives H1 and H2:[8]

H1: *Happenstance*: No multiverse exists, the value of Λ is just another physical parameter that has that value by happenstance, and we were just lucky that this value allowed life to come into being.

H2: *Intention:* The universe was in some sense meant to allow life to come into existence.

H2 could be understood in various ways. For example, the now fashionable idea (in some circles) that the universe might be a computer simulation would be one context where that would be the case.

Choosing between these options is a philosophical rather than scientific issue, as no scientific data can distinguish between them; this is essentially the issue of underdetermination of scientific theory building on the basis of the available empirical data. One is choosing between hypotheses that one regards as a satisfactory explanation – that is, one that supports a predetermined worldview with particular choice criteria; there is 'no alternative' only if one a priori excludes two of these three possible hypotheses. A major issue here is which data will be taken into account in making that decision – only scientifically measurable data versus data from a broader swathe of life (Ellis, 2017).

2. The *meta-inductive argument* (MIA: Dawid, 2013) states that other comparable theories in the research programme were successful later on, so we can assume this one will be, too. This argument is rather weak, as many counterexamples can be cited where MIA has not worked:

- The $SU(5)$ grand unified theory
- The Einstein static universe (assumed to be a good model of the universe from 1917 to 1931)
- The Einstein–de Sitter universe (assumed to be a good model of the later expansion of the universe from the 1960s to the 1980s)
- The assumption that there are no large-scale motions or structures in the universe (only overcome by observationalists against great resistance)

A meta-analysis of the MIA (MMIA) suggests that this theory is not generally true.

3. The *argument of unexpected explanatory coherence* (UEA: Dawid, 2013) explains something without having been developed to that end, so it has achieve-

---

[8] The further alternative H4, *Inevitability*; It's the only possible outcome; it is difficult to sustain, given that physics could in principle be different (as strongly suggested by the string landscape theory); or there might in principle be no physical laws at all.

ments that were not foreseen at the time of its construction. In a sense, this is true for the multiverse argument, for in principle it can explain anything whatsoever (particularly Tegmark's version IV). This is a vacuous victory, however, as the theory does not explain anything specific. UEA needs to be sharpened to 'explain something specific', at which point the strongest claim is made regarding the value of Λ, as discussed earlier.

It is apparent that these proposals are not overwhelmingly strong, in the multiverse case. But in any case, the key point is that NAA, MIA, and UEA are *philosophical* principles. They cannot be tested by any scientific experiments or observation. Hence, insofar as they work, they are philosophical arguments for the multiverse, rather physics arguments. They do imply multiverses are worth investigating, but do not imply more than that – specifically, that multiverses actually exist.

There is a suggestion that *inference to the best explanation* (IBE), put in a Bayesian perspective, will make these arguments more solid (Dawid, 2013). Bayes' theorem states;[9].

$$P(H|E) = P(H) * P(E|H)/P(E) \tag{15.6}$$

where $P(H)$ is the initial degree of belief in the hypothesis H (the prior probability), $P(H|E)$ is the degree of belief after having accounted for new evidence E (the posterior probability), $P(E|H)$ is the likelihood of data E being measured if H is true, and $P(E)$ is the initial probability that the evidence itself is true. There are several issues here:

1. The prior probability $P(H)$ can be freely chosen to fit one's philosophical predilictions, and will give results according to those choices.
2. The probability $P(E|H)$ in the multiverse case is, in essence, the universe probability measure, when runs into all of the problems alluded to in Section 4.3.
3. To fit in the usual scientific framework, the new evidence E must be a new measurement, not just a revision of theory; but it is sometimes used where E just represents an updating of theory, as in Dawid's use to support the NAA (Dawid, 2013, p. 70). It is not a new measurement, but in Dawid's view is not just a revision of theory, either. Instead, his claim is that the evidence is a meta-level observation of the kind 'no alternatives have been found' or 'in a relevant sense, comparable scientific theories had a tendency to be predictively successful in the past.' From my perspective, that is not an observation, but rather theory revision. No new observational data have been used in this Bayesian updating. I suggest that the Duhem–Quine thesis, Lakatos's theories, and Bayes' theorem don't change the essentials embodied in the usual scientific view: that there is a need for observational or experimental data to test any extrapolation from the

---

[9] www.gaussianwaves.com/2013/10/bayes-theorem/.

known to the unknown, so as to change it from a provisional hypothesis to a well-established theory, not just a revision of theory. That is the core issue at stake (Rovelli, 2016).

15.5.4 Falsifiability

Getting down to brass tacks, the question for multiverse enthusiasts is:

- What data/observation would lead you to abandon the multiverse model?
  If the answer to this question is 'none', then the theory is dogma, not science. Because multiverses can explain anything, the answer is, indeed, likely to be 'none'. If this is not the case, multiverse proponents need to state clearly what those data are – and stick to the implied commitment. Conversely, the question for multiverse skeptics is:
- What data/observation would lead you to accep the multiverse model?
  If the answer is 'none', then, again the opposition is dogma, not science. If this is not the case, multiverse opponents need to state clearly what those data are – and stick to the implied commitment.

15.6 The Existence of Physically Existing Infinities

A particular warning sign regarding the claims made for existence of multiverses arises from the assertion that they involve an infinite number of bubbles, each containing an infinite number of galaxies (Susskind, 2005; Guth, 2007; Tegmark, 2014). An example can be found in *Many Worlds in One: The Search for Other Universes* by Alex Vilenkin (2007), which is described by the publishers as follows:

He goes on to posit that our universe is but one of an infinite series, many of them populated by our 'clones.' Vilenkin is well aware of the implications of this assertion: 'countless identical civilizations [to ours] are scattered in the infinite expanse of the cosmos. With humankind reduced to absolute cosmic insignificance, our descent from the center of the world is now complete.'

What such statements forget is that infinity is an unattainable state, rather than a very large number; it is forever beyond what can be experienced or realised (Ellis, 2018). David Hilbert stated this as follows: 'The infinite is nowhere to be found in reality, no matter what experiences, observations, and knowledge are appealed to.' (Hilbert 1925).

Thus, while the universe expands toward an infinite size in the far future, it never actually attains that state at any finite time, for it is unattainable; the scale factor of the universe is unlimited in the future, but is never truly infinite. Again, any claim that there is an infinite number of galaxies in the universe is not a scientific statement, if science involves testability by either observation or experiment, for the visual horizon restricts us to observing only a finite number of galaxies. Supposing we could see indefinitely far (for example, if gravity weakened at early times so that expansion was linear and hence there was no visual horizon), we could still never prove the existence of an infinite number of galaxies, because we could not count them even if we could see them. No matter how many galaxies we had counted, say out to $10^{1000,000,000,000,000,000,000}$ Hubble radii, we would still not have taken the first step toward infinity – for that is the nature of infinity.

An interesting argument for the coming into being of spatial infinities is given by Freivogel et al. (2005). Because of the nature of hyperbolic geometry, if tunnelling from one vacuum to another took place at a point P, then a hypersurface at constant proper time ε from P along all future directed timelike geodesics – for example, 1 ms proper time measured from P – is an infinite spatial surface of constant negative curvature, no matter how small ε is. Hence one can claim such an infinite spatial surface comes into being instantaneously, implying an infinite amount of matter also comes into being.

This argument is faulty (Ellis & Stoeger, 2010). It is not true if one remembers that the origin of such physical processes cannot be an exact point: It has to be a domain of finite size (no physical process can take place precisely at a point, which has zero dimensions). In fact, the time to reach that hypersurface from a point a spatial distance $\varepsilon > 0$ gets longer and longer the farther up the spatial surface one goes, and it takes an eternity of time for a physical process at the origin to reach the future infinity of this spacelike surface, no matter how small the distance ε (representing the nucleus of the physical action) is. This is just another example of how 0 and ∞ are dual to each other (Ellis, 2018).

The conclusion is that any claims of existence of an infinite number of any physical entities whatever should be treated with scepticism, and this applies in particular to multiverse theories. A recent example can be found in *New Scientist* of 18 January 2017:

Mapping the multiverse: How far away is your parallel self? There seems to be an infinity of invisible worlds lurking out there. Now we're starting to get a handle on where they are, and what it might take to reach them.

Any such claims are certainly not testable science. Note again, we are concerned with *really existing* multiverses, not potential or hypothetical multiverses. We can't prove they are there (or not), and there is no way we can reach them. We should stop misleading the public about these issues. It discredits genuine science.

15.7 Conclusion

There are a few specific cases where the multiverse idea might be tested or rejected. Apart from these cases, the various arguments for claiming that the multiverse is an established scientific result (for example, those based on the small value of the cosmological constant) do so by in one way or another weakening the usual standards for a theory to be considered scientific. This approach should be employed only with extreme care, else it will be a giant step backward to the time before Galileo and the establishment of the bases of successful scientific theories. The kinds of arguments given are sound proposals for considering the possibility seriously and pursuing its consequences further (Rovelli, 2016), but they are not adequate for claiming the theory is a well-established scientific result. Having a well-developed mathematical theory is not sufficient.

References

Ade, P. A. R., et al. (The Planck collaboration) (2015). Planck 2015 results. XVI. Isotropy and statistics of the CMB. *A&A* **594**, A16, [arXiv:1506.07135].

Aguirre, A. (2005). "On making predictions in a multiverse: Conundrums, dangers, and coincidences. In *Universe or Multiverse?*, ed. B. Carr (Cambridge: Cambridge University Press), [arXiv:astro-ph/0506519v1].

Aguirre, A., & M. C. Johnson. (2011). A status report on the observability of cosmic bubble collisions. *Reports on Progress in Physics* **74**, 074901, [arXiv 0908.4105].

Banks, T. (2012). The top 10^{500} reasons not to believe in the landscape. [arXiv:1208.5715].

Barrow, J., & F. Tipler. (1988). *The Anthropic Cosmological Principle* (Oxford: Oxford University Press)

Bondi, H. (1960). *Cosmology* (Cambridge: Cambridge University Press).

Carr, B. (2005). *Universe or Multiverse?* (Cambridge: Cambridge University Press).

Carroll, S. M. (2001). The cosmological constant. *Living Rev. Rel.* **4**:1. [arXiv:astro-ph/0004075v2].

Carroll, S. (2014). Edge essay: What scientific ideas are ready for retirement?: Falsifiability. www.edge.org/response-detail/25322.

Cornish, N. J., D. N. Spergel, & G. D. Starkman. (1998). Measuring the topology of the universe. *Proc. Nat. Acad. Sci.* **95**: 82–88.

Davies, P. (2007). *The Goldilocks Enigma* (London: Penguin Books).

Dawid, R. (2013). *String Theory and the Scientific Method* (Cambridge: Cambridge University Press).

Dawid, R., S. Hartmann, and J. Sprenger. (2015). The No Alternatives Argument. *British Journal for the Philosophy of Science* 66(1), 213–34.

Deutsch, D. (1997). *The Fabric of Reality* (New York: Viking Adult).

Dodelson, S. (2003). *Modern Cosmology* (Cambridge: Academic Press).

Durrer, R. (2008). *The Cosmic Microwave Background* (Cambridge: Cambridge University Press).

Efstathiou, G. (1995). An anthropic argument for a cosmological constant. *MNRAS* **274**:L73–6.

Ellis, G. F. R. (1975). Cosmology and verifiability. *QJRAS* **16**:245–64.

Ellis, G. F. R. (1980). Limits to verification in cosmology. *Ann. NY Acad. Sci.* **336**:130–60.

Ellis, G. F. R. (2006). Issues in the philosophy of cosmology. In. *Handbook in Philosophy of Physics*, ed. J. Butterfield & J. Earman (Amsterdam: Elsevier). [arXiv:**astro-**ph/0602280v2].

Ellis, G. F. R. (2017). The domain of cosmology and the testing of cosmological theories. In *The Philosophy of Cosmology*, ed. K. Chamchan et al. (Cambridge: Cambridge University Press).

Ellis, G. F. R., R. Maartens, & M. A. H. MacCallum (2012). *Relativistic Cosmology* (Cambridge: Cambridge University Press).

Ellis, G. F., K. A. Meissner, & H. Nicolai, (2018). The physics of infinity. *Nat. Phys.* **14**:770.

Ellis, G. F. R., & G. Schreiber. (1986). Observational and dynamical properties of small universes. *Phys. Lett.* **A115**:97–107.

Ellis, G. F. R., & J. Silk. (2014). Scientific method: Defend the integrity of physics. *Nature* **516**:321–3.

Ellis, G. F. R., & W. Stoeger. (2010). A note on infinities in eternal inflation. *Gen. Rel. Grav.* **41**:1475–84. [arXiv:1001.4590].

Ellis, G. F. R., & J.-P. Uzan. (2015). Causal structures in cosmology. *Comptes Rendus Physique* **16**:928–47.

Feeney, S. M., M. C. Johnson, D. J. Mortlock, & H. V. Peiris. (2011). First observational tests of eternal inflation: Analysis methods and WMAP 7-year results. *Phys. Rev. D* **84**:043507.

Freivogel, B., M. Kleban, M. R. Martinez, & L. Susskind. (2006). Observational consequences of a landscape. *JHEP* **0603**:039.

Greene, G. (2011). *The Hidden Reality: Parallel Universes and the Deep Laws of the Cosmos* (New York: Vintage).

Guth, A. (2007). Eternal inflation and its implications. *J. Phys. A: Math. Theor.* **40**:6811. [arXiv:hep-th/0702178].

Hilbert, H. (1925). On the infinite. English translation in *Philosophy of Mathematics: Selected Readings*, ed. P. Benacerraf & H. Putnam (Cambridge: Cambridge University Press). (https://math.dartmouth.edu/~matc/Readers/HowManyAngels/Philosophy/Philosophy.html).

Kachru, S., R. Kallosh, A. Linde, & S. P. Trivedi. (2003). De Sitter Vacua in String Theory. *Phys. Rev.D* **68**:046005. [arXiv:hep-th/0301240v2].

Kleban, M. (2011). Cosmic bubbles collisions. [arXiv:1107.2593].

Kragh, H. 2013. The criteria of science, cosmology and the lessons of history. in M Heller, B Brozek & L Kurek (eds), Between Philosophy and Science. Copernicus Center Press, Krakow, pp. 149–172.

Lachieze-Rey, M., & J. P. Luminet. (1995). Cosmic topology. *Phys. Rep.* **254**:135–214.

Levy, J. (2009). *Newton's Notebook: The Life, Times, and Discoveries of Sir Isaac Newton* (Cape Town: Zebra Press).

Lewis, G. F., & L. A. Barnes. (2016). *A Fortunate Universe: Life in a Finely Tuned Cosmos* (Cambridge: Cambridge University Press).

Linde, A. (1982). A new inflationary universe scenario: A possible solution of the horizon, flatness, homogeneity, isotropy and primordial monopole problems. *Phys. Lett. B* **108**:389–93.

Linde, A., & M. Noorbala. (2010). Measure problem for eternal and non-eternal inflation. *JCAP* **1009**:008. [arXiv:1006.2170].

Luminet, J.-P., J. R. Weeks, A. Riazuelo, R. Lehoucq, & J.-P. Uzan. (2003). Dodecahedral space topology as an explanation for weak wide-angle temperature correlations in the cosmic microwave background. *Nature* **425**:593–5.

Martin, J., C. Ringeval, & V. Vennin. (2014). Encyclopaedia inflationaris. *Phys. Dark Univ.* 5–6, 75–235. [arXiv:1303.3787].

Mukhanov, V. (2005). *Physical Foundations of Cosmology* (Cambridge: Cambridge University Press).

Musgrave A., & C. Pigden. (2016). Imre Lakatos. In *The Stanford Encyclopedia of Philosophy*, ed. Edward N. Zalta. https://plato.stanford.edu/archives/win2016/entries/lakatos/.

Penrose, R. (2011). *Cycles of Time: An Extraordinary New View of the Universe* (New York: Random House).

Peter, P., & J.-P. Uzan. (2013). *Primordial Cosmology* (Oxford: Oxford University Press).

Rees, M. (2001). *Just Six Numbers: The Deep Forces That Shape the Universe* (New York: Basic Books).

Rees, M. (2017). *Our Cosmic Habitat* (Princeton, NJ: Princeton University Press).

Rovelli, C. (2016). The danger of non-empirical confirmation. [http://arxiv.org/abs/1609.01966].

Smolin, L. (1997). *The Life of the Cosmos* (Oxford: Oxford University Press).

Starkman, G. D. (2006). Why anthropic reasoning cannot predict lambda. *Phys. Rev. Lett.* **97**:201301. [arXiv:astro-ph/0607227v2].

Susskind, L. (2005). *The Cosmic Landscape: String Theory and the Illusion of Intelligent Design* (Boston: Little, Brown and Company).

Tegmark, M. (2003b). Parallel universes: Not just a staple of science fiction, other universes are a direct implication of cosmological observations. *Sci. Am.* **288**(5):40–51.

Tegmark, M. (2003b). Parallel universes. In *Science and Ultimate Reality: From Quantum to Cosmos*, ed. J. D. Barrow, P. C. W. Davies, & C. L. Harper (Cambridge: Cambridge University Press). [astro-ph/0302131].

Tegmark, M. (2014). *Our Mathematical Universe: My Quest for the Ultimate Nature of Reality.* (New York: Knopf Doubleday Publishing Group).

Thornton, S. (2016). Karl Popper. In *The Stanford Encyclopedia of Philosophy*, ed. Edward N. Zalta. https://plato.stanford.edu/archives/win2016/entries/popper/.

Uzan, J.-P. (2010). Varying constants, gravitation and cosmology. *Living Rev. Relat.* [arXiv:1009.5514].

Vilenkin, A. (2007). *Many Worlds in One: The Search for Other Universes* (New York: Hill and Wang).

Weinberg, S. (1987). Anthropic bound on the cosmological constant. *Phys. Rev. Lett.* **59**:2607.

Weinberg, S. (1989). The cosmological constant. *Rev. Mod. Phys.* **61**:1.

Weinberg, S. (2001). The cosmological constant problems. In *Sources and detection of dark matter and dark energy in the universe* (pp. 18–26). Berlin: Springer, Berlin.

16

Beyond Falsifiability: Normal Science in a Multiverse

SEAN M. CARROLL

16.1 Introduction

The universe seems to be larger than the part we can see. The Big Bang happened a finite time ago, and light moves at a finite speed, so there is only a finite amount of space accessible to our observations. In principle, space could be a compact manifold with a size comparable to that of our observable region, but there is no evidence for such a possibility, and it seems likely that space extends much farther than what we see [1]. It is natural, then, for scientists to wonder what the universe is like beyond our observable part of it.

Or is it?

Two basic options for the ultra-large-scale universe are worth distinguishing. The first is uniformity. The universe we see is approximately homogeneous and isotropic on large scales. A simple possibility is that this uniformity extends throughout all of space, which might itself be finite or infinite. But a second possibility is that conditions beyond our horizon eventually start looking very different. At a prosaic level, the average density of matter, or the relative abundances of ordinary matter and dark matter, could vary from place to place. More dramatically, but consistent with many ideas about fundamental physics, even the local laws of physics could appear different in different regions – the number and masses of particles, forms and strengths of interactions between them, or even the number of macroscopic dimensions of space. The case where there exist many such regions of space, with noticeably different local conditions, has been dubbed the "cosmological multiverse," even if space itself is connected. (It would be equally valid but less poetic to simply say "a universe where things are very different from place to place on ultra-large scales.")

For a number of reasons, both cosmologists and non-cosmologists have paid a lot of attention to the prospect of a cosmological multiverse in recent years [2, 3]. At the same time, however, there has been a backlash. A number of highly respected scientists have objected strongly to the idea, in large part due to their conviction that what happens outside the universe we can possibly observe simply shouldn't *matter* [4–7]. The job of science, in this view, is to account for what we observe, not to speculate about what we don't. There is a real worry that the multiverse represents imagination allowed to roam unfettered from empirical observation, unable to be tested by conventional means.

In its strongest from, the objection argues that the very idea of an unobservable multiverse shouldn't count as science at all, often appealing to Karl Popper's dictum that a theory should be falsifiable to be considered scientific. At the same time, proponents of the multiverse (and its partner in crime, the anthropic principle) sometimes argue that while multiverse cosmologies are definitely part of science, they represent a new *kind* of science, "a deep change of paradigm that revolutionizes our understanding of nature and opens new fields of possible scientific thought" [8]. (Similar conceptual issues are relevant for other kinds of theories, including string theory and the many-worlds of Everettian quantum mechanics, but to keep things focused here I will discuss only the cosmological multiverse.)

In this chapter, I stake out a judicious middle position. Multiverse models are scientific in an utterly conventional sense; they describe definite physical situations, and they are ultimately judged on their ability to provide an explanation for data collected in observations and experiments. But they are perfectly ordinary science, so the ways in which we evaluate the multiverse as a scientific hypothesis are precisely the ways in which hypotheses have always been judged (cf. [9]).

The point is not that we are changing the nature of science by allowing unfalsifiable hypotheses into our purview. Instead, the point is that "falsifiability" was never the way that scientific theories were judged (although scientists have often talked as if it were). While the multiverse is a standard scientific hypothesis, it does highlight interesting and nontrivial issues in the methodology and epistemology of science. The best outcome of current controversies over the multiverse and related ideas (other than the hopeful prospect of finding the correct description of nature) is if working scientists are nudged toward accepting a somewhat more nuanced and accurate picture of scientific practice.

16.2 Falsifiability and Its Discontents

Falsifiability arises as an answer to the "demarcation problem": What is science and what is not? The criterion, introduced by Karl Popper [10], states that truly scientific theories are ones that stick their necks out, offering specific predictions we might imagine testing by "possible or conceivable observations" [11]. If a theory could, in principle, be proved false by the appropriate experimental results, it qualifies as scientific; if not, it doesn't. Indeed, the hope was that such a principle could fully capture the logic of scientific discovery: We imagine all possible falsifiable theories, set about falsifying all the ones that turn out to be false, and what remains standing will be the truth.

Popper offered an alternative to the intuitive idea that we garner support for ideas by verifying or confirming them. In particular, he was concerned that theories such as the psychoanalysis of Freud and Adler, or Marxist historical analysis, made no definite predictions; no matter what evidence was obtained from patients or from history, one could come up with a story within the appropriate theory that seemed to fit all of the evidence. Falsifiability was meant as a corrective to the claims of such theories to scientific status.

On the face of it, the case of the multiverse seems quite different from the theories with which Popper was directly concerned. There is no doubt that any particular multiverse scenario makes very definite claims about what is true. Such claims could conceivably be falsified, if we allow ourselves to count as "conceivable" observations made outside our light cone. (We can't actually make such observations in practice, but we can conceive of them.) So whatever one's stance toward the multiverse, its potential problems are of a different sort than those raised (in Popper's view) by psychoanalysis or Marxist history.

More broadly, falsifiability doesn't actually work as a solution to the demarcation problem, for reasons that have been discussed at great length by philosophers of science. In contrast with the romantic ideal of a killer experiment that shuts the door on a theory's prospects, most scientific discoveries have a more ambiguous impact, especially at first. When the motions of the outer planets were found to be anomalous, no one declared that Newtonian gravity had been falsified; instead, they posited that the anomalies were evidence of the existence of a new planet and eventually discovered Neptune. When a similar thing happened with Mercury, they tried the same trick, but this time the correct conclusion was entirely different: Newtonian gravity actually had been falsified, as we understood once general relativity came on the scene. When Einstein realized that general relativity predicted an expanding or contracting universe, he felt free to alter his theory by adding a cosmological constant, to bring it in line with the understanding at the time that the universe was static. When the OPERA experiment discovered neutrinos that apparently traveled faster than light, very few physicists thought that special relativity had been falsified. Individual experiments are rarely probative.

In any of these cases, it is a simple matter to explain why an overly naive version of falsifiability failed to capture the reality of the scientific process: Experimental results can be wrong or misinterpreted, and theories can be tweaked to fit the data. This is the point. Science proceeds via an ongoing dialogue between theory and experiment, searching for the best possible understanding, rather than cleanly lopping off falsified theories one by one. (Popper himself thought Marxism had started out scientific, but had become unfalsifiable over time as its predictions failed to come true.)

While philosophers of science have long since moved past falsifiability as a simple solution to the demarcation problem, many scientists have seized on it with gusto, going so far as to argue that falsifiability is manifestly a central part of the definition of science. In the words of philosopher Alex Broadbent:

It is remarkable and interesting that Popper remains extremely popular among natural scientists, despite almost universal agreement among philosophers that – notwithstanding his ingenuity and philosophical prowess – his central claims are false. [12]

It is precisely challenging cases such as the multiverse that require us to be a bit more careful and accurate in our understanding of the scientific process.

Our goal here, however, is not to look into what precisely Karl Popper might have said or thought. Even if the most naive reading of the falsifiability criterion doesn't successfully separate science from non-science, it does seem to get at something important about the nature of a scientific theory. If we want to have a defensible position on how to view the multiverse hypothesis, it would be helpful to figure out what is useful about falsifiability and how the essence of the idea can be adapted to realistic situations.

Falsifiability gets at two of the (potentially many) important aspects of science:

- **Definiteness.** A good scientific theory says something specific and inflexible about how nature works. It shouldn't be possible to take any conceivable set of facts and claim that they are compatible with the theory.
- **Empiricism.** The ultimate purpose of a theory is to account for what we observe. No theory should be judged to be correct solely on the basis of its beauty or reasonableness or other qualities that can be established without getting out of our armchairs and looking at the world.

The falsifiability criterion attempts to operationalize these characteristics by insisting that a theory say something definite that can then be experimentally tested. The multiverse is an example of a theory that says something definite (there are other regions of space far away where conditions are very different), but in such a way that this prediction cannot be directly tested, now or at any time in the future.

But there is a crucial difference between "makes a specific prediction, but one we can't observe" and "has no explanatory impact on what we do observe." Even if the multiverse itself is unobservable, its existence may well change how we account for features of the universe we *can* observe. Allowing for such a situation is an important way in which the most naive imposition of a falsifiability criterion would fail to do justice to the reality of science as it is practiced.

To try to characterize what falsifiability gets at, we can distinguish between different ways of construing what it might mean for a theory to be falsifiable. Levels of falsifiability, from least to most, might include the following:

1. There is no conceivable empirical test, in principle or in practice or in our imaginations, that could return a result that is incompatible with the theory. This would be Popper's view of the theories of Freud, Adler, and Marx.
2. There exist tests that are imaginable, but it is impossible for us to ever conduct them. This is the situation we could be in with the multiverse, where we can conceive of looking beyond our cosmological horizon, but can't actually do so.
3. There exist tests that are possible to perform within the laws of physics, but are hopelessly impractical. We can contemplate building a particle accelerator the size of our galaxy, but it's not something that will happen no matter how far technology progresses; likewise, we can imagine decoherent branches of an Everettian wave function recohering, but the timescales over which that becomes likely are much longer than the age of the universe.
4. There exist tests that can be performed, but which will cover only a certain subset of parameter space for the theory. Models of dynamical dark energy (as opposed to a non-dynamical cosmological constant) can be parameterized by the rate at which the energy density evolves; this rate can be constrained closer and closer to zero, but it's always conceivable that the actual variation is less than any particular observational limit yet doesn't strictly vanish.
5. There exist doable, definitive tests that could falsify a theory. For example, general relativity would be falsified if large masses were shown not to deflect passing light.

Most scientists would probably agree that theories in category 1 aren't really scientific in any practical sense, while those in category 5 certainly are (at least as far as testability is concerned). But the purpose of the preceding list isn't to figure out where to definitively draw the line between "scientific" and "non-scientific" theories; for the purposes of scientific practice, in any event, it is not clear what use such a distinction actually has. Nevertheless, this list helps us to appreciate the sharp divide between category 1 and categories 2–5. That difference can be summarized by pointing out that any theory in categories 2–5. might reasonably be *true*: It could provide a correct description of nature, to some degree of accuracy and within some domain of applicability. Theories in category 1, by contrast, don't really have a chance of being true; they don't successfully distinguish true things from not-true things, regardless of their testability by empirical means. Such theories are unambiguously not helpful to the scientific enterprise.

It would be wrong, therefore, to lump together categories 1–4 in opposition to category 5, while glossing over the distinction between category 1 and categories 2–5. The former distinction is epistemological: It is a statement about how difficult it is to figure out whether what the theories are saying is actually true. The latter distinction is metaphysical There is a real sense in which theories in category 1 aren't actually saying anything about the world.

The multiverse as a general paradigm – regions far outside our observable part of the universe, where local conditions or even low-energy laws of physics could be very different – is in category 2. The universe could be like that, and there is nothing unscientific about admitting so. The fact that it might be difficult to gather empirical evidence for or against such a possibility doesn't change the fact that it might be true.

Another way to emphasize that the important dividing line is between categories 1 and 2, not between categories 4 and 5, is to note that some specific multiverse models are actually in category 4, where a certain region of parameter space allows the theory to be explicitly falsified. For the moment, let's put aside the general idea of "the universe looking very different outside our observable region," and focus on the most popular realization of this idea in contemporary cosmology, known as "false-vacuum eternal inflation" [13]. In this kind of scenario, bubbles of lower-energy configurations appear via quantum tunneling within a region of space undergoing inflation; these bubbles grow at nearly the speed of light, and can contain within them distinct "universes" with potentially different local laws of physics. The distribution of bubbles depends on details of physics at high energies – details about which we currently have next to no firm ideas. Nevertheless, we can parameterize our ignorance in terms the nucleation rate of such bubbles and the energy density within them. If the rate is sufficiently high, this model makes a falsifiable prediction: the existence of circular features in the anisotropy of the cosmic microwave background, remnants of when other bubbles literally bumped into our own. Cosmologists have searched for such a signature, allowing us to put quantitative limits on the parameters of false-vacuum eternal inflation models [14, 15].

This particular version of the multiverse, in other words, is indubitably falsifiable for certain parameter values. It would be strange, indeed, if the status of an idea as scientific versus unscientific depended on the parameters used. One might suggest that models with such tunable parameters should be labeled unscientific for *any* parameter values – even the testable ones – on the grounds that such theories can wriggle out of being falsified through a clever choice of parameters. In actual scientific practice, though, that's not how we behave. In the dark-energy example given previously, the density of dark matter is imagined to vary at some slow rate; if the rate is slow enough, we will never be able to distinguish it from a truly constant vacuum energy. No one claims that therefore the idea of dynamical dark energy is unscientific, nor should they. Instead, we test the theory where we can, and our interest in the idea gradually declines as it is squeezed into ever-small regions of parameter space.

In the end, "what counts as science" is less important than how science gets done. Science is about what is true. Any proposal about how the world truly is should be able to be judged by the methods of science, even if the answer is "we don't know and never will." Excluding a possible way the world could be on the basis of a philosophical predisposition is contrary to the spirit of science. The real question is, How should scientific practice accommodate this possibility?

16.3 Abduction and Inference

While we don't currently have a final and complete understanding of how scientific theories are evaluated, we can do a little bit better than the falsifiability criterion. The multiverse provides an especially interesting test case for attempts to explain how science progresses in the real world.

One way of thinking about the scientific method is as an example of "abduction," or inference to the best explanation, which is to be contrasted with the logical techniques of deduction and induction [16]. Deduction reasons perfectly from a given set of axioms; induction attempts to discern patterns in repeated observations. Abduction, by contrast, takes a given set of data and attempts to find the most likely explanation for them. If you eat spoiled food, and you know that eating spoiled food usually makes you sick, by deduction you can conclude that you are likely to get sick. Abduction lets you go the other way around: If you get sick, and there are no other obvious reasons why, you might conclude that you probably ate some spoiled food. This isn't a logically necessary conclusion, but it's the best explanation given the context. Science works along analogous lines.

This raises the knotty question of what makes an explanation "the best." Thomas Kuhn, after his influential book *The Structure of Scientific Revolutions* led many people to think of him as a relativist when it came to scientific claims, attempted to correct this misimpression by offering a list of criteria that scientists use in practice to judge one theory better than another one: accuracy, consistency, broad scope, simplicity, and fruitfulness [17]. "Accuracy" (fitting the data) is one of these criteria, but by no means the sole one. Any working scientist can think of cases where each of these concepts has been invoked in favor of one theory or another. But there is no unambiguous algorithm according to which we can feed in these criteria, a list of theories, and a set of data, and expect the best theory to pop out. The way in which we judge scientific theories is inescapably reflective, messy, and human. That's the reality of how science is actually done; it's a matter of judgment, not of drawing bright lines between truth and falsity or science and non-science. Fortunately, in typical cases the accumulation of evidence eventually leaves only one viable theory in the eyes of most reasonable observers.

This kind of abductive reasoning can be cast in quantitative terms by turning to Bayesian inference. In this procedure, we imagine an exhaustive set of theories T_i (one of which might be a catch-all "some theory we haven't thought of yet"), and to each we assign a prior probability that the theory is true, $P(T_i)$. Then we collect some new data, D. Bayes's theorem tells us $P(T_i|D)$, the posterior probability that we should assign to each theory in light of the new data, via the formula

$$P(T_i|D) = \frac{P(D|T_i)P(T_i)}{P(D)}. \tag{16.1}$$

Here, $P(D|T_i)$ is the likelihood of obtaining those data if the theory were true, and the normalization factor $P(D) = \sum_i P(D|T_i)P(T_i)$ is simply the total probability of obtaining the data under the weighted set of all theories.

Bayes' theorem is a mathematical theorem; it is necessarily true under the appropriate assumptions. The additional substantive claim is that this theorem provides the correct model for how we think about our credences as scientific theories, and how we update those credences in light of new information.

The role of priors is crucial. Our priors depend on considerations such as Kuhn's criteria (aside from "accuracy," which is a posterior notion). Consider two theories of gravitation. Theory 1 is precisely Einstein's general relativity. Theory 2 is also general relativity, but with the additional stipulation that Newton's constant will suddenly change sign in the year 2100, so that gravity becomes repulsive rather than attractive.[1] Every sensible scientist will assign different credences to these two theories, even though the empirical support for them is exactly the same and they are equally falsifiable. The second theory would be considered extremely unlikely, as it is unnecessarily complicated without gaining any increase in consistency, scope, or fruitfulness over conventional general relativity.

How does the shift from an overly simplistic "falsification" paradigm to one based on abduction and Bayesian inference affect our attitude toward the multiverse? Rather than simply pointing out that the multiverse cannot be directly observed (at least for some parameter values) and therefore can't be falsified, and therefore isn't science, we should ask whether a multiverse scenario might provide the best explanation for the data we do observe. We should then attempt to quantify this notion of "best" by assigning priors to different possibilities and updating them in the face of new data – just as we do for any other scientific theory.

---

[1] Any similarities to Nelson Goodman's invented property "grue," meaning "green" before some specified future date and "blue" afterward [18], are entirely intentional.

Put this way, it becomes clear that there isn't that much qualitatively special about the multiverse hypothesis. It's an idea about nature that could be true or false, and to which we can assign some prior probability, which then gets updated as new information comes in.

Indeed, this is precisely what happened in 1998, when observations indicated that the universe is accelerating, something that is most easily explained by the existence of a nonzero cosmological constant (vacuum energy) [19–21]. Before this discovery was made, it was known that the cosmological constant had to be much smaller than its supposedly natural value at the Planck scale, but it wasn't known whether the constant was exactly zero or just very close. (For convenience, I am not distinguishing here between a truly constant vacuum energy and some kind of dynamical dark energy.) At that point we could distinguish between four broad theories of the cosmological constant (each of which might include multiple more specific theories):

1. There is a mechanism that sets the cosmological constant equal to zero once and for all.
2. There is a mechanism that sets the cosmological constant to some small nonzero value.
3. There is no mechanism that sets the cosmological constant to a small number, but it nevertheless is just by random chance.
4. The cosmological constant takes on different values in different parts of a multiverse, and is small in our observable region for anthropic reasons. (If the vacuum energy is too large and positive, galaxies never form; if it is too large and negative, the universe rapidly recollapses. The vacuum energy must be small in magnitude relative to the Planck scale to allow for the existence of life.)

Within the community of theoretical physicists, there was (and is) very little support for option 3, and pre-1998 a substantial prior was attached to option 1. The reasoning why option 1 was given a higher probability than option 2 was essentially that it's difficult to think of any mechanism that makes the vacuum energy very small, but in the set of all not-yet-proposed mechanisms that might someday be invented, it's easier to conceive of dynamics that set the cosmological constant all the way to zero (perhaps due to some unknown symmetry) than it is to imagine a mechanism that puts it just below the threshold of what had then been observed, but no lower. In a 1987 article, Steven Weinberg [22] argued that under the multiverse scenario 4, we should actually expect the cosmological constant to be nonzero and near the observational bounds that existed at that time. In other words, the likelihood of observing a nonzero cosmological constant was relatively large under option 4 (and presumably under option 3, though that was thought to be unlikely from the start), somewhat smaller under option 2 (since a dynamical mechanism might prefer numbers very close to zero), and would be vanishing under option 1.

Ten years later, when the evidence came in for an accelerating universe, scientists quite rightly updated the probabilities they attached to these theoretical alternatives. By Bayes' theorem, finding a nonzero cosmological constant decreased their credence in the popular theory 1, while increasing it in the other options. In particular, this experimental result led directly to an increase in the credence assigned by working physicists to the multiverse hypothesis 4 [23]. We didn't observe other universes directly, but we observed something about our universe that boosted the likelihood that a multiverse provides the best explanation for what we see.

The multiverse, therefore, is a case of science as usual: We evaluate it on the basis of how likely it is to be true, given what we know on the basis of what we actually have observed. But it is not only examples of literal new data that can cause our credences in a theory to change. The multiverse hypothesis reminds us of how better understanding, as well as actual experimental or observational input, can serve as "data" for the purposes of Bayesian inference.

Dawid, writing in the context of string theory, has highlighted the role of considerations other than simply fitting the data in evaluating theories [24]. He uses the term "non-empirical confirmation," which has led to some marketing problems in the scientific community. In epistemology, "confirmation" is used to refer to any evidence or considerations that increase our credence in a theory; it does not correspond exactly to the everyday-language notion of "proving the truth of." Even "non-empirical" is a bit misleading, as such considerations are always used in addition to empirical data, not in place of it.

It is obviously true that our credences in scientific theories are affected by things other than collecting new data via observation and experiment. Choosing priors is one pertinent example, but an important role is also played by simply understanding a theory better (which is the job of theoretical physicists). When Einstein calculated the precession of Mercury in general relativity and found that it correctly accounted for the known discrepancy with Newtonian gravity, there is no doubt that his credence in his theory increased substantially, as it should have. No new *data* were collected; the situation concerning Mercury's orbit was already known. Instead, an improved theoretical understanding changed the degree of belief it made sense to have in that particular model. When Gerard 't Hooft showed that gauge theories with spontaneous symmetry breaking were renormalizable [25], the credence in such models increased dramatically in the minds of high-energy physicists, and justifiably so. Again, this change came not in response to new data, but following a better understanding of the theory.

Although these cases are sharp and relatively uncontroversial examples of credences changing as a result of improved theoretical understanding, most changes occur more gradually. In the case of the multiverse, physicists have been greatly influenced by the realization that a multiverse need not be simply posited as a logical possibility, but is actually a *prediction* of theories that are attractive for entirely different reasons: inflation and string theory.

It was realized early in the development of inflationary cosmology that inflation is often (depending on the model) eternal: Once begun, inflation ends in some regions but continues in others, leading to the creation of different regions of space with potentially very different physical conditions [26–28]. String theory, meanwhile, was originally hoped to give a unique set of predictions for observable quantities, but more recently appears to be able to support an extremely large number of metastable vacuum states, each of which corresponds to different low-energy physics [29–31]. (This realization was actually spurred in part by the discovery of a nonzero cosmological constant – a rare instance where progress in string theory was catalyzed by experimental data.) Taken together, these theories seem to predict that a multiverse is quite likely; correspondingly, anyone who assigns substantial credences to inflation and string theory would naturally assign substantial credence to the multiverse.

Once again, this is an example of science proceeding as usual. Ideas are not evaluated in a vacuum; we ask how they fit in with other ideas we have reason to believe might play a role in the ultimate correct theory of the universe. What theories predict and how one theory fits together with another one are perfectly respectable considerations to take into account when we assign credences to them. Moreover, an improved understanding of these issues qualifies as "data" in the context of Bayes' theorem.

Of course, this is not meant to suggest that this kind of data could serve as a *replacement* for the good old-fashioned evidence we collect via telescopes and microscopes. Our priors for different ideas are naturally affected by non-empirical ideas such as elegance and beauty, and those may even affect our likelihoods, such as when we realize that there is an unanticipated hidden beauty underlying an existing theory. Nevertheless, an ability to account for the data is always the overwhelmingly important quality a scientific theory must have. Our credence in general relativity increased due to a pencil-and-paper calculation by Einstein, but the impact of that calculation rested on the fact that it helped account for something we observed about the world, the precession of Mercury's perihelion. Many physicists increased their credence in the multiverse when they realized that it was predicted by inflation and string theory, but inflation and string theory are ultimately interesting because they purport to explain features of observational reality – the large-scale structure of the cosmos as well as the existence of gravitation and gauge theories. A correct accounting for the multitude of influences that shape our credences concerning scientific hypotheses is in no sense a repudiation of empiricism; it is simply an acknowledgment of the way it works in the real world.

16.4 The Unavoidable Unobservable

The best reason for classifying the multiverse as a straightforwardly scientific theory is that we don't have any choice. This is the case for any hypothesis that satisfies two criteria:

- It might be true.
- Whether or not it is true affects how we understand what we observe.

Consider a relevant example: attempting to solve the cosmological constant problem (understanding why the observed vacuum energy is so much smaller than the Planckian value we might naively expect). Given that we don't currently know the answer to this problem, it might be the case that some dynamical mechanism awaits discovery, which will provide a simple formula that uniquely relates the observed cosmological constant to other measurable constants of nature. Or it might be the case that no such mechanism exists. As a working theoretical physicist interested in solving this problem, one cannot spend equal amounts of research effort contemplating every possible idea, as the number of ideas is extremely large. We use our taste, lessons from experience, and what we know about the rest of physics to help guide us in (we hope) productive directions.

In this case, whether there is a multiverse in which the local cosmological constant takes on different values in different regions is an inescapable consideration. If there is, then the need to find a unique dynamical mechanism is severely diminished, since no such mechanism can seemingly exist (given that the vacuum energy is different from place to place) and that there is a plausible alternative (anthropic selection). If there is no multiverse, the interest in a unique dynamical mechanism is correspondingly enhanced. Since our credence in the multiverse should never be exactly 0 or 1, individual scientists will naturally have different predilections toward pursuing different theoretical approaches, which is a sign of a healthy scientific community. But it makes no sense to say that the possible existence of a multiverse is irrelevant to how we deal with the problem.

The universe beyond what we can see might continue uniformly forever, or it might feature a heterogeneous collection of regions with different local conditions. Since direct observations cannot reveal which is the case, both alternatives are equally "scientific." To the extent that which alternative might be true affects our scientific practice, keeping both possibilities in mind is unavoidable. The only unscientific move would be to reject one or the other hypotheses a priori on the basis of an invented methodological principle.

Of course, this doesn't rule out the special challenges posed by the multiverse. At a technical level, we have the measure problem: Given an infinite multiverse, how do we calculate the relative probabilities of different local conditions? Skeptics will sometimes say that since everything happens somewhere in the multiverse, it is impossible to make even probabilistic predictions. Neither of these two clauses is necessarily correct; even if a multiverse is infinitely big, it does not follow that everything happens, and even if everything happens, it does not follow that there are no rules for the relative frequencies with which things happen. After all, probabilistic theories such as statistical mechanics and quantum mechanics also require a measure, and in the case of cosmology several alternatives have been proposed [32–34]. For inflation in particular, it seems sensible to imagine that the measure is strongly peaked on trajectories resembling traditional single-universe inflationary scenarios. Given that eternal inflation is common in realistic inflationary potentials, some provisional assumption along these lines is necessary to license the interpretation of cosmological observables in terms of inflationary parameters (as most working cosmologists are more than willing to do).

But the question remains: Even if there is a correct measure of the multiverse, how will we ever *know*? It seems hard to imagine doing experiments to provide an answer. Rather, we are stuck with the more indirect kinds of reasoning mentioned earlier, especially how any particular proposal fits in with other physical theories. (Presumably in this case that would involve some understanding of quantum gravity and the emergence of spacetime, which doesn't yet exist but might someday be forthcoming.) And while such reasoning is necessary, it is unlikely to be definitive. In Bayesian terms, we can nudge our credences upward or downward, but it is very hard on the basis of indirect evidence alone to send those credences so close to 0 or 1 that the question can be considered definitively answered.

That, in a nutshell, is the biggest challenge posed by the prospect of the multiverse. It is not that the theory is unscientific, or that it is impossible to evaluate it. Rather, the challenge is that evaluating the theory is *hard*. It is entirely conceivable that we could make significant progress in understanding quantum gravity and inflationary cosmology, yet still be left with a situation where a century in the future our credence in the multiverse is still somewhere between 0.2 and 0.8.

Of course, the difficulty that we human beings might have in deciding whether a theory is correct should be of little relevance to the seriousness with which we contemplate it. There really might be a multiverse out there, whether we like it or not. To the extent that such a possibility affects how we attempt to explain features of the universe we do see, we need to treat this hypothesis just as we would any other scientific idea. Nobody ever said science was going to be easy.

Acknowledgments

This research is funded in part by the Walter Burke Institute for Theoretical Physics at Caltech and by DOE grant DE-SC0011632.

References

[1] P. M. Vaudrevange, G. D. Starkman, N. J. Cornish, and D. N. Spergel. (2012). "Constraints on the Topology of the Universe: Extension to General Geometries." *Phys. Rev. D* **86**, 083526; [arXiv:1206.2939 [astro-ph.CO]].

[2] S. Weinberg. (2005). "Living in the multiverse." In B. Carr, ed., *Universe or Multiverse?*, pp. 29–42 (Cambridge: Cambridge University Press) [hep-th/0511037].

[3] H. Kragh. (2009). "Contemporary History of Cosmology and the Controversy over the Multiverse." *Ann. Sci.*, **66**, 529.

[4] G. Ellis and J. Silk. (2014), "Scientific Method: Defend the Integrity of Physics." *Nature News*, **516**(7531), 321.

[5] L. Smolin. (2015). "You Think There's a Multiverse? Get Real." *New Scientist*, **225**(3004), 24–25.

[6] Ijjas, A., Steinhardt, P. J., & Loeb, A. (2017). POP goes the universe. Scientific American, **316**(2), 32–39.

[7] G. Ellis. (2017). "Physics on Edge." *Inference Review*, **3**, http://inference-review.com/article/physics-on-edge.

[8] Barrau, A. (2007). Physics in the multiverse: An Introductory review. *CERN Cour.*, *47*(arXiv: 0711.4460), 13–17.

[9] D. Harlow. (2017). "Business as Usual." *Inference Review*, **3**, http://inference-review.com/article/business-as-usual.

[10] K. Popper. (1959). *The Logic of Scientific Discovery*, translation of *Logik der Forschung* (London: Hutchinson).

[11] K. Popper. (1962). *Conjectures and Refutations: The Growth of Scientific Knowledge* (New York: Basic Books).

[12] A. Broadbent. (2016). *Philosophy for Graduate Students: Metaphysics and Epistemology* (London: Routledge).

[13] M. Kleban. (2011). "Cosmic Bubble Collisions." *Class. Quant. Grav.* **28**, 204008; doi:10.1088/0264-9381/28/20/204008 [arXiv:1107.2593 [astro-ph.CO]].

[14] M. Kleban, T. S. Levi, and K. Sigurdson. (2013). "Observing the Multiverse with Cosmic Wakes." *Phys. Rev. D* **87**(4), 041301; doi:10.1103/PhysRevD.87.041301 [arXiv:1109.3473 [astro-ph.CO]].

[15] J. Braden, M. C. Johnson, H. V. Peiris, and A. Aguirre. (2017). "Constraining Cosmological Ultralarge Scale Structure Using Numerical Relativity," *Phys. Rev. D* **96**(2), 023541; doi:10.1103/PhysRevD.96.023541 [arXiv:1604.04001 [astro-ph.CO]].

[16] I. Douven. (2017, Summer). "Abduction." In E. N. Zalta, ed., *The Stanford Encyclopedia of Philosophy* https://plato.stanford.edu/archives/sum2017/entries/ abduction/.

[17] T. S. Kuhn. (1977). "Objectivity, Value Judgment, and Theory Choice." *The Essential Tension* (Chicago: University of Chicago Press), pp. 320–39.

[18] N. Goodman. (1995). *Fact, Fiction, and Forecast* (Cambridge, MA: Harvard University Press).

[19] A. G. Riess et al. (1998). "Observational Evidence from Supernovae for an Accelerating Universe and a Cosmological Constant." *Astron. J.* **116**, 1009; doi:10.1086/300499 [astro-ph/9805201].

[20] S. Perlmutter et al. (1999). "Measurements of Omega and Lambda from 42 High Redshift Supernovae." *Astrophys. J.* **517**, 565; doi:10.1086/307221 [astro-ph/9812133].

[21] S. M. Carroll. (2001). "The Cosmological Constant." *Living Rev. Rel.* **4**, 1; doi:10.12942/lrr-2001-1 [astro-ph/0004075].

[22] S. Weinberg. (1987). "Anthropic Bound on the Cosmological Constant." *Phys. Rev. Lett.* **59**, 2607.

[23] J. Polchinski. (2017). "Memories of a Theoretical Physicist." arXiv:1708.09093 [physics.hist-ph].

[24] R. Dawid. (2013). *String Theory and the Scientific Method* (Cambridge: Cambridge University Press).

[25] G. 't Hooft. (1971). "Renormalizable Lagrangians for Massive Yang–Mills Fields." *Nucl. Phys. B* **35**, 167; doi:10.1016/0550-3213(71)90139-8.

[26] P. J. Steinhardt. (1983). "Natural Inflation." in *The Very Early Universe*, Proceedings of the Nuffield Workshop, Cambridge, 21 June–9 July, 1982, ed: G. W. Gibbons, S. W. Hawking, and S. T. C. Siklos (Cambridge: Cambridge University Press), pp. 251–66.

[27] A. Vilenkin. (1983). "The Birth of Inflationary Universes." *Phys. Rev. D* **27**, 2848–55.

[28] A. H. Guth. (2007). "Eternal Inflation and Its Implications." *J. Phys. A* **40**, 6811; doi:10.1088/1751-8113/40/25/S25 [hep-th/0702178 [HEP-TH]].

[29] R. Bousso and J. Polchinski. (2000). "Quantization of Four Form Fluxes and Dynamical Neutralization of the Cosmological Constant." *JHEP* **0006**, 006; doi:10.1088/1126-6708/2000/06/006 [hep-th/0004134].

[30] S. Kachru, R. Kallosh, A. D. Linde, and S. P. Trivedi. (2003). "De Sitter Vacua in String Theory," *Phys. Rev. D* **68**, 046005; doi:10.1103/PhysRevD.68.046005 [hep-th/0301240].

[31] L. Susskind. (2003). "The Anthropic Landscape of String Theory." In B. Carr, ed., *Universe or Multiverse?*, pp. 247–66 (Cambridge: Cambridge University Press [hep-th/0302219]).

[32] A. De Simone, A. H. Guth, A. D. Linde, M. Noorbala, M. P. Salem, and A. Vilenkin. "Boltzmann Brains and the Scale-Factor Cutoff Measure of the Multiverse." *Phys. Rev. D* **82**, 063520; [arXiv:0808.3778 [hep-th]].

[33] B. Freivogel. (2011). "Making Predictions in the Multiverse." *Class. Quant. Grav.* **28**, 204007; [arXiv:1105.0244 [hep-th]].

[34] M. P. Salem. (2012). "Bubble Collisions and Measures of the Multiverse." *JCAP* **1201**, 021; [arXiv:1108.0040 [hep-th]].

17

Gaining Access to the Early Universe

CHRIS SMEENK

17.1 Introduction

Particle physics and cosmology each have a Standard Model consistent with an astonishing array of observations and experimental results. Both models are, to some extent, victims of their own success: There are few clear empirical anomalies that could serve as signposts guiding physicists' next steps. The reach of experiments and observations barely extends past the domains they cover, providing few glimpses of the undiscovered country of novel phenomena. Although theoretical accounts of this territory have been offered, the almost complete lack of empirical access to this domain makes it difficult to determine whether these accounts are reliable. Research programs such as string theory and eternal inflation have been successful in one sense – winning widespread acceptance in the relevant communities – without having the record of correct novel predictions that is often taken as a prerequisite for empirical success. Trust in these proposals is seen as either justified by a reasonable extension of scientific methodology, in light of changed circumstances, or a sign that physicists have ventured into a worrisome new phase of "post-empirical" – or even "post-scientific"– inquiry.

Debates regarding the status of such theories reflect a fundamental disagreement regarding what constitutes success, and how to establish it. A commitment to empiricism is often taken to imply that how well a theory fits the data is the only relevant factor in assessing its truth, or to what degree current evidence confirms it.[1] Competing accounts of confirmation take a broader view, allowing for what is sometimes called "indirect" confirmation, based on factors other than compatibility with the data.

---

[1] There is a contrast between the way (most) philosophers and scientists use the term "confirm." Here I will follow the philosophers: Evidence in favor of a theory "confirms" it, even if it only leads to an incremental boost in confidence. With this usage, confirmation admits degrees, whereas scientists often use "confirm" as a success term – applied only to cases of extremely strong evidence. (Readers who adopt the latter usage should substitute "incrementally confirm" for "confirm" throughout.)

One motivation for these broader views is that scientists arguably do take a variety of factors over and above successful predictions into account when assessing theories; they demand that theories do much more than merely fit the data. For example, the theory should be compatible with other relevant theories. Direct evidence for one theory may extend outward to support several related theories.[2] A more contentious issue is whether features such as elegance should be taken into consideration. An advocate of "elegance" owes us at least an account of how to assess theoretical elegance, along with an argument that questions of elegance are not best left to tailors, as Einstein quipped. The challenge is to provide such a defense, without tacitly assuming that the world has an order or structure that we will find beautiful. One central question in debates about what I will call "supra-empirical" physics regards how much weight scientists should give to a theory's provenance.[3] Suppose that a particular research strategy has led to successful theories when it has been employed in the past. To what extent should evidence for the reliability of the strategy, in the form of its past successes, boost confidence in the strategy's latest "output"? There is no reason to expect a strategy that has been reliable in the past to falter just as it reaches past accessible domains. When confirmation is considered as a descriptive claim, it is clear that scientists often take factors like these into account in theory assessment; the more delicate question is whether they have abandoned empiricism in doing so.

A second motivation for a broader account of confirmation starts from the observation that mere compatibility with the data is too weak to justify the confidence we have in successful theories. A narrow construal of evidential support leaves us without sufficient reason to eliminate rival theories. Underdetermination is the claim that for a theory T_0 supported by a given body of evidence E, there are rival theories $\{T_1, T_2, \dots\}$, incompatible with T_0, that are also compatible with E.[4] Without an effective way to limit underdetermination, we would have no grounds to trust T_0's claims about a given domain, or its extension to as yet unobserved domains, over the competing claims of its rivals. Scientists would only be justified

---

[2] Laudan and Leplin (1991) argue that theories can be confirmed by direct evidence for a seemingly unrelated theory, if both are themselves consequences of a more general theory.

[3] I am using "supra-empirical" as an alternative to "post-empirical," because the latter implies that empirical evaluation is no longer relevant. This does not need to be the case for an account of indirect confirmation, and it is not for Dawid's view. This debate regards physical theories that cantilever out into new domains from some empirical foundation. "Supra-" is intended to emphasize that further direct tests of these aspects of the theory are impossible. Yet evidence is still relevant; for example, further evidence that shores up the foundation can increase confidence in speculative extensions. My thanks to Richard Dawid for comments on this point.

[4] There are a number of variations on this theme. For example, we have a temporary or transient form of underdetermination if E is taken to be evidence available *at a given time* (or within a restricted domain) and permanent underdetermination if E includes *all available evidence*. Our focus here is on transient underdetermination. I am assuming that theories make claims about the world that extend beyond what is reflected in the evidence, and that the account given by rival theories differs. An instrumentalist would deny that this contrast makes sense, instead regarding some parts of the theories as "mere instruments," not to be interpreted literally.

in accepting low-level generalizations of the data, about which all the competitors agree. But we clearly do have sufficient evidence to justify, for example, reliance on Maxwell's equations within the domain of weak, slowly varying electromagnetic fields. At a minimum, we need an account of theory assessment that clarifies the contrast between the appropriate level of credibility in mature, reliable theories versus theories that, however promising, remain speculative.

Ideally, these two motivations would dovetail, leading to an account of theory assessment that both describes how scientists have, in fact, evaluated theories (at least when they are at their best) and makes the case that these considerations are a reliable guide to true theories, compatible with the empiricist requirement that substantive knowledge of the world is grounded in experience.

Dawid (2013) proposes such an account, based on the idea that three arguments working in conjunction can be used to effectively limit the scope of underdetermination. The first of these, the "meta-inductive argument," concerns the provenance of a theory: Our credence in a theory should be enhanced if it is the outcome of a successful research strategy. The second regards the failure to find alternative theories for a given domain ("no alternatives"), and the third, a theory's ability to explain things it was not designed to account for ("unexpected explanatory coherence"). Dawid gives two different defenses of this account: a formal analysis within the framework of Bayesian confirmation theory and a historical analysis tracing the role of these arguments through several case studies. More strikingly, he applies this account to the supra-empirical phase of physics and makes the case that we *could* still have strong support for theories, focusing in particular on string theory.[5]

Clearly, Dawid is correct that a satisfactory account of theory assessment should explain how phyisicists have effectively limited underdetermination. But when it comes to extending historically successful strategies to our unfortunate current state, the devil is in the details of how these arguments are characterized. In this chapter, I propose an alternative characterization of the ways in which physicists have limited underdetermination, with an emphasis on the context of application. There is a family resemblance between the arguments that I see as playing a role in historical cases and Dawid's analysis. However, there is a striking contrast between the historical cases and their extensions and contemporary fundamental physics. In the historical cases, several issues regarding the content of these arguments (e.g., What are the viable alternatives? What counts as explanatory coherence?) can be settled relatively straightforwardly; by comparison, in supra-empirical physics, the lack of concrete applications leaves the same issues open to vigorous debate.

---

[5] He is more cautious regarding whether we *do* have strong support; he elucidates a pattern of argumentation that (arguably) could support string theory, while acknowledging that some of the assumptions that such an argument must make may turn out to be false.

This chapter begins by clarifying what it means to trust theory. We typically take mature theories to deserve a strong sense of trust, as we rely on them to guide practical actions in a variety of ways. Justifying this level of trust requires limiting the scope of underdetermination. In Section 17.3, I argue in favor of taking the content of physical theories to be reflected in the possibilities they introduce for measurements of theoretical quantities. Given this account of content, the threat of underdetermination is limited to two types of cases. First, a given theory may be compatible with different extensions to new domains; second, a rival research tradition may reject the theory's claim to have achieved stable, convergent measurements of theoretical quantities outright. I consider strategies that have been used historically in response to both types of underdetermination in Section 17.4, arguing that these do not extend to supra-empirical physics. Finally, Section 17.5 shows how these problems play out in eternal inflation.

17.2 Why Trust Theory?

There is an ambiguity in asking whether scientists should "trust" theories. Trust comes in various forms, and one contrast is particularly relevant here. When a theory is introduced, the evidence that can be marshalled in its support is usually provisional. Scientists often "trust theory," nonetheless, in the sense that they develop an understanding of some domain of phenomena based on the theory. The theory may allow scientists to gain access to new phenomena as well as provide guidance in the search for particularly telling types of evidence. This sense of "trust" or acceptance is heuristic and pragmatic: How much can be gained by assuming that the theory holds? This question is distinct from an assessment of a theory's credibility: How likely is it that the theory is true? This is not to say that credibility is irrelevant to the pragmatic choice; it would be unusual for scientists to accept a theory to which they assign very low credibility. Scientists who assign the same credibility may nonetheless make different choices about whether to accept a theory as a basis for further work – reflecting their risk tolerance, training, potential for fruitful work in the area, and so on.

Trust as provisional acceptance differs from the trust granted to theories that we take as reliably representing nature. Our trust in Newton's gravitational theory is reflected in the confident assertion that, despite its flaws, errors that arise from using it to describe the motion of bodies can be made arbitrarily small for sufficiently weak fields and low relative velocities.[6] The evidence available to Newton, and

---

[6] Even though Newton's theory provides only an approximate description of the gravitational interaction, in light of general relativity, the errors introduced by using it are negligible in this domain (at least, with this crude characterization filled out appropriately); see also the end of Section 17.3 in this chapter.

even to Laplace, was not sufficient to justify this level of trust. But Newtonian ideas became the warp and woof of celestial mechanics. The law of gravity has undergone ongoing tests through the development of a dynamical account of the solar system, incorporating a steadily increasing body of physical details and matching ever more precise and comprehensive observations. Newton's characterization of gravity is but one example of a scientific claim that we plausibly regard as permanent, in that it will be included, at least as an approximation valid in some domain, in any future scientific description of the world. It is difficult, albeit not impossible, to imagine a way in which all of the evidence acquired in its favor could simply unravel. To reject it completely, we would need to explain how a line of research based on a false assumption had nonetheless appeared to make steady progress. Our trust in using Newtonian gravity for a variety of practical purposes rests on our confidence in its permanence.

Just as in personal relationships, developing such deep and abiding trust is a long-term achievement. In the case of scientific theories, such trust should be based on a detailed quantitative comparsion between the actual world and the possible worlds described by a theory. Typically, the comparison proceeds based on accepting a theory provisionally.[7] The theory guides the choice of experiments or observations that will be particularly revealing, and also plays the role of a tool used to extend scientists' reach, making more detailed assessment possible. As Laplace remarked, for example, Newton's law of gravity provided a tool for the study of celestial dynamics that was as essential as the telescope.

One concrete way theories play this role is in underwriting theory-mediated measurements.[8] Physical theories typically achieve clarity and conceptual economy by introducing various quantities that are not directly observable. They then owe us an account of how these theoretical quantities are revealed in properties that are accessible to us – that is, an account of how these quantities can be measured. Schematically, what is required is an explanation of what kind of system can be used as a measurement apparatus, with functional dependencies linking some readily accessible property of the apparatus (the "pointer-variable") to the theoretical quantity of interest. The functional dependencies linking the pointer-variable and target quantity follow from the dynamical description of the combined system (measuring device plus target). Further questions about the utility and reliability of the measuring apparatus can be answered based on this dynamical description (e.g., In what domains is a particular type of device a reliable way of measuring the target quantity?) As a simple illustration, according to Newtonian gravity, local surface

---

[7] This is one of the main themes of Kuhn's work; see, in particular, Kuhn (1961).

[8] Here I am following the account of measurements in Newton's physics developed by Harper and Smith (1995) and by Smith (2002); see also Chang (2004).

gravity (the target quantity) can be reliably measured by a pendulum (whose length is the pointer-variable). The domain in which pendulums of different types (small-arc circular, cycloidal, and so on) provide reliable measurements is determined by the theory.

There is clearly a great deal to be gained by accepting a theory. Theory-mediated measurements make it possible for a given body of data to constrain and inform a richer description of the phenomena. Yet there is also a risk to accepting a theory. The new quantities introduced may merely fit a given body of data, without accurately representing the phenomena. How should we avoid the fruitless line of research that would result from accepting such a theory? One response would be to demand a higher evidential threshhold for acceptance. This strikes me as misguided as a general policy, given the points made previously, in addition to being descriptively inaccurate (see also Kuhn, 1961). Scientists do often accept theories on the basis of remarkably weak evidence, with the expectation that subsequent work will reveal any mistakes. In most cases the only way to reveal mistakes or limitations is through a long-term assessment of the theory's potential, which is guided by the theory itself. The crucial question is whether the theory is subject to further critical evaluation, or if it instead shapes further inquiry in a way that shields it from ongoing scrutiny. This risk is particularly salient when accepting a theory that grants exclusive access to some set of novel quantities, in the sense that all measurements of a target quantity (or set of such quantities) rely on a single theory. If the interpretation of available data depends on the very theory that the data are being used to evaluate, there is an obvious risk of circularity.

One further clarification regarding what scientists accept or trust will be essential in what follows. In my view, the object of trust or acceptance is not restricted to theories, because what is used as a tool to guide further work may be more specific or more general than what philosophers usually call "theories." The functional dependencies supporting inferences from pointer-variable readings to target quantities may follow from a theory, but this need not be the case. In many fields, empirically discovered regularities, not yet integrated into a general theory, are used to gain access to new quantities.[9] In other cases, one example of which is discussed in this chapter, quite general background assumptions (shared by many theories) may suffice for some inferences from the data.

---

[9] Theories may sometimes support the idea that such a relationship should exist, while leaving the details to be fixed by observation; in other cases, empirically stable regularities may be discovered that cannot be accounted for theoretically. Understanding how these regularities fit into a larger theory makes it possible to answer a number of further questions, such as those regarding reliability and domain of applicability. But a regularity may support useful inferences from the data without being successfully integrated. Reliability may be established empirically through repeated successful applications, even if there are open theoretical questions as to why the regularity holds. I owe recognition of this point to conversations with George Smith.

On the one hand, scientists routinely make judgments about how evidence bears on claims that are more specific, or more general, than theories. For example, astronomers in the eighteenth century carefully analyzed which parts of Newton's characterization of gravity were actually supported by successful descriptions of orbital motion. This success bears directly on the dependence of gravitational force with distance, but the same cannot be said for the further claim that gravitational attraction is proprotional to the mass of the attracting body.[10] On the other hand, scientists often have compelling arguments for principles that stand above specific theories, such as symmetry principles in contemporary physics.

Philosophical accounts that treat questions of evidence as a relation between a collection of data and a monolithic "theory T" hide this complexity from view (see also Wilson, 2006). A more fine-grained analysis of which principles are actually relevant for specific lines of evidential reasoning makes it possible to see whether a set of results rely on independent principles. The evidence provided by convergence of independent lines of reasoning would be overlooked if the results are regarded as all simply following from "theory T." In addition, in some cases principles more general than a theory are sufficient to support inferences from the data, as we will see later when considering eliminative programs.

These considerations lead to a sharper formulation of the question of trust: How have scientists successfully made the transition from a risky, pragmatic choice to accept a theory, to taking it as a permanent contribution to our knowledge of the world and a reliable guide for action? And what role does the differential evaluation of components of a theory, or more general principles it satisfies, play in that transition? A theory is worthy of trust in the stronger sense to the extent that it is not merely compatible with the data, but is the only viable characterization of the law-like relationships that hold in the relevant domain. In the next section, I turn to the challenge that apparently arises when establishing trust due to the underdetermination of theory by evidence.

17.3 Underdetermination

One of Dawid's (2013) main themes is that trust, in the stronger sense, can be earned only through an effective response to underdetermination. While I agree on this point, the details of how scientists have effectively limited underdetermination

---

[10] As Euler and others pointed out, purely gravitational interactions among celestial bodies cannot distinguish between the contributions of the gravitational constant G and the mass of the attracting body M to the overall attractive force. The mass can be assigned freely if the value of G is allowed to vary, unless the third law is taken to apply to this interaction. (Euler, among others, regarded this extension of the third law to action between non-contiguous bodies as not sufficiently supported by direct evidence.) The strongest evidence that the force of gravity is proportional to M comes from experimental measurements establishing that G is a universal constant. See Smith (2007) for a detailed assessment of this issue.

in the past matter a great deal in trying to evaluate whether they can succeed, using similar approaches, in a context where theories cannot be developed and assessed through concrete application to accessible phenomena.

The first task is to clarify the nature of underdetermination and the threat it poses. There is a striking contrast between the treatments of underdetermination by philosophers and physicists: Philosophers, following the pronouncements of Quine and others, hold that *many* rival theories fare equally well with respect to some body of evidence. By contrast, physicists tend to emphasize how challenging it is to find even *one* theory compatible with the evidence (as emphasized by Norton, 1993). The contrast reflects differing views about the empirical content of theories and about what success with respect to some body of data implies.

Philosophers who regard rival theories as plentiful usually adopt a narrow conception of empirical content: It is limited to a set of "observational claims," or to an "empirical substructure" (part of a theoretical model). By contrast, I endorse Wilson's (1980) proposal that a theory's content should be characterized in terms of the possibilities it introduces for theory-mediated measurements. Rather than thinking of the evidence as exhausted by a set of observational claims, this view assumes that the evidence consists of a characterization of systems that qualify as measuring devices for specific quantities, along with the results of measurements performed using these devices (schematically, pointer-variable readings).

There are several reasons to prefer an understanding of empirical content along these lines. Perhaps the most compelling is that the dynamical description of measurements makes it possible to assess their stability and reliability, which are clearly both essential to scientific practice.[11] Measuring devices can often be thought of as amplifying some feature of a target system, so that it is registered in an accessible pointer-variable. When does the device reliably amplify the target quantity, and to what level of precision can it be trusted? Answers to these and related questions depend on the theoretical description of the functional dependence holding between the target quantity and the pointer-variable. Suppose that a community of scientists has established a reliable means of measuring a target quantity, at some level of precision, that is *stable* in the sense that repeated measurements in varying circumstances yield consistent results. This success lends support to the theoretical account of measurement, including its counterfactual implications (e.g., to what extent we could still use the measuring device confidently in a new setting). This extends beyond what would be included on a narrow construal of content.

Consider, for example, the use of the length of a pendulum (ℓ) of a given period to measure the variation of surface gravity with latitude [$g(\theta)$]. Using a narrow

---

[11] This is not the only reason to adopt an alternative conception of content; for further related discussions, see Bogen and Woodward (1988) and Roberts (2008).

construal of content, a rival theory would need to recover a set of data points (roughly, ordered pairs $\{\ell, \theta\}$); with the proposed alternative, by contrast, a rival theory would need to not just recover these results but also account for the reliability and limits of using pendulums for these measurements. Any rival theory recovering content in this stronger sense would agree on the fragment of Newtonian dynamics needed to establish the functional dependencies holding between ℓ and g.

Theories of broad scope typically provide multiple ways to measure theoretical quantities, making it possible to triangulate on them using independent methods. Continuing with the example, Newton argued that the same force causes the motion of pendulums and the Moon, because they both give consistent measurements, in this general sense, of the Earth's gravitational field. A further expectation of successful theories is that the multiple measurements they make possible will converge on a consistent assignment of theoretical quantities.[12] Insofar as each distinctive measurement method employs a different fragment of the theory, they provide independent constraints. The convergence of a set of independent measurements on a common value for the target quantity is a particularly powerful reply to a skeptic who holds that the theory succeeds by simply "fitting" parameters to match the data.

A crucial further aspect of assessment regards how the theory fares in response to the pressure of systematically improving standards of measurement precision. As noted earlier, a theory is often accepted to gain access to a new domain of phenomena. The initial applications of a theory typically start with simple models of the measurement interaction as well as the system being studied. As standards of precision improve, discrepancies between observations and these simple models are used to guide the development of more detailed models. A theory is successful to the extent that it can be consistently applied to develop successively more detailed and precise models of the relevant phenomena, without abandoning its core principles. The development of models incorporating further physical details to account for discrepancies also provides more opportunities for corroboration – for example, astronomical observations of Nepture, which was introduced to resolve discrepancies in Uranus's orbit.

Returning to the question of underdetermination, a theory that is successful in the sense of supporting stable, convergent measurements of the theoretical quantities it introduces, and that is consistently applied as measurement precision improves, leaves little room for rival theories covering the same domain. According to the proposal I have been considering, the content of the theory includes all the elements

---

[12] Harper (2012) argues that Newton introduced an "ideal of empirical success" along these lines. See Chapters 4 and 6 for a particularly thorough discussion of the comparison between pendulum results and the lunar orbit (called the "moon test"), which defends attributing this account to Newton.

needed to account for theory-mediated measurements of the relevant set of target quantities. If we construct a rival theory that captures the same content, as Wilson (1980) argues, we will recover very nearly the same theory – a reformulation rather than a true rival.[13]

There are two important qualifications to this uniqueness claim.[14] First, there may be "locally indinstinguishable" rival theories that share the same content in this sense, to some specified level of precision within a specified domain, yet differ in other domains. For example, by construction, general relativity reduces to Newtonian gravity in an appropriate limit. Within the domains to which Newtonian theory applies, general relativity is a locally indistinguishable rival.[15] Even though Newton's theory provides only an approximate description of the gravitational interaction, in light of general relativity, the errors introduced by using it are negligible in this domain. The existence of rivals in this sense does not undercut trust in the theory. If general relativity is true, then the Newtonian results hold as good approximations within the relevant domain. To undermine the reliability of Newtonian gravity in the relevant sense, we would need to discover a rival theory that substantively differs from it *even within this domain*.

The limits of the domain within which we can trust a given theory often become clear only retrospectively. The existence of locally indistinguishable rivals reflects the ineliminable risk associated with inductive generalizations. Newtonian gravity adequately describes the law-like relations revealed in solar system motions, but it would be a mistake to take these laws describing weak-field effects as characterizing the gravitational interaction more generally.[16] Of course, it is extremely difficult to assess to what extent the evidence available at a given stage of inquiry is parochial or limited in this sense. These limits are often uncovered by extrapolating the theory boldly and determining where it begins to lose fidelity.

Second, "rival research traditions" may differ sharply from an existing theory. Although this idea is admittedly somewhat vague, for my purposes the defining characteristic of such a rival is that it rejects a theory's claims to have achieved stable, convergent measurements of theoretical quantities. In other words, the initial theory failed to establish that there are real phenomena that must be preserved. From the standpoint of a rival research tradition, the apparent successes have to

---

[13] Wilson (1980) makes the case that classical mechanics is essentially the only theory compatible with the phenomena, in this rich sense, within its domain of applicability.

[14] There is a third qualification regarding the scope of this approach: My primary focus is on physical theories, which are distinctive in usually requiring an explicit account of measurement of novel quantities on their own terms. It is not clear how much of this approach carries over to areas of science that treat measurement in different ways – for example, if access to quantities is mediated by theories from another field.

[15] It is also an "unconceived alternative" in the sense discussed by Stanford (2006), because it was not explicitly formulated as an alternative when Newton's theory was originally accepted.

[16] See Smith (2014), especially §3, for a discussion of the relationship between Newtonian theory and general relativity, which I draw on here.

be explained away as misleading – the result of systematic errors, coincidences, or something else along these lines. Such explanations undercut the rationale for preserving the content of the theory in the sense defended previously. All that needs to be preserved is the raw data, although it may not have much intrinsic interest. The rival research tradition also has to provide some account of how to understand the earlier data, without recovering the theory-mediated measurements and associated commitments. Such rival research traditions typically pursue different aims for inquiry in a given area, prioritizing an altogether different set of problems and suggesting alternative approaches to resolving them.

17.4 Eliminating Rivals

To what extent have physicists been able to eliminate rival theories, in these two different senses? An answer to this question effectively rebuts a skeptic, who objects that the apparent success of a theory may merely reflect its flexibility, and is an essential part of establishing trust in the stronger sense. I will briefly discuss two concrete cases from the history of physics exemplifying successful strategies.[17] This is not intended to exhaust the approaches physicists have used in eliminating rival theories, but I expect that two cases will be sufficient to raise questions regarding whether similar approaches can be extended to supra-empirical physics.

First, physicists in many areas have pursued eliminative programs that proceed by constructing a space of possible theories, and then winnowing this down to a small subset – or even a single theory – compatible with the evidence.[18] In the first step, the parametrized possibility space includes known alternative theories as well as merely possible competitors that have not yet been explicitly formulated. This allows comparison between existing theories and an entire class of rival theories. In the second step, the implications of different types of data follow from assumptions held in common among all these competing theories. It is then possible to use data to constrain the space of allowed theories, ideally eliminating almost all the possibilities, without needing to perform calculations for each theory. Success in this program supports a local version of what Dawid calls the "no-alternatives argument," by unambiguously identifying the "best" theory among a clearly articulated space of competitors.

---

[17] In both cases, I am isolating one line of argument that has been used to make the case for the theory discussed; obviously, there is much more to be said about evidence in favor of general relatively or quantum electrodynamics than I have the space to discuss here.

[18] Dorling (1973), Earman (1992, Chapter 7), and Norton (1993) all emphasize the importance of eliminative inferences in physics; it also plays a central role in Kitcher's (1993) account of scientific progress.

To take one prominent example, the "parametrized post-Newtonian" (PPN) formalism is a systematic framework designed to allow observations to choose among general relativity (GR) and various competing accounts of gravity. It is relatively straightforward to compare competing theories in the post-Newtonian limit, in which the differences among spacetime metrics for a broad class of competing theories can be characterized in terms of 10 coefficients. The PPN formalism represents a "possible gravitational theory" with a point in a 10-dimensional parameter space, although distinct theories may correspond to the same point.

Two main claims justify taking this parameter space as delimiting the relevant alternatives. First, there is broad agreement regarding the domain of "gravitational" phenomena and how facts about this domain should be handled. Candidate theories are all expected to account for facts about this domain based primarily on the dynamics of the gravitational interaction. The second more substantive argument limits consideration to "metric theories" of gravity, in which gravity is treated as an effect of spacetime curvature (Will, 2014). The eliminative program thus begins with a carefully circumscribed set of possible theories. Over the last half-century, experimental tests have constrained all 10 PPN parameters to be very close to the values for GR (summarized in Will, 2014). At an abstract level, at least, these tests are relatively straightforward: Based on an account of some phenomena valid for all metric theories, including explicit dependence on the PPN parameters, observations are used to determine the best values of the parameters using standard statistical techniques. This is possible because the structure shared by metric theories is sufficient to link observations to the PPN parameters.

This argument for GR is local in the sense that it applies to this specific regime. This focus enables the construction of parametrized possibility space, and identification of common principles needed to link observations to parameters. Yet due to its local character, this argument leads to a limited conclusion: GR, or any locally indistinguishable rival, outperforms alternative metric theories of gravity in accounting for solar system dynamics. The map from specific PPN parameters to dynamical theories of gravity is one to many, however, so these results will not distinguish among metric theories that share the same limiting behavior in this regime. Because of this degeneracy, a thorough eliminative program requires local tests in a variety of regimes. GR and competing dynamical theories of gravity specify links among these various regimes, and the tests can be complementary if they have different degeneracies. Recent gravitational wave observations have provided an opportunity to pursue tests of GR in the strong-field regime, providing a powerful complement to the solar-system tests.

This case illustrates that tests in several distinct regimes can effectively eliminate locally indistinguishable rivals. Tests of gravitational effects in the strong-field regime and in cosmology promise to differentiate among theories that match GR's success in describing the solar system. This strategy is based on explicitly defining the space of possible theories in terms of common principles similar to those that hold in GR, along with assumptions regarding what domain of phenomena a gravitational theory should explain. These assumptions define a research tradition in gravitational physics. Our assessment of GR as a permanent contribution to science – a characterization of the gravitational interaction that will at least be recovered as a limiting case of some future theory – is based on accepting this research tradition. The argument in favor of this acceptance itself has a different character, based on assessment of how much progress has been made in developing increasingly detailed models of phenomena in light of a growing body of observational and experimental results.

Second, the convergence of theory-mediated measurements sometimes takes a particularly sharp form: multiple measurements of a single fundamental constant. Perrin's (1923) famous case in favor of the existence of atoms was based on 13 independent measurements of Avogadro's number. Physicists have often emphasized the "overdetermination of constants" (borrowing Norton's (2000) term) as a reply to skeptical worries that a theory fits the data merely because of its flexibility. The strength of this argument depends on two claims. As the number and diversity of ways of determining the value of a fundamental constant increase, the odds of attributing the agreement to systematic error associated with each measurement should decrease. The chance that the various experimental measurements would agree, even if the theory were fundamentally false, is also expected to decrease.

Consider, for example, the evidence in favor of quantum electrodynamics (QED) based on agreeing measurements of the fine-structure constant α. Remarkable levels of precision have been reached in low-energy tests of QED: State-of-the-art measurements of the anomalous magnetic moment of the electron (a_{e^-}) achieve a precision to better than one part in 10^{12} (Gabrielse et al., 2006). For these measurements, Gabrielse and his collaborators have effectively created something that is as close to a "pure QED system" as possible – in effect, a single-electron cyclotron. The system can be described without needing to worry about the complicated structure of protons, as would be required to perform spectroscopic measurements of comparable precision. QED determines a theoretical value of a_{e^-}, through a perturbative expansion in terms of the fine structure constant α. Schwinger's (1948) success in calculating corrections to Dirac's value for the magnetic moment of the electron initially inspired confidence in QED. Seventy years later, the quantitative comparison of QED and the world that has been achieved is simply astonishing: Theoretical calculations have now been carried out to 10th order in α, including a total of 12,672 Feynman diagrams (Aoyama et al., 2015). These results lead to a consistency check (rather than a direct prediction of a_{e^-}): Given an independent determination of α, the computed value of a_{e^-} can be compared with the precision measurements.[19]

QED underwrites numerous experimental measurements of α. Alongside high-precision measurements of a single electron in a Penning trap, the value of the fine-structure constant α can be determined based on atomic recoil experiments, spectroscopic measurements, the quantum Hall effect, the AC Josephson effect, and various scattering amplitudes (see, for example, Kinoshita, 1996). The agreement among these determinations is required to hold, insofar as QED correctly describes electromagnetism in these domains. The fact that they coincide within the bounds of experimental error provides powerful evidence in favor of QED: It provides a coherent account of a wide range of phenomena at an astonishing level of precision. As the number of theory-mediated measurements increases, along with their precision, it becomes more difficult to preserve the connections between these diverse experimental situations in a theory that truly differs from QED in the relevant domains.

The strength of the conclusion of overdetermination arguments depends on whether this agreement would be expected on other grounds. What is the probability that these various determinations of α would agree within experimental error if QED were false? The argument thus relies implicitly on an assessment of the space of competing theories.

Dawid (2013) advocates a "new paradigm of theory assessment" in which arguments to limit underdetermination play a central role. There is a family resemblance between the arguments Dawid identifies and the strategies just described. The "no-alternatives argument" (NAA) holds that a search for an alternatives to a given theory T can provide evidence for T if it fails to turn up a better alternative. Successfully implementing an eliminative program provides a precise version of this idea, by determining which theory among a space of competitors fares best with respect to the evidence. A theory's unified description of diverse phenomena, which were not taken into account when the theory was initially formulated, is an example of "unexpected explanatory coherence" (UEC). Several of the phenomena that now provide precision constraints on α, based on QED, played no role in the historical development of the theory, and hence provide examples of unexpected coherence. Finally, I noted earlier that theories are successful insofar as they are applied consistently even as the scope, detail, and precision of available

---

[19] The resulting comparison does not straightforwardly vindicate QED: There is a discrepancy between the QED calculation and observations. Agreement can be recovered by including hadronic loop contributions and contributions from the weak interaction. In other words, the low-energy precision tests have reached such remarkable precision that it is actually the Standard Model being tested, rather than QED alone.

observations increases. This idea presupposes that there is a distinction between maintaining the same strategy – maintaining core principles, while changing or adding details – through applciations. Dawid's third, "meta-inductive" argument (MIA) assumes a similar notion of constancy of research strategy: The provenance of a theory, as the output of a research strategy with a good track record, also provides evidence in its favor. Dawid takes a "good track record" to include, in particular, successful novel predictions.[20] The appealing idea is that these arguments have played an essential role in inquiry in the past, but – unlike more direct forms of empirical confirmation – they can continue to guide inquiry even as we depart the realm of the accessible.

This position, appealing though it may be for other reasons, is not compatible with the characterization of responses to underdetermination given previously. I have emphasized the local character of two arguments that have been important in eliminating rival theories. Any eliminative program has to start somewhere, with an explicit choice regarding which type of theories qualify as legitimate competitors. Similarly, the overdetermination of constants argument exemplified by QED requires some assessment of the space of rival theories. The assessment of relevant competing theories is based on what I have called, loosely, a research tradition. In the case of the PPN framework, the competing theories are all similar to GR: metric theories satisfying the Einstein equivalence principle, taking similar phenomena as their explanatory target. Following through on the eliminative program makes it possible to eliminate nearby rival theories in favor of GR. The assumptions defining this framework are motivated by the success of a long research tradition in gravitational physics.

This acknowledgment of the limitations of these strategies does not undercut trust in theories, however, because there is more to the story. The assessment of rival research traditions has a different character than these local responses to underdetermination. This kind of argument addresses a different concern. Once a theory is accepted as the basis for inquiry, much of the subsequent reasoning is heavily theory-dependent. How can the resulting line of inquiry be compared with a rival approach that shares little common ground?

Without minimizing the difficulties in doing so, in my view competing research traditions can usually be evaluated effectively based on concrete applications. In many cases, a theory-dependent line of reasoning leads to striking claims that can be evaluated by entirely independent methods. The validity and value of these claims is obvious, even without contentious decisions regarding how to characterize

---

[20] The language of "research strategies" is my own, whereas Dawid formulates his position in terms of a broader category, "research fields." There is an interesting question here regarding the appropriate unit of analysis, but I do not have the space to pursue it here – and it is not important for what follows.

the phenomena or how to explain them. An example discussed by Smith (2014) illustrates the general point beautifully. The Hill–Brown lunar theory, published in 1919, is an enormously complicated Newtonian description incorporating roughly 1400 different physical sources of perturbations. Brown discovered a discrepancy between this theoretical description and observed motion that he called the "Great Empirical Term." Jones (1939) argued that this discrepancy should be attributed to a fluctuation in the Earth's rotation with a 225-year period, which has subsequently been confirmed through a variety of independent methods. The agreement of these different precision measurements of the Earth's rotation is a striking result regardless of how the domain of gravitational phenomena is delineated.

Results of this sort can be decisive when assessing competing research traditions, and they provide compelling evidence that a research tradition has accurately identified law-like dependencies and fundamental quantities within a given domain. The application of a theory to an accessible domain of phenomena is essential to making this case. The challenge for supra-empirical physics is to find similarly compelling arguments when the fruits of successful applications are not available.

17.5 Eternal Inflation

Inflationary cosmology is often discussed, alongside string theory, as an example of "post-empirical" physics. The simplest models of inflation account for a variety of observed features of the early universe as the consequence of the dynamical evolution of a scalar field, which drove a transient phase of exponential expansion. In this section, I briefly consider how the line of thought developed previously applies in this case: To what extent can inflation be subjected to ongoing scrutiny if it is accepted, and is there a way to effectively eliminate rival theories?[21]

Observations have led to a remarkably simple picture of the early universe, which is well described by a flat Friedmann–Lemaître–Robertson–Walker (FLRW) model, with Gaussian, adiabatic, linear, nearly scale-invariant density pertubations. Inflation has remained the most widely accepted explanation of why the early universe has these features, even as three decades of increasingly precise observations of cosmic microwave background radiation (CMBR) have successfully ruled out competing proposals.[22] The details of inflation are not directly accessible to experiments, given the features of the "inflaton" field in the simplest models. In

---

[21] This line of argument is similar in some ways to recent prominent criticisms of eternal inflation due to Ijjas, Steinhardt, and Loeb (see Ijjas et al., 2014). Many of these points have been raised earlier, by critics including George Ellis, but these publications have brought renewed attention. For example, Ijjas et al. (2017), in a *Scientific American* article presenting their arguments at a more accessible level, led proponents of inflation to defend their theory in a letter to the editor signed by 33 prominent physicists and cosmologists.

[22] Observations of acoustic peaks lend support to the idea that the fluctuations are primordial, as they are in inflation, contrasting with predictions from competing models of structure formation based on active sources

principle an "inflaton" field has implications for accelerator physics, but in practice we will clearly never build an accelerator to probe its properties. Cosmological observations are thus the primary source of evidence for the inflaton.

Inflation might qualify as a "post-experimental," yet still observational science, if not for a further step. Many cosmologists hold that inflation is "generically eternal," in the sense that inflation produces a an ensemble of "pocket universes," quasi-isolated regions in which inflation has come to an end, within a still-inflating background (see, for example, Aguirre, 2007). The mechanism leading to this multiverse structure is typically assumed to produce variation in low-energy physics among the different pocket universes. This version of inflation, in particular, exemplifies the risk identified earlier: It provides a framework for further inquiry that shields basic theoretical assumptions from substantive tests.

For the sake of contrast, consider first a view of inflation that takes observations to constrain the properties of the inflaton field – in effect, determining the form of the Lagrangian for the inflaton field. Observations of the CMB and large-scale structure would give theory-mediated measurements of the Lagrangian in two main ways. Primordial fluctuations have implications for features of the inflaton field at a time well before the end of inflation.[23] The energy density of other matter fields is rapidly diluted during inflation, so most inflationary models require a phase of "reheating," during which the inflaton field decays into other particle species such that the post-inflationary state is compatible with a "hot" Big Bang. Observations can also be used to determine features of the inflaton Lagrangian, such as interaction terms with other fields, based on an account of reheating.

The best-case scenario for this kind of approach is obtained when the inflaton is identified within a specific particle physics model. The parameters appearing in the Lagrangian would then be constrained by cosmological data related to the details of inflation, along with experimental data relevant to the particle physics model. The promise of obtaining constraints from cosmology alongside particle physics was surely one of inflation's most appealing features. Agreement of theory-mediated measurements of the properties of the inflaton field from such strikingly different domains would be an astonishing example of unification.

Fulfilling this promise has proved an elusive task. No canonical candidate for the inflaton field has been widely accepted. There are now a wide variety of inflationary models, and the prospects for establishing a tighter link to particle physics appear

---

for fluctuations (such as topological defects). (See, for example, Durrer et al. [2002] for a review of structure formation via topological defects, and its contrasting predictions for CMB anisotropies.)

[23] Inflation generates scalar and tensor perturbations whose physical properties depend on the features of the effective potential $V(\phi)$ at horizon exit, with $\frac{k}{R} \approx H$. Perturbations relevant to CMB observations typically crossed the horizon at approximately 60 e-foldings before the end of inflation, whereas those reentering the horizon now were produced a few e-foldings later.

bleak. This certainly makes the prospects for observationally constraining the properties of the inflaton field more daunting. Even though cosmological observations can be used to constrain properties of the inflaton field, it is much harder to make the case that they are independent and that they provide consistent constraints on the same underlying feature.[24] Furthermore, in part because of the lack of a canonical model, inflation has not led to the identification of robust physical features of the early universe that can be tested independently. There is no analog of the discovery of Neptune based on the comparison of inflationary models with observations – that is, a result that is compelling independent of the research tradition that led to it.

In sum, on this first approach it is challenging to see how to make a more detailed empirical case for inflation, primarily due to the inaccessibility of the relevant energy scales. The compatibility of the observed state of the universe with inflation is a significant success. To date, however, this success has not been followed up with the development of a more detailed account of how inflation transpired. The worry is that the progress inflation has made is analogous to that in Ptolemaic astronomy: progress in developing a more precise model with better parameter fits, without leading to a more accurate model.

The situation is much worse if inflation leads to eternal inflation (EI). Accepting EI undermines the observational program of attempting to constrain and fix the features of the inflaton field.[25] Developing strong evidence, or eliminating rival theories, relies on the exactness of a theory, and EI is anything but exact.

The central question can be put quite simply: What do we gain by accepting that EI is true? Advocates of EI often claim that we can make statistical "predictions" about, for example, the value of fundamental parameters such as Λ. Several further questions need to be answered in the course of making these predictions. The first regard a physical characterization of the ensemble generated by inflation: How is the ensemble of pocket universes characterized? (For example, does a given fundamental constant or other aspect of low energy physics vary among the members of the ensemble?) How do we count different types of pocket universes in the ensemble – in other words, what is the measure over the ensemble? A distinct type of question regards the selection effect associated with the presence of observers like us. What part of the ensemble includes observers? What are the necessary conditions for observers like us? None of these questions have clear answers. But even if they are all satisfactorily resolved, the further assumptions needed to derive

---

[24] The second difficulty reflects the fact that the observations constrain different aspects of the inflaton Lagrangian: different parts of the effective potential $V(\phi)$ or interaction terms.

[25] Ijjas et al. (2014) argue that many of the defenders of inflation fail to acknowledge the full impact of shifting to this "post-modern" version of the theory (see also Smeenk, 2014, 2017).

"predictions" put substantial distance between the basic physical description of inflation and our assessment of it.

The appeal to anthropic selection effects is particularly problematic. Consider the "prediction" we obtain from EI for the value of some parameter α_i. One part of this calculation is based on evaluating the range of values of this parameter that are compatible with our existence. (Imagine varying α_i to see what range of values is compatible with the existence of certain physical systems that are necessary conditions for our presence, such as gravitationally bound systems at various scales.) The EI proponents see this calculation as clarifying which part of the ensemble we might inhabit, and they modulate their probabilistic predictions in light of the selection effect. Why not take the calculation of the range of values α_i as in itself explaining why we observe the value that we do? This calculation would then "screen off" the alleged explanatory value of the multiverse. Perhaps the multiverse can be said to explain how it is possible that α_i has the value that it does. A much cheaper explanation is available, which simply takes the parameter value as a contingent feature of the universe. More importantly, this line of argument does not provide information about the kinds of law-like relations among parameters or features of the universe that would support ongoing scrutiny of the framework.

Acknowledgments

I have benefied from questions and discussions with several people following talks based on this material, including Peter Achinstein, Andy Albrecht, Radin Darshati, George Ellis, Helge Kragh, and Kyle Stanford. It is a pleasure to thank Richard Dawid for discussions as well as comments on an earlier draft of the chapter. The title alludes to a paper by George Smith, whose influence will be apparent to those familiar with his work. Finally, I gratefully acknowledge that work on this chapter was supported by a grant from the John Templeton Foundation. The views expressed here are those of the author and are not necessarily endorsed by the John Templeton Foundation.

References

Aguirre, Anthony. (2007). Eternal inflation, past and future. arXiv preprint arXiv:0712.0571.

Aoyama, Tatsumi, Hayakawa, Masashi, Kinoshita, Toichiro, and Nio, Makiko. (2015). Tenth-order electron anomalous magnetic moment: Contribution of diagrams without closed lepton loops. *Physical Review D* 91, 033006.

Bogen, James, and Woodward, James. (1988). Saving the phenomena. *Philosophical Review* 97, 303–52.

Chang, Hasok. (2004). *Inventing Temperature: Measurement and Scientific Progress.* Oxford: Oxford University Press.

Dawid, Richard. (2013). *String Theory and the Scientific Method.* Cambridge: Cambridge University Press.

Dorling, Jon. (1973). Demonstrative induction: Its significant role in the history of physics. *Philosophy of Science* 40, 360–72.

Durrer, Ruth, Kunz, Martin, and Melchiorri, Alessandro. (2002). Cosmic structure formation with topological defects. *Physics Reports* 364, 1–81.

Earman, John. (1992). *Bayes or Bust?* Cambridge, MA: MIT Press.

Gabrielse, G., Hanneke, D., Kinoshita, T., Nio, M., and Odom, B. (2006). New determination of the fine structure constant from the electron *G* value and QED. *Physical Review Letters* 97, 030802.

Harper, W. L. (2012). *Isaac Newton's Scientific Method: Turning Data into Evidence about Gravity and Cosmology.* Oxford: Oxford University Press.

Harper, W. L., and Smith, George E. (1995). Newton's new way of inquiry. In *The Creation of Ideas in Physics*, ed. Jarrett Leplin, Vol. 55 of *University of Western Ontario Series in Philosophy of Science.* Dordrecht: Springer Verlag, 113–66.

Ijjas, Anna, Steinhardt, Paul J., and Loeb, Abraham. (2014). Inflationary schism. *Physics Letters B* 736, 142–46.

   (2017). Pop goes the universe. *Scientific American* 316, 32–9.

Jones, H., Spencer. (1939). The rotation of the earth, and the secular accelerations of the sun, moon and planets. *Monthly Notices of the Royal Astronomical Society* 99, 541–58.

Kinoshita, Toichiro. (1996). The fine structure constant. *Reports on Progress in Physics* 59, 1459.

Kitcher, Philip. (1993). *The Advancement of Science.* Oxford: Oxford University Press.

Kuhn, Thomas S. (1961). The function of measurement in modern physical science. *Isis* 52, 161–93.

Laudan, Larry, and Leplin, Jarrett. (1991). Empirical equivalence and underdetermination. *Journal of Philosophy* 88, 449–72.

Norton, John D. (1993). The determination of theory by evidence: The case for quantum discontinuity, 1900–1915. *Synthese* 97, 1–31.

   (2000). How we know about electrons. In *After Popper, Kuhn and Feyerabend: Recent Issues in Theories of Scientific Method*, ed. R. Nola and H. Sankey. Dordrecht: Kluwer Academic Publishers, 67–97.

Perrin, Jean. (1923). *Atoms*, trans. D. L. Hammick. New York: Van Nostrand.

Roberts, John T. (2008). *The Law-Governed Universe.* Oxford: Oxford University Press.

Schwinger, Julian. (1948) On quantum-electrodynamics and the magnetic moment of the electron. *Physical Review* 73, 416.

Smeenk, Chris. (2014). Predictability crisis in early universe cosmology. *Studies in History and Philosophy of Science Part B: Studies in History and Philosophy of Modern Physics* 46, 122–33.

   (2017). Testing inflation. In *The Philosophy of Cosmology*, ed. Khalil Chamcham, John D. Barrow, Simon Saunders, and Joseph Silk. Cambridge: Cambridge University Press, 206–27.

Smith, George E. (2002). The methodology of the *Principia*. In *Cambridge Companion to Newton*, ed. I. B. Cohen and George E. Smith. Cambridge: Cambridge University Press, 138–73.

   (2007). The question of mass in Newton's law of gravity. Manuscript of talk given in Leiden and Austin.

(2014). Closing the loop. In *Newton and Empiricism*, ed. Zvi Biener and Eric Schliesser. Oxford: Oxford University Press, 262–351.

Stanford, P. Kyle. (2006). *Exceeding Our Grasp: Science, History, and the Problem of Unconceived Alternatives*. Oxford: Oxford University Press.

Will, Clifford M. (2014). The confrontation between general relativity and experiment. *Living Reviews in Relativity* 17, 4.

Wilson, Mark. (1980). The observational uniqueness of some theories. *Journal of Philosophy* 77, 208–33.

(2006). *Wandering Significance: An Essay on Conceptual Behavior*. Oxford: Oxford University Press.

Part IV

Prospects for Confirmation in String Theory

18

String Theory to the Rescue

JOSEPH POLCHINSKI

The search for a theory of quantum gravity faces two great challenges: the incredibly small scales of the Planck length and time, and the possibility that the observed constants of nature are in part the result of random processes. A priori, one might have expected these to be insuperable obstacles. However, clues from observed physics and the discovery of string theory raise the hope that the unification of quantum mechanics and general relativity is within reach.

18.1 The Planck Scale

The question of how we are to understand quantum gravity is vitally important, and in this chapter I look at it in a broad way. I begin with the first, and perhaps the most important, calculation in this subject. This is Planck's 1899 use of dimensional analysis, combining the speed of light c, the gravitational constant G, and his own constant $\hbar$, to identify a fundamental length scale in nature [1],

$$l_P = \sqrt{\hbar G/c^3} = 1.6 \times 10^{-33} \text{ cm.} \qquad (18.1)$$

Planck had a strong sense of the importance of what he had done:

These necessarily retain their meaning for all times and for all civilizations, even extraterrestrial and non-human ones, and can therefore be designated as *natural units*.

I want to ask the question:

In light of this calculation, what would Planck have predicted for the state of fundamental physics today, nearly 120 years later?

Of course, prediction in science is hard, but the incredibly small scale of the Planck length is a powerful fact. In Planck's time, it had taken approximately 300 years to move *four* orders of magnitude, from the 10^{-4} cm achievable by optical microscopes to the 10^{-8} cm of the atomic scale. Now he was arguing that to reach

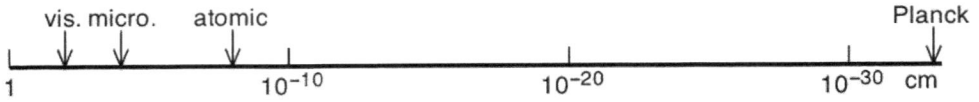

Figure 18.1 The gulf to be crossed.

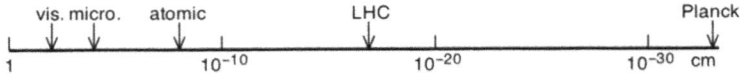

Figure 18.2 Progress to date.

the natural length scale of nature one would have to improve by a further *25* orders of magnitude (Figure 18.1).

The first thing that Planck might have predicted is a meeting to explore the relevant issues. He could anticipate that this natural scale would remain inaccessible to direct observation for a very long time, perhaps forever. So we would be confronted with the problem of how we are to proceed.

It is interesting to consider the situation if Planck's calculation had worked out to 10^{-17} cm. This would still have seemed remote to him, nine orders of magnitude past the atomic scale. But, in fact, this is the length scale that the Large Hadron Collider (LHC) is reaching today (Figure 18.2). If this had been the Planck length, we would not be sitting around here whining about falsifiability, but instead anticipating a wonderland of experimental quantum gravity.

I am a practical physicist; I don't use expressions like "post-empirical." But physics is driven by numbers and scales, and the fact that the Planck length is 10^{-33} cm and not 10^{-17} cm very much affects the tools that I can use and the rate at which I can expect to make progress. Theories in the making are judged in a different way than finished theories. Moreover, because the Planck scale is so remote, this first period is likely to be much longer than we are used to. For those who say that science has changed, what has changed is the magnitude of what we are trying to do, of what we need to do given what nature has presented us with.

So how are we to proceed? After moving four orders of magnitude in 300 years, and then nine orders in a little more than 100 years, we might continue to progress incrementally, and hope that some sort of scientific Moore's law will carry us the remaining 16 orders of magnitude in time. Certainly, there is no shortage of effort going into the next power of 10, and there are many indirect probes of higher energy physics, though almost all of these have been negative. Even so, we have reason to hope that we are in a position to leap to the answer.

There have been times when theory has been able to leap a smaller gap: from Maxwell to light; from Dirac to antiparticles; from the Standard Model to the gluon,

W, *Z*, top, and Higgs. The gap here is vastly larger than any we have encountered before, but at the same time our theoretical sophistication has grown, and we have great theories – quantum mechanics and relativity – to build on.

There is a danger of defining science too rigidly, so that one might decree that any discussion of the physics at 10^{-33} cm is unscientific because it is beyond the reach of direct observation. This was the attitude for a long period after Planck, lingering even when I was a graduate student. Such a perspective makes science too weak – that is, to decree that some aspects of the natural world are outside of its domain. We have to make the effort; it may prove to be fruitless, or premature, but I will argue that the situation is not so bad.

Let me continue my Planckian prediction game. One scenario Planck might have imagined is that his calculation was overly pessimistic. Dimensional analysis works only if one has identified all the relevant constants, and perhaps he had the wrong ones. But the subsequent history seems to have borne out his choice. Each of his constants, c, G, and $\hbar$, was about to launch a scientific revolution – special relativity, general relativity, and quantum mechanics – and these revolutions remain the center of our understanding of space, time, matter, and even reality. Further, various ways in which quantum gravity might have manifested at longer distances – large extra dimensions, low-scale strings, and violations of Lorentz invariance (though this is problematic, as I will discuss later) – have thus far failed to appear. Thus, we don't seem to be working with the optimistic scenario.

At the opposite extreme, Planck might have imagined that quantum gravity would largely be ignored, for how can science progress without observation? Of course, given the importance of the problem, physicists would continue to dabble in it from time to time, and a few would devote themselves to it. In fact, this is not the present situation, either.

What has actually happened is something that would have been difficult to predict. A large fraction of the theoretical physics community is working on quantum gravity, not with the enthusiasm that they would have had in the optimistic scenario, but certainly with far more enthusiasm and a greater sense of progress than in the pessimistic scenario. Something has happened, and that something is the discovery of string theory.

Of course, what I have described so far is a sociological phenomenon, albeit the reflection of a scientific one. But let me first analyze the sociology a bit further. One might think that my claim is circular: "Of course, string theorists work on string theory." But science does not fall into such narrow silos. Indeed, it is striking that, with few exceptions, those who have contributed major ideas to string theory have also made major contributions to other areas of science, including particle physics [2–10], quantum field theory [11–16], mathematics [17–21], condensed matter physics [22–26], general relativity [27–30], cosmology [31–34], nuclear physics [35–38], and most recently quantum information theory [39, 40].[1]

There is a rather silly criticism of string theory [41], that string theorists are not enough like Einstein. Such a perspective focuses on Einstein's philosophical bent. But there was more to Einstein: In addition to his work on relativity, Einstein explained the photoelectric effect in terms of the quantization of light and Brownian motion in terms of atoms, predicted stimulated emission, and understood the heat capacity of matter, among many other insights. It seems that there is a unity to physics, so that the ability to perceive new principles is not limited to narrow areas.

As the references cited in the preceding paragraph attest, the scientists working on string theory are theoretical physicists in a broad sense (and many of those working on other approaches to quantum gravity suffer for a lack of this viewpoint). Their willingness to engage in such research also helps to explain how string theory is able to remain vital even with limited data. In particular, I am happy to see that many young people appear as authors in the references list.

18.2 String Theory

I am arguing that the discovery of a theory like strings is a surprise, something that Planck could not have anticipated. To see why, I will enumerate five features: the solution to the short distance problem, the uniqueness of the dynamics, the unification of physics and geometry, the duality between gauge fields and strings, and new insights into the quantum mechanics of black holes.

18.2.1 The Short Distance Problem

The first question to ask is: What do our existing theories predict for physics at shorter distances?

If we simply take Einstein's theory and feed it into the path integral, applying the standard recipe for quantizing a classical theory, we get infinities. Again this is implicit in Planck's calculation (1): Setting $\hbar = c = 1$, the gravitational coupling G has units of length (cm) squared. If we probe the physics at some length scale d, the dimensionless coupling is G/d^2, growing stronger as $d \to 0$. In the quantum theory, there will be virtual effects running over all values of d, giving a divergent result even for observations on longer distance scales. The typical observation would be the scattering of gravitons, or more geometrically the measurement of quantum corrections to the solutions to Einstein's equations. In the language of particle

---

[1] I felt it important to illustrate this point with examples, but I have certainly omitted other scientists and papers of equal importance. Most of these examples involve physicists who are generally identified as string theorists contributing in other areas, but some (e.g., [9, 10, 35, 40]) are physicists from other areas employing ideas from string theory, usually to the enrichment of both fields, while others are a combination of the two (e.g., [19, 20, 23, 25, 26, 28, 29, 33, 36]).

physics, this is the problem of nonrenormalizability; more geometrically, it is the problem of spacetime foam, in which quantum fluctuations tear spacetime apart at short distances.

This problem is not unique to gravity. The Fermi theory of the weak interaction also has a coupling constant (G_F) with units of cm^2. This was a key clue that allowed theorists to predict the W, Z, and Higgs, and most of their detailed properties, before experiments gave any direct sign that they even existed – one of the gaps that has successfully been leapt.

To see why this clue is so powerful, imagine that we try to develop a new theory that solves this problem, by smearing out the interaction so that it is not so strong at short distances. Because of special relativity, smearing in space implies smearing in time, so we lose causality or unitarity and our theory does not make sense. It would be easy to cure the problem, and to find many candidate theories of the weak interaction or of gravity, if we were willing to give up Lorentz invariance. But Lorentz invariance has been confirmed to very high accuracy. Even if it were broken only at very short distances, virtual effects would transmit large breaking to observed quantities [42]. Thus, this is the first thing that Planck might not have anticipated – that our theories of physics would become so tightly intertwined as to allow few ways forward.

In the case of the weak interaction, introducing the W, Z, and Higgs produced the necessary smearing in a physically sensible way. In the case of gravity, the problem is more challenging. Rather than just adding a few particles, it seems necessary to change the nature of all particles, from points to strings. At least, that was the first thing that worked.

There are other ideas out there, but I have reservations about most or all of them (for example, many violate Lorentz invariance). In this chapter, however, I am trying to explain not why theorists work on one theory of quantum gravity rather than another, but rather why they work on quantum gravity at all. I am arguing that string theory is successful on an absolute scale, not a relative one. This requires positive arguments, not negative ones, and so we move on to the next positive.

18.2.2 Uniqueness of Dynamics

A remarkable feature of string theory is that the dynamics, the equation of motion, is completely fixed by general principle. This is consistent with the overall direction of fundamental theory, describing the vast range of phenomena that we see in terms of fewer and fewer underlying principles. Uniqueness would seem to be the natural endpoint to this process, but such theories are truly rare.

General relativity for a time seemed to have this property, with the equivalence principle determining the form of the Einstein equation. Of course, the cosmological constant was a confusion, and the equation does not describe matter, but more importantly the restriction to terms linear in the curvature was artificial. The equivalence principle allows terms quadratic in the curvature, and cubic, and so on. Indeed, even if one tries to omit them, they will be generated by quantum effects [43]. The reason that we do not usually discuss such terms is that their effects are minuscule for dimensional reasons – Planck's calculation again – but as a matter of principle they will be there.

In quantum field theory, we also have many choices: gauge symmetries, field content, and couplings, including higher derivative terms. There seems to be nothing else like string theory.[2] Its existence is a surprising discovery, and here I emphasize the word "discovery": It is a mathematical-physical structure that exists, and we are discovering it.

Indeed, when I assert that the equations of string theory are fully determined by general principle, I must admit that we do not yet know the full form of the equations, or the ultimate principle underlying them. String theory was discovered in an incomplete and approximate form, and our understanding of it has gradually deepened. In fact, string theory is no longer about the strings; our best understanding would put the holographic principle closer to the center. Even so, the uniqueness was visible even in the first form: The condition of world-sheet conformal invariance $T^a_a = 0$ implies Einstein's equation plus matter, and the coupling constant becomes a field. This property continues to hold in anti-de Sitter/conformal field theory (AdS/CFT).

18.2.3 Physics from Geometry

In general relativity, gravity comes from the curvature of spacetime. Unity of physics would suggest that the other interactions arise in this way as well. But we have used up the evident part of spacetime in accounting for gravity, so there must be more to it. The uniqueness of string theory actually forces more on us, extra dimensions plus dual structures such as branes. It seems that there is about the right amount of structure to describe the physics that we see. In particular, our laws of physics arise from the geometry of the extra dimensions. Understanding this geometry ties string theory to some of the most interesting questions in modern mathematics; it has also shed new light on them, such as mirror symmetry.

Coming from a background in quantum field theory and particle physics, I find this is less remarkable than most of the other features on my list, but it accounts for some of the great interest in string theory. The physicists who work on string theory come from a great variety of scientific backgrounds (consider the references list [2–40]) and would be expected to have very different intuitions as to what a fundamental theory should look like – yet string theory appeals to all of them.

---

[2] Supergravity comes closest, but its symmetries still allow an infinite number of higher derivative terms, and the concept does not include matter.

18.2.4 Duality between Gauge Fields and Strings

A surprising discovery in recent decades about the structure of quantum physics is that duality is a common property. The word "quantize" suggests a one-to-one correspondence: Go from the classical theory to the quantum theory by quantizing (for example, with the path integral), and go back again by taking the classical limit. For a long time, thinking about quantum field theory in particular proceeded in this way. Now, however, we know that many rich quantum theories have multiple classical limits, and that these can look quite different from one another.

Most remarkable is the duality between gauge theory and gravity [16] (presaged in [44–46]). A theory of quantum gravity, with all its puzzles, can be obtained by quantizing a gauge theory, something that we are very familiar with. Even quantum chromodynamics (QCD), the theory of the strong interaction, contains quantum gravity, albeit in an unfamiliar regime of high curvature. The gravitational theory obtained by this duality is restricted to spacetimes with special boundary conditions, but within this boundary one can have rich dynamics, including the graviton scattering discussed earlier, change of the topology of spacetime, or the black hole thought experiments to be discussed later. Of course, extending this theory to more interesting spacetimes is a key research direction.

Moreover, when quantum gravity emerges in this way, so does the rest of string theory.[3] Indeed, our most complete description of string theory results from this process. It reproduces the approximate forms that we know in various limits, and fills in the parameter space between them. Moreover, the strings themselves are emergent: The starting point is no longer a theory of strings, and the principle by which the whole theory emerges is now best understood as holography.

Recognizing a common origin for diverse phenomena has been one of the ongoing successes of science. Maxwell's understanding of light in terms of electricity and magnetism was one of the great steps forward. The duality between gravity and gauge theory is equal to this in intellectual magnitude and surprise.

18.2.5 Quantum Mechanics of Black Holes

Nonrenormalizability is the immediate problem that one encounters when one tries to quantize gravity, but other challenges are present as well. Quantum mechanics and general relativity seem to give different pictures of the black hole. In general relativity, the black hole is hairless. In quantum theory, it has a temperature and entropy, and so should have a statistical mechanical description in terms of a

---

[3] To be precise, if we have a theory of quantum gravity in AdS, it defines formally a field theory (CFT) on its boundary. But for all CFT's for which we have an independent construction (e.g., via a Lagrangian), the emergent theory is string theory.

definite number (the exponential of the Bekenstein–Hawking entropy) of micro-scopic states. This is the entropy puzzle: What are these states? Beyond this is the information paradox: Black hole evaporation seems to destroy information. To be precise, pure states appear to evolve into mixed states [47], in contradiction to the ordinary laws of quantum mechanics. This is problematic, but so are the alternatives.

Strominger and Vafa showed that the microstates could be understood in string theory, in agreement with the Bekenstein–Hawking count [48]. Moreover, gauge/gravity duality shows that black hole evaporation must take place within the ordinary framework of quantum mechanics, contradicting Hawking. This ability to shed light on difficult decades-old puzzles that are far from its original motivation is another success of the theory.

The black hole puzzles largely drove the discovery of gauge/gravity duality, forcing theorists to better elucidate the relation between the dynamics of black branes and D-branes. It seems that these puzzles will continue to be a fruitful source of insight and new ideas [49].

18.2.6 Summary

Of the five properties I have discussed, I regard three (discussed in Sections 18.2.1, 18.2.3, and 18.2.5) as very important, and two (Sections 18.2.2 and 18.2.4) as remarkable. Finding all of them in a single theory tells me that we have been fortunate. In spite of the great gulf that faces us, we have learned enough from what we can see to figure out what lies on the other side.

18.3 The Multiverse

Two of the positive features on my list – physics from geometry, and uniqueness of dynamics – come with a dark side. If the physics that we see is determined by the geometry of spacetime, what determines that geometry? It should solve Einstein's equation, or something like it. But how many solutions does Einstein's equation have that look something like our universe, with four large dimensions and the rest small and compact? This question was first addressed by Calabi and Yau, who provided the framework to show that the number is large, perhaps billions, even if we restrict ourselves to spaces with only the metric and no other fields.[4] A more complete count is very indefinite, with the number 10^{500} often used as a stand-in (this arises as a small integer raised to a large power coming from

---

[4] To be precise, each of these has moduli, so the spaces are actually a continuous family of solutions (formally the number of solutions is billions times infinity). Realistic solutions [33, 50] do not have moduli.

topology [31, 51, 52]).[5] Even Einstein encountered a small landscape, the radius of the Kaluza–Klein theory. If he had known about the nuclear forces, he would been led to the billion or so solutions of Calabi and Yau.

Thus, the laws that we see do not follow uniquely from the dynamics, but depend also on the specific solution. This is the price we pay for getting physics from geometry and for having unique dynamical laws. Of course, one of the central features of physics is that simple equations can have many and complicated solutions. But we did not expect, or want, our observed physical laws to depend on the solution.

So what determines which solution, and laws, we find around us? One might think that this outcome depends on the initial conditions, so we need to develop a theory of these. Indeed, we do need a theory of the initial conditions, but it will probably do little to solve our problem, because quantum mechanics and relativity conspire to hide the past. If we start in any solution with a positive vacuum energy, it will inflate. Quantum mechanics will cause small regions to tunnel into any of the other vacua, and the process can repeat indefinitely. We end up with a vast universe, with all possible solutions realized in one place or another. This has come to be called the multiverse. The things that we are trying to explain, the observed laws of nature, then do not follow from the fundamental theory but rather vary from place to place; in other words, they are environmental.

In fact, there is reason to believe that we live in just such a universe [55]. The Standard Model vacuum is a rich place, with zero point energies, Higgs fields, quark condensates, color fluctuations, and so on. Why, then, is there not an enormous vacuum energy? Virtually everyone interested in fundamental theory has wracked their brain over this question. For very general reasons, all known theories that predict a definite value of the cosmological constant give either an enormous value or exactly zero. In the latter case, this is accompanied by unbroken supersymmetry, which we do not have. Theories that do not predict a definite value are of two types: Either they have free parameters, so that the cosmological constant can be set to any value at all (fine-tuning), or they have a multiverse. The former case seems unsatisfactory; ultimately, we expect that there will be a theory with definite dynamics, and we will be in one of the other cases. In the latter case, different values of the cosmological constant will be realized in different regions. However, most of these regions will be boring, without interesting structure. For structure to develop requires many degrees of freedom, large volume, and long times; these will be realized only in the rare regions where the cosmological constant is small. This problem was analyzed by Weinberg [56], who sharpened arguments made by others [57–60]. Thus, the only

---

[5] A recent paper [53] suggests a much larger number The early paper [54] gave the number 10^{1500}, but it is not clear to what extent these represent discrete points in a much smaller set of moduli spaces.

known class of theories that have definite dynamics and are consistent with our observations of the universe are those that give a multiverse.

Before 1997, it was widely assumed that the cosmological constant was exactly zero for some reason (perhaps a symmetry) that had not yet been discovered. The only framework that predicted otherwise was the multiverse [56], since the cosmological constant need only be small for structure to develop, and an exact zero value would be a set of measure zero. Thus, the discovery of the vacuum energy in 1997 came as a great surprise to all but two classes of people: those who had paid attention to the data without theoretical prejudice (because there had long been evidence of a cosmological constant) and those who knew how hard the problem was and had paid attention to Weinberg's prediction. Speaking as one of the latter, my attitude toward the vacuum energy was not so much expectation as fear, and the hope that the evidence would go away, because a vacuum energy would bring the multiverse into physics.

It is often said that the multiverse is not predictive, but in a very real sense the exact opposite is true. In the last half-century, there has been no more surprising, or more important, discovery about the fundamental nature of our universe than the vacuum energy. This was the prediction most worth making, and only the multiverse made it. Of course, this is a special circumstance coming from the extreme discrepancy between other theories and the observed value, and for many other quantities it is unpredictive.

However, we do not get to decide how predictive the laws of nature are – how much is random or environmental and how much is fixed. It is something that we have to discover. Of course, if the answer is that we live in a less predictable universe, it will be much more difficult to know that this is right. But we can figure it out.

To conclude this section, I will make a quasi-Bayesian estimate of the likelihood that there is a multiverse. To establish a prior, I note that a multiverse is easy to make: It requires quantum mechanics and general relativity, and it requires that the building blocks of spacetime can exist in many metastable states. We do not know if this last assumption is true. It is true for the building blocks of ordinary matter, and it seems to be a natural corollary to getting physics from geometry. So I will start with a prior of 50%. I will first update this with the fact that the observed cosmological constant is not enormous. Now, if I consider only known theories, this pushes the odds of a multiverse close to 100%. But I have to allow for the possibility that the correct theory is still undiscovered, so I will be conservative and reduce the no-multiverse probability by a factor of 2, to 25%. The second update is that the vacuum energy is not exactly zero. By the same (conservative) logic, I reduce the no-multiverse probability to 12%. The final update is the fact that our outstanding candidate for a theory of quantum gravity, string theory, most likely predicts a multiverse.[6] But again I will be conservative and take only a factor of 2. So this is my estimate for the likelihood that the multiverse exists: 94%.[7]

Occam may be respected in some circles where Bayes is not, so let me also express this in terms of his razor, as in [61]. Occam charged for assumptions. As a physicist, it seems that physics from geometry and the uniqueness of dynamics are a minimal count to the razor. Yet, it has always amazed me that general relativity does not charge for volume and quantum mechanics does not charge for branches: These are no count against the razor. The way that physics creates a rich universe from simple laws is that simple assumptions lead to rich dynamics, and the multiverse may be more of the same.

This is not to say that the multiverse is on the same footing as the Higgs or the Big Bang. A probability of 94% is two sigma; two sigma effects do go away (though I factored in the look-elsewhere effect; otherwise, I would get a number much closer to 1). The standard for the Higgs discovery was five sigma, 99.9999%.

18.4 The Next 120 Years

I started by talking about the great barrier posed by the Planck scale. If the physical laws that we directly see are environmental rather than fixed, then we face a second barrier that is just as great. You can disagree with my 94% estimate, but there is no rational argument that a multiverse does not exist, or even that it is unlikely.

Supposing that both barriers do exist, one might think that progress will be impossible. Indeed, if one takes an overly rigid definition of science, then science will be useless by definition. But one should not be so pessimistic – or so dismissive of science. We have perhaps had two pieces of bad luck, but also one of good luck: String theory exists, and we have found it.

So how to go forward? Experimentalists must explore all avenues, and theorists should follow their own instincts. Other authors will talk about the ongoing program to connect string vacua to observation. For me, the challenge is clear: to complete string theory, the same way that Einstein finished general relativity, using the kind of tools that have brought us to our current understanding. Obviously it will take longer, given that string theory encompasses the quantum as well as the relativity. It may require concepts that we do not yet suspect. Eventually, though, we will succeed. This subject is currently in an exciting state, with new ideas from quantum information theory [39, 40] and the black hole information problem [49 et seq.].

---

[6] With a few exceptions, those who argue in the opposite direction are indulging in wishful thinking. By the way, I count the prediction of a multiverse as a sixth unexpected success of string theory: It accounts for the small but nonzero cosmological constant.

[7] For those who find this calculation amusing, I ask you: How many of you expected a nonzero cosmological constant in 1997? If not, perhaps you should be a bit more humble. Like Dick Cheney pontificating on Iraq, past performance is an indicator of future results.

So what is my prediction for 2131? It is that we will have figured out the theory of quantum gravity, and that it will be built upon string theory. What and how much it will be able to predict I cannot say; how much can be predicted is something for us to discover. Nevertheless, unification has always led to unexpected insights and predictions, so I will be optimistic and predict that when we have crossed the greatest gap and completed the greatest unification, we will know that we are correct and the result will be wonderful. Of course, I am working hard to reduce the interval: I want to know the answer.

Acknowledgments

I am grateful to Richard Dawid and his collaborators Radin Dardashti, George Ellis, Dieter Lust, Joseph Silk, and Karim Thébault, and to the Munich Center for Mathematical Philosophy, LMU Munich, and the Arnold Sommerfeld Center for Theoretical Physics, LMU Munich, for sponsoring the meeting "Why Trust a Theory?: Reconsidering Scientic Methodology in Light of Modern Physics," held in Munich on December 7–9, 2015. I thank Raphael Bousso, Lars Brink, Andre Linde, Dave Morrison, and Bill Zajc for communications. The work of J.P. was supported by NSF Grant PHY11-25915 (academic year) and PHY13-16748 (summer).

References

[1] M. Planck, "About irreversible radiation processes," *S.-B. Preuss. Akad. Wiss.*, S.479–80 (1899).

[2] A. A. Migdal and A. M. Polyakov, "Spontaneous breakdown of strong interaction symmetry and the absence of massless particles," *Sov. Phys. JETP* **24**, 91 (1967) [*Zh. Eksp. Teor. Fiz.* **51**, 135 (1966)].

[3] D. J. Gross and F. Wilczek, "Ultraviolet behavior of nonabelian gauge theories," *Phys. Rev. Lett.* **30**, 1343 (1973).

[4] L. Susskind, "Dynamics of spontaneous symmetry breaking in the Weinberg–Salam theory," *Phys. Rev. D* **20**, 2619 (1979).

[5] M. Gell-Mann, P. Ramond, and R. Slansky, "Complex spinors and unified theories," *Conf. Proc. C* **790927**, 315 (1979) [arXiv:1306.4669 [hep-th]].

[6] R. J. Crewther, P. Di Vecchia, G. Veneziano, and E. Witten, "Chiral estimate of the electric dipole moment of the neutron in quantum chromodynamics," *Phys. Lett. B* **88**, 123 (1979) [*Phys. Lett. B* **91**, 487 (1980)].

[7] M. Dine, W. Fischler, and M. Srednicki, "A simple solution to the strong CP problem with a harmless axion," *Phys. Lett. B* **104**, 199 (1981).

[8] M. W. Goodman and E. Witten, "Detectability of certain dark matter candidates," *Phys. Rev. D* **31**, 3059 (1985).

[9] N. Arkani-Hamed, S. Dimopoulos, and G. R. Dvali, "The hierarchy problem and new dimensions at a millimeter," *Phys. Lett. B* **429**, 263 (1998) [hep-ph/9803315].

[10] L. Randall and R. Sundrum, "A large mass hierarchy from a small extra dimension," *Phys. Rev. Lett.* **83**, 3370 (1999) [hep-ph/9905221].

[11] A. M. Polyakov, "Particle spectrum in the quantum field theory," *JETP Lett.* **20**, 194 (1974) [*Pisma Zh. Eksp. Teor. Fiz.* **20**, 430 (1974)].

[12] T. Banks and A. Zaks, "On the phase structure of vector-like gauge theories with massless fermions," *Nucl. Phys. B* **196**, 189 (1982).

[13] J. Polchinski, "Renormalization and effective Lagrangians," *Nucl. Phys. B* **231**, 269 (1984).

[14] E. Witten, "Quantum field theory and the Jones polynomial," *Commun. Math. Phys.* **121**, 351 (1989).

[15] N. Seiberg, "Electric–magnetic duality in supersymmetric nonAbelian gauge theories," *Nucl. Phys. B* **435**, 129 (1995) [hep-th/9411149].

[16] J. M. Maldacena, "The Large N limit of superconformal field theories and supergravity," *Int. J. Theor. Phys.* **38**, 1113 (1999) [*Adv. Theor. Math. Phys.* **2**, 231 (1998)] [hep-th/9711200].

[17] P. Candelas, X. C. De La Ossa, P. S. Green, and L. Parkes, "A pair of Calabi–Yau manifolds as an exactly soluble superconformal theory," *Nucl. Phys. B* **359**, 21 (1991).

[18] E. Witten, "Monopoles and four manifolds," *Math. Res. Lett.* **1**, 769 (1994) [hep-th/9411102].

[19] A. Strominger, S. T. Yau, and E. Zaslow, "Mirror symmetry is T duality," *Nucl. Phys. B* **479**, 243 (1996) [hep-th/9606040].

[20] A. Connes, M. R. Douglas, and A. S. Schwarz, "Noncommutative geometry and matrix theory: Compactification on tori," *JHEP* **9802**, 003 (1998) [hep-th/9711162].

[21] H. Ooguri, A. Strominger, and C. Vafa, "Black hole attractors and the topological string," *Phys. Rev. D* **70**, 106007 (2004) [hep-th/0405146].

[22] D. Friedan, Z. A. Qiu, and S. H. Shenker, "Conformal invariance, unitarity and two-dimensional critical exponents," *Phys. Rev. Lett.* **52**, 1575 (1984).

[23] G. W. Moore and N. Read, "Nonabelions in the fractional quantum Hall effect," *Nucl. Phys. B* **360**, 362 (1991).

[24] J. Polchinski, "Effective field theory and the Fermi surface," in *Boulder 1992, Proceedings: Recent directions in particle theory*, 235–74 [hep-th/9210046].

[25] S. Ryu and T. Takayanagi, "Holographic derivation of entanglement entropy from AdS/CFT," *Phys. Rev. Lett.* **96**, 181602 (2006) [hep-th/0603001].

[26] S. A. Hartnoll, P. K. Kovtun, M. Muller, and S. Sachdev, "Theory of the Nernst effect near quantum phase transitions in condensed matter, and in dyonic black holes," *Phys. Rev. B* **76**, 144502 (2007) [arXiv:0706.3215 [cond-mat.str-el]].

[27] E. Witten, "A simple proof of the positive energy theorem," *Commun. Math. Phys.* **80**, 381 (1981).

[28] D. Garfinkle, G. T. Horowitz, and A. Strominger, "Charged black holes in string theory," *Phys. Rev. D* **43**, 3140 (1991) [*Phys. Rev. D* **45**, 3888 (1992)].

[29] T. Damour and A. M. Polyakov, "String theory and gravity," *Gen. Rel. Grav.* **26**, 1171 (1994) [gr-qc/9411069].

[30] G. W. Gibbons, "Born–Infeld particles and Dirichlet p-branes," *Nucl. Phys. B* **514**, 603 (1998) doi:10.1016/S0550-3213(97)00795-5 [hep-th/9709027].

[31] R. Bousso and J. Polchinski, "Quantization of four form fluxes and dynamical neutralization of the cosmological constant," *JHEP* **0006**, 006 (2000) [hep-th/0004134].

[32] J. M. Maldacena, "Non-Gaussian features of primordial fluctuations in single field inflationary models," *JHEP* **0305**, 013 (2003) [astro-ph/0210603].

[33] S. Kachru, R. Kallosh, A. D. Linde, and S. P. Trivedi, "De Sitter vacua in string theory," *Phys. Rev. D* **68**, 046005 (2003) [hep-th/0301240].

[34] M. Alishahiha, E. Silverstein, and D. Tong, "DBI in the sky," *Phys. Rev. D* **70**, 123505 (2004) [hep-th/0404084].

[35] P. Kovtun, D. T. Son, and A. O. Starinets, "Viscosity in strongly interacting quantum field theories from black hole physics," *Phys. Rev. Lett.* **94**, 111601 (2005) [hep-th/0405231].

[36] H. Liu, K. Rajagopal, and U. A. Wiedemann, "Calculating the jet quenching parameter from AdS/CFT," *Phys. Rev. Lett.* **97**, 182301 (2006) [hep-ph/0605178].

[37] S. Bhattacharyya, V. E. Hubeny, S. Minwalla, and M. Rangamani, "Nonlinear fluid dynamics from gravity," *JHEP* **0802**, 045 (2008) doi:10.1088/1126-6708/2008/02/045 [arXiv:0712.2456 [hep-th]].

[38] S. S. Gubser, "Symmetry constraints on generalizations of Bjorken flow," *Phys. Rev. D* **82**, 085027 (2010) [arXiv:1006.0006 [hep-th]].

[39] F. Pastawski, B. Yoshida, D. Harlow, and J. Preskill, "Holographic quantum error-correcting codes: Toy models for the bulk/boundary correspondence," *JHEP* **1506**, 149 (2015) [arXiv:1503.06237 [hep-th]].

[40] A. Kitaev, "Hidden correlations in the Hawking radiation and thermal noise," Breakthrough Prize Fundamental Physics Symposium November 10, 2014, KITP seminar February 12, 2015; "A simple model of quantum holography (part 1)," KITP seminar April 7, 2015.

[41] L. Smolin, *The Trouble with Physics,* ch. 18, "Seers and Craftspeople," Boston: Houghton Mifflin Harcourt (2006).

[42] J. Collins, A. Perez, D. Sudarsky, L. Urrutia, and H. Vucetich, "Lorentz invariance and quantum gravity: An additional fine-tuning problem?," *Phys. Rev. Lett.* **93**, 191301 (2004) [gr-qc/0403053].

[43] J. F. Donoghue, "General relativity as an effective field theory: The leading quantum corrections," *Phys. Rev. D* **50**, 3874 (1994) [gr-qc/9405057].

[44] G. 't Hooft, "A planar diagram theory for strong interactions," *Nucl. Phys. B* **72**, 461 (1974). doi:10.1016/0550-3213(74)90154-0

[45] A. M. Polyakov, "Gauge fields and strings," *Contemp. Concepts Phys.* **3**, 1 (1987).

[46] T. Banks, W. Fischler, S. H. Shenker, and L. Susskind, "M theory as a matrix model: A conjecture," *Phys. Rev. D* **55**, 5112 (1997) [hep-th/9610043].

[47] S. W. Hawking, "Breakdown of predictability in gravitational collapse," *Phys. Rev. D* **14**, 2460 (1976).

[48] A. Strominger and C. Vafa, "Microscopic origin of the Bekenstein–Hawking entropy," *Phys. Lett. B* **379**, 99 (1996) [hep-th/9601029].

[49] A. Almheiri, D. Marolf, J. Polchinski, and J. Sully, "Black holes: Complementarity or firewalls?," *JHEP* **1302**, 062 (2013) [arXiv:1207.3123 [hep-th]].

[50] E. Silverstein, "(A)dS backgrounds from asymmetric orientifolds" [hep-th/0106209].

[51] L. Susskind, "The anthropic landscape of string theory," In Bernard Carr, ed., *Universe or multiverse?*, 247–66 [hep-th/0302219].

[52] M. R. Douglas and S. Kachru, "Flux compactification," *Rev. Mod. Phys.* **79**, 733 (2007) [hep-th/0610102].

[53] W. Taylor and Y. N. Wang, "The F-theory geometry with most flux vacua" [arXiv:1511.03209 [hep-th]].

[54] W. Lerche, D. Lust, and A. N. Schellekens, "Chiral four-dimensional heterotic strings from selfdual lattices," *Nucl. Phys. B* **287**, 477 (1987).

[55] A. Linde, "A brief history of the multiverse" [arXiv:1512.01203 [hep-th]].

[56] S. Weinberg, "The cosmological constant problem," *Rev. Mod. Phys.* **61**, 1 (1989).

[57] P. C. W. Davies and S. D. Unwin, "Why is the cosmological constant so small," *Proc. Roy. Soc. Lond.*, **377**, 147 (1981).

[58] A. D. Linde, "The inflationary universe," *Rept. Prog. Phys.* **47**, 925 (1984), doi:10.1088/0034-4885/47/8/002.

[59] A. D. Sakharov, "Cosmological transitions with a change in metric signature," *Sov. Phys. JETP* **60**, 214 (1984) [*Zh. Eksp. Teor. Fiz.* **87**, 375 (1984)] [*Sov. Phys. Usp.* **34**, 409 (1991)].

[60] T. Banks, "T C P, quantum gravity, the cosmological constant and all that . . . ," *Nucl. Phys. B* **249**, 332 (1985).

[61] A. Vilenkin and M. Tegmark, www.scientificamerican.com/article/multiverse-the-case-for-parallel-universe/.

19

Why Trust a Theory? Some Further Remarks

JOSEPH POLCHINSKI

19.1 Introduction

The meeting "Why Trust a Theory? Reconsidering Scientific Methodology in Light of Modern Physics," which took place at the Ludwig Maximilian University in Munich in December, 2015, was for me a great opportunity to think in a broad way about where we stand in the search for a theory of fundamental physics.

In this follow-up discussion, I have two goals. The first is to expand on some of the ideas presented in the previous Chapter, and to emphasize some aspects of the discussion that I believe need more attention. As the only scientific representative of the multiverse at that meeting, a major goal for me was to explain why I believe with a fairly high degree of probability that this is the nature of our universe. But an equally important goal was to explain why, contrary to what many believe, string theory has been a tremendous success and remains a great and inspiring hope for our ultimate success in understanding fundamental physics. For natural reasons, certain aspects of the multiverse discussion have gotten the greatest initial reaction. I will explain why some things that have been most widely discussed are actually the least important, while other things are much more central. The second goal is to respond to some of the other speakers, as I was unexpectedly unable to attend the actual meeting.

To summarize the main theme of Chapter 18, nature has presented us with two great challenges. The first is that the Planck length – the fundamental length scale at which quantum mechanics and relativity come together – is incredibly small. The second is the possibility that the constants of nature that we are working so hard to measure at the Large Hadron Collider (LHC) and elsewhere are environmental. If so, then they are very different from the constants of nature that we have usually dealt with, such as those in the theories of the weak or strong interaction. There, from a very small number of inputs from the Weinberg–Salam theory or quantum chromodynamics (QCD), one can predict with very high precision hundreds of measurable properties of elementary particles. This is "normal physics." But if some of the constants of nature, like the masses of the quarks and leptons, are environmental, it means that these are much more loosely connected to the fundamental mathematics that underlies physics. There is an essential randomness to them, not unlike the weather. This is not "normal physics," and it is not what anyone wants – but unfortunately there is significant evidence that this is the case.

These difficulties seem often to be presented as failures of string theory, as though trying a different idea will somehow make the Planck length larger or force that environmental property not to hold. In reality, these are challenges for humankind as a whole, at least for those of us who care about the fundamental laws of nature. My conclusion is that these challenges are not insuperable, that we will succeed in figuring things out, and that string theory is a large part of the answer.

In this chapter, both Sections 19.2 and 19.3 are largely concerned with the multiverse. In Section 19.2, I explain some statistical issues that have attracted wide attention. In Section 19.3, I review my own long scientific interaction with the idea of a multiverse, because it provides a good framework for explaining how and why we have come to this idea. In Section 19.4, I briefly discuss some critics of string theory.

19.2 It's Not about the Bayes. It's about the Physics.

In section 18.3 of Chapter 18, I made an important distinction between the 5-sigma or 99.9999% probability for the Higgs boson to exist, and a 2-sigma or 94% probability for the multiverse to exist. For a particle physicist, the difference between such numbers is a familiar thing. However, I have seen that for members of the public, for many cynics, and even for some philosophers, this distinction is not so obvious, and so I will explain it further. The 5-sigma standard for the Higgs has been widely discussed in many blogs and other public sources, so I will not expand on it here. It is the meaning of the 2-sigma that I want to explain.

I am coming at this issue as an active scientist trying to solve a scientific problem – in my case, the theory of quantum gravity. Every day I need to make my best estimate of where to put my effort. This, of course, is not unique to the problem that I am working on. For example, every experimental or theoretical particle physicist, when confronted with evidence for some new particle or dark matter observation, wants to know how strong is the evidence – that is, what is the sigma – the physicist will know how much effort to apply to the problem.

There is inevitably a significant amount of personal taste and judgment, and there is no right answer at this level for a "correct" value of sigma. Science works better if different people try different assumptions. For my own problem, it is essential to factor in which ideas I and others have already tried. This has been criticized in some quarters, and indeed it might not be the proper approach for a discovery measurement like the Higgs. But for my purpose it is obviously necessary.

In the end, my estimate simply came down to combining four factors of 2. This led to the probability 94% that the multiverse exists. If I had quoted this value in binary, probability 0.1111, the scientific content would have been exactly the same but it would have led to much less merriment.

My estimate uses a kind of analysis known as Bayesian statistics. There has recently been much discussion of this method, partly in response to the Munich meeting. So let me state very clearly: Bayesian analysis is not the point. It is not even 1% of the point. Every word spent on this subject is a wasted word; in fact, it has negative value because it distracts attention from the real point, which is the physics. I blame myself for this kerfuffle, because I chose to frame the problem in this way. So let me summarize what the relevant physics is. In the next section, I give a longer discussion of some of the ideas and my own personal history with them.

The main issue is the vacuum energy, the cosmological constant. This goes back a long way – to Pauli, who argued in 1920 that the zero point energy of the radiation field would cause the radius of the universe to be no larger than the distance to the Moon [1]. In the ensuing century, there has been essentially no progress on this problem using "normal" science. Indeed, the difficulty has been sharpened: The vacuum energy is too large by a factor of 10^{120}, though this can be reduced to 10^{60} in theories with low energy supersymmetry.

So normal science has a problem: It gives a universe that does not look like ours. But if the cosmological constant is environmental – if it depends on space or time – then suddenly we get a universe like ours, where observers find themselves in a large universe with small curvature.

The environmental theory comes with a great surprise. In the context of normal science, it was almost universally assumed that the actual value of the cosmological constant was exactly zero, and this was what we were trying to explain. But in the environmental theory there is no reason for it to be zero, and in fact the value zero would be infinitely unlikely. The prediction is just that the cosmological constant be small enough (more on this later). Of course, this is what was discovered in 1998 [2, 3], to the great surprise of many – but not those who had been following Weinberg's environmental approach.

Finally, string theory, which I credit as an enormous success for reasons reviewed in section 18.2 of Chapter 18, leads to environmental physics, the string landscape.

In all, there are four arguments here for the multiverse: the failure of conventional methods in explicating why the cosmological constant is not large, the success of the environmental theories in doing so, the successful prediction of the nonzero cosmological constant, and the string landscape. This is the physics, with no reference to Bayes.

Returning to Bayes, I chose to package these four numbers each as a factor of 2, a nice mnemonic for the four kinds of evidence, leading to the 94% probability. Perhaps it was silly to do so, but expressing things in terms of numbers is a good practice, provided this information is approached in the correct spirit. That is, we are not talking about discovery mode like the Higgs, but rather trying to make a best estimate of our current state of knowledge as guidance to how to move forward. Of course, those who choose to belittle others' efforts will always do so, but these are not the ones to listen to.

Obviously there is no precision to our number. Andrei Linde would argue for a larger number [4], taking into account additional cosmological evidence beyond the cosmological constant. I would have come up with a larger number if I had given more weight to my long history of trying to understand the cosmological constant by normal methods, using both my own ideas and those of others. However, I thought it important to factor in possible unknown unknowns. While I believe that the road to quantum gravity will go through string theory, new and game-changing concepts may be encountered along the way.

19.3 The Multiverse and Me

I have sometimes seen the assertion that the multiverse is largely the resort of older theorists who, having run out of ideas or time, are willing to give up the "usual" process of science and use the multiverse as an excuse [5]. In my own experience, the opposite is true. My own interaction with the multiverse began at the age of 34, and I think that this is much more typical. In this section, I review my own personal associations with the multiverse, because I believe that it is instructive in a number of ways. I also discuss some of the experiences of two scientists with whom I have been closely associated, Steven Weinberg and Raphael Bousso. Other important perspectives would include those of Andrei Linde [4], Tom Banks [6] (though he is no longer an advocate of these ideas), Bert Schellekens [7], and Lenny Susskind [8].

I began to think about the problem of the cosmological constant as a postdoctoral student, around 1980. The then-new idea of inflation had emphasized the importance of the vacuum energy, and of the connections between particle physics and cosmology more generally. The presumed goal was to understand why the value of the cosmological constant was exactly zero. I went through the usual attempts and tried to find some new wrinkles, a process that began with Pauli and has certainly been repeated by countless others over the years. I failed to explain why the cosmological constant would vanish.[1] I added this to my list of the most important questions to keep in mind, believing that we would one day solve it and that it would be a crucial clue in understanding fundamental physics.

I eventually got a good opportunity to review my attempts at solving the problem [14]. Bousso has presented a similar review [15]. Gerard 't Hooft had proposed for many years to write a review, "100 Solutions to the Cosmological Constant Problem, and Why They Are Wrong." He never did so, but decided that it would be a good exercise for his graduate student [16].

In 1988, Weinberg similarly set out to review the possible solutions to the cosmological constant problem, but he approached it in a particularly broad and original way [17]. Weinberg is a great scientist who has two qualities that I find distinctively striking, and that are special to him. As science progresses, it often happens that certain assumptions become widely accepted without really being properly inspected. Eventually someone asks the question, "What if B is true instead of A?" The ensuing investigation may reveal that B is a perfectly good possibility, and a new set of ideas emerges. Weinberg is not the only person who asks such questions, but when he does his approach is both distinctive and powerful. Most of us will say, "Let us see what happens if we make a model in which B is true instead of A. But this looks hard, so we'll also assume C and D so we can get an answer." Weinberg does not take these shortcuts. As much as possible, he tries to ask, "If B is true instead of A, then what is the most general set of consequences that one can draw?" Generally one cannot make progress without some additional assumptions, but Weinberg's effort is always to make these minimal, and most importantly to state as precisely as possible any such extra assumptions being made. This approach takes great clarity of mind, so as to formulate such a general question in a way that can be usefully answered, and great integrity so as not to take shortcuts. It is an approach that Weinberg has applied many times, usually in more ordinary areas of science, to great effect.

The second and related quality is that he takes an unusually broad and independent point of view as to what the possibilities for B might be. One example is asymptotic safety, the possibility that the short distance problem of quantum gravity is solved not by new physics like string theory, but rather by a nontrivial UV fixed point of the renormalization group [18]. A second example, the one to which Weinberg turned in his 1988 review, is the possibility that the cosmologi-

---

[1] As an aside, there is a certain collection of ideas that incorporate some aspects of the environmental approach, in that the value of the cosmological constant can change in time or in different branches of the wavefunction [6, 9–12]. However, for these the cosmological constant changes in a rather systematic way, which is amenable to "normal science." By contrast, in the environmental approach the calculation of the cosmological constant is a problem of a particularly hard sort, NP (nondeterministic polynomial time) hard [13].

cal constant might be environmental, varying from place to place. Other theorists had already begun to explore this idea [6, 19–21], but Weinberg brought to it his unique strength. He argued that most places will be uninteresting, with space either expanding so rapidly as to be essentially empty, or contracting to a singularity long before anything interesting can happen. But there will be a sweet spot in between, where spacetime looks much like what we see around us. Here galaxies would form, and life would presumably follow by the usual processes, and eventually creatures would form who could do cosmological measurements. Moreover, because this can happen only in the sweet spot, theorists will find a value for the cosmological constant that is extremely small, just as we find.

I saw Weinberg's work at first hand, because at the time I was his colleague at the University of Texas at Austin. The effect of his analysis on me was profoundly upsetting. If correct, he had solved the problem that I had failed to solve, but he had done so in a way that took away some of our most important tools. I had hoped to make progress in fundamental physics by finding the symmetry or dynamics that would explain why the cosmological constant is exactly zero. But now it was no longer zero. Even worse, it is not a number that we can calculate at all, because it takes different values in different places, with a vast number of possible values running over different solutions.

So I very much wanted Weinberg to be wrong, and I hoped that the cosmological constant would turn out to be exactly zero. Unfortunately, there was already some evidence that it was not. There was the age problem, the fact that some stars seemed to be older than the universe. And there was the missing mass, the fact that the total amount of matter that could be found seemed to account for only 30% of the energy budget of the universe. Neither of these problems requires the cosmological constant, but the simplest way to make it all balance would be if the cosmological constant were nonzero. There was also a sign that Weinberg might be wrong: His estimate for the cosmological constant came out a factor of 10^2 or 10^3 too large. But missing by 10^3 is a lot better than missing by 10^{60}, and one could see that by changing some of the details of Weinberg's argument one might end up with a much closer number. So I spent the next 10 years hoping that the age problem and the missing mass problem would go away, but they did not. Finally, in 1998, the cosmological constant was confirmed as the correct answer.

Fortunately, string theorists had something much more interesting to think about in 1998, so the cosmological constant, important as it was, sat on the back burner for a while. The second superstring revolution began in 1995. Over a period of four years, we discovered dualities of quantum field theories, dualities of string theories, duality between quantum field theories and string theories (that is, anti-de Sitter/conformal field theory [AdS/CFT]), D-branes, matrix theory, and a quantitative understanding of black hole entropy. Our understanding of the basic structure of fundamental theory had moved forward in a way that seems almost miraculous. One cannot compare this to something as profound as the discovery of quantum mechanics or calculus, but there are few points in the history of science when our depth of understanding have moved forward so rapidly, and I feel fortunate to have been part of it.

The capabilities that we had before and after the second superstring revolution are much like what the rest of the world experienced before and after the World Wide Web. If I had to choose only one of these, I know which it would be.

Much of the initial analysis during the second superstring revolution focused on states with some amount of supersymmetry, because these new tools that we were learning are most tractable in this setting. But all of these ideas apply to the full quantum theory, states of all kinds. By 1999, the principles had been understood, and it was time to think about the implications for more physical systems.

Fortunately, Raphael Bousso was then visiting from Stanford University, and both he and I had a strong sense that string theory as now understood provided the microscopic theory that was needed for Weinberg's environmental theory to work. In his spirit of minimal assumptions, he had been able to formulate the problem in a way that did not require knowing the actual microscopic theory. We had the perfect combination of skills, comprising my understanding of the dynamics of strings and Raphael's understanding of cosmology. We understood why many attempts to find a microscopic model of Weinberg's theory had failed, and we could see how it all worked in string theory. We were quickly able to make a simple but convincing model that showed string theory was very likely to lead to exactly the kind of environmental theory that Weinberg had anticipated [22]. I say "quickly," but there was a significant delay due to my own personal misgivings with the landscape, something that I will discuss later in this section. But we published our model, and other string theorists began to show in detail that this is actually how string theory does work [23, 24].

My misgivings with Weinberg's solution had not ended, but in fact had gotten worse. Planck had taught us that because his scale was so small, the possible measurements of fundamental physics would be very limited. Now Weinberg, with Raphael's and my own connivance, was saying that even these few things were actually largely random numbers. We had gotten far, but it seemed that we would never get farther. The Bayesian estimate suggested that there was a 94% probability that our clues are just random numbers. What, then, could I work on? Shortly before the 1998 discovery of the cosmological constant, I told Sean Carroll that if the cosmological constant were found, he could have my office, because physics would be over.

In fact, even when Raphael and I were writing our paper, I had pushed him to remove some of the most anthropic parts of the discussion. It was not that I thought that they were wrong – we were in perfect agreement about the physics – but I felt that it was too discouraging. In effect, we would be telling the experimentalists, who were spending billions of dollars and countless human-years of brilliant scientists, that they were essentially measuring random numbers, and I did not want to be the bearer of bad news. But for the most part Raphael's point of view prevailed: We both agreed clearly on what the science was saying, and in the end one must be true to the science. I am not by nature a radical, but seem to have become one both with the multiverse and with the black hole firewall [25] just by following ideas to their logical conclusions. In both cases it helped to have brilliant and bold young collaborators, Raphael in the one case and Don Marolf in the other, who drove me to suppress my natural conservatism.

Still my anxiety grew, until eventually I needed serious help. So you can say quite literally that the multiverse drove me to a psychiatrist.[2]

In the end, I have come to a much more positive point of view, as expressed in Chapter 18. We are looking after all for the theory of everything. Actually, this is an expression that I dislike, and never use, because "everything" is both arrogant and ill defined. But I will use it on this one occasion to refer to the set of ideas that we are thinking about in quantum gravity, plus possibly some more that will turn out to be part of the story. There will not be a unique route to the theory of everything. If one route becomes more difficult, another will work. It will likely require clues from many directions – particle physics, cosmology, certainly string theory, and likely other areas as well – but we will succeed.

19.4 Some Critics

19.4.1 Some Comments on Ellis, Silk, and Dawid

In late 2014, George Ellis and Joseph Silk published in *Nature* a criticism of string theory and of the multiverse – indeed, an argument that it is not even science [27]. They argued that the issue was so important that a conference should be convened to discuss it. I am very grateful that they followed through on this suggestion, because it is indeed a vital issue, and it has given me a venue to present a much more positive and inspiring point of view. I wish that I had done so long ago.

I would like to make a few comments on some specifics of the arguments made by Ellis and Silk. One of the arguments that underlies their discussion is that physics is simply getting too hard, so we will never get the answer. Indeed, this is a logical possibility, and I worry about it. But the lesson of Planck is that we are working on

---

[2] In truth, I should have gotten help for anxiety sooner, and for more general reasons. One should not be reluctant to seek help.

a problem that is very hard, so we should think about this on a very long timescale. We have been working on Planck's problem for nearly 120 years and have made enormous progress. It is far too early to say that we have reached the end of our rope. It is a race between finding the key clues and running out of data and ideas, and I very much expect that we (humankind) will win.

A second point of disagreement is the claim of Ellis and Silk that unimodular gravity is a satisfactory alternative theory of the cosmological constant [27]. Unimodular gravity is Einstein's theory with one fewer equation, the one with the cosmological constant. But the value of the cosmological constant matters, because it affects physics, so we need some way to get that last piece. In unimodular gravity, effectively the value of the cosmological constant is assumed to be measured, and whatever value we get is taken as the missing piece. One gives up on trying to ever understand the particular value, which one finds to be around 7×10^{-27} kg m^3. Why is the value that one finds not the zero that Einstein expected? Why is it not the 10^{60} larger that Pauli expected? These points simply cannot be explained. This does not seem to me a satisfactory way for theory to work. Moreover, it seems to me that it is *less* predictive than the multiverse. Why the cosmological constant is not enormous cannot be explained in unimodular gravity, nor can it be explained why it has a small nonzero value. In contrast, the multiverse explains both. So I would think that Ellis and Silk would regard the multiverse as a better scientific theory, but they have a different take.

However, like Ellis and Silk, I am troubled by the expression "non-empirical theory confirmation." Perhaps I have not understood clearly what is meant by this. If someone is interpreting the rules in such a way that string theory is already at 5-sigma level, discovery, I cannot agree. How could I, when we do not even know the full form of the theory? And what is the rush? Again, long ago Planck taught us that it will take a long time to solve difficult problems.

By comparison, I think Dawid means something different, and more consistent with how I see things [28]. In particular, he seems to be using the same kind of evidence that I argue for in Chapter 18. Coming back to the theme of sigma values, I counted five kinds of evidence in section 19.2 and one more in section 19.3. If I just make the same crude counting of a factor of 2 for each, I end up with probability 98.5% for string theory to be correct. But here, where most of the focus is on quantum theory rather than the multiverse, I am on grounds that I understand better, and I would assign a higher level of confidence, something greater than 3-sigma.

Although these statistical games are fun, they are not the real point. There is some fascinating physics to explore right now, this is what we want to be doing, and it may transmute our understanding in ways that we cannot anticipate.

Acknowledgments

I think Raphael Bousso, Melody McLaren, and Bill Zajc for advice and extensive proofreading. I also thank C. Brown and L. Mezincescu for their typo-finding and comments. The work of J.P. was supported by NSF Grant PHY11-25915 (academic year) and PHY13-16748 (summer).

References

[1] N. Straumann, The history of the cosmological constant problem, [gr-qc/0208027].

[2] A. G. Riess et al. [Supernova Search Team Collaboration], Observational evidence from supernovae for an accelerating universe and a cosmological constant, *Astron. J.* **116**, 1009 (1998), doi:10.1086/300499 [astro-ph/9805201].

[3] S. Perlmutter et al. [Supernova Cosmology Project Collaboration], Measurements of omega and lambda from 42 high redshift supernovae, *Astrophys. J.* **517**, 565 (1999), doi:10.1086/307221 [astro-ph/9812133].

[4] A. Linde, A brief history of the multiverse, arXiv:1512.01203 [hep-th].

[5] https://4gravitons.wordpress.com/2016/01/01/who-needs-non-empirical-confirmation/.

[6] T. Banks, T C P, quantum gravity, the cosmological constant and all that . . . , *Nucl. Phys. B* **249**, 332 (1985), doi:10.1016/0550-3213(85)90020-3.

[7] A. N. Schellekens, Life at the interface of particle physics and string theory," *Rev. Mod. Phys.* **85**(4), 1491 (2013), doi:10.1103/RevModPhys.85.1491 [arXiv:1306.5083 [hep-ph]].

[8] L. Susskind, The anthropic landscape of string theory, in Bernard Carr, ed., *Universe or multiverse?*, 247–66 [hep-th/0302219].

[9] S. W. Hawking, "The cosmological constant is probably zero," *Phys. Lett. B* **134**, 403 (1984), doi:10.1016/0370-2693(84)91370-4.

[10] L. F. Abbott, "A mechanism for reducing the value of the cosmological constant, *Phys. Lett. B* **150**, 427 (1985), doi:10.1016/0370-2693(85)90459-9.

[11] J. D. Brown and C. Teitelboim, Neutralization of the cosmological constant by membrane creation, *Nucl. Phys. B* **297**, 787 (1988), doi:10.1016/0550-3213(88)90559-7.

[12] S. R. Coleman, "Why there is nothing rather than something: A theory of the cosmological constant, *Nucl. Phys. B* **310**, 643 (1988), doi:10.1016/0550-3213(88)90097-1.

[13] F. Denef and M. R. Douglas, Computational complexity of the landscape. I., *Annals Phys.* **322**, 1096 (2007), doi:10.1016/j.aop.2006.07.013 [hep-th/0602072].

[14] J. Polchinski, The cosmological constant and the string landscape, hep-th/0603249.

[15] R. Bousso, TASI lectures on the cosmological constant, *Gen. Rel. Grav.* **40**, 607 (2008), doi:10.1007/s10714-007-0557-5 [arXiv:0708.4231 [hep-th]].

[16] S. Nobbenhuis, Categorizing different approaches to the cosmological constant problem, *Found. Phys.* **36**, 613 (2006), doi:10.1007/s10701-005-9042-8 [gr-qc/0411093].

[17] S. Weinberg, The cosmological constant problem, *Rev. Mod. Phys.* **61**, 1 (1989), doi:10.1103/RevModPhys.61.1.

[18] S. Weinberg, Ultraviolet divergences in quantum theories of gravitation, in S. W. Hawking and W. Israel, eds., *General Relativity: An Einstein centenary survey*, 790–831 (Cambridge: Cambridge University Press, 1979).

[19] P. C. W. Davies and S. D. Unwin, Why is the cosmological constant so small, *Proc. Roy. Soc. Lond.*, **377**, 147 (1981).

[20] A. D. Linde, The inflationary universe, *Rept. Prog. Phys.* **47**, 925 (1984), doi:10.1088/0034-4885/47/8/002.

[21] A. D. Sakharov, Cosmological transitions with a change in metric signature, *Sov. Phys. JETP* **60**, 214 (1984) [*Zh. Eksp. Teor. Fiz.* **87**, 375 (1984)] [*Sov. Phys. Usp.* **34**, 409 (1991)].

[22] R. Bousso and J. Polchinski, Quantization of four form fluxes and dynamical neutralization of the cosmological constant, *JHEP* **0006**, 006 (2000), [hep-th/0004134].

[23] E. Silverstein, (A)dS backgrounds from asymmetric orientifolds, [hep-th/0106209].

[24] S. Kachru, R. Kallosh, A. D. Linde, and S. P. Trivedi, De Sitter vacua in string theory, *Phys. Rev. D* **68**, 046005 (2003) [hep-th/0301240].

[25] A. Almheiri, D. Marolf, J. Polchinski, and J. Sully, "Black holes: Complementarity or firewalls?, *JHEP* **1302**, 062 (2013), doi:10.1007/JHEP02(2013)062 [arXiv:1207.3123 [hep-th]].

[26] G. Ellis and J. Silk, Scientific method: Defend the integrity of physics, *Nature* 516, 321–3, doi:10.1038/516321a.

[27] G. F. R. Ellis, The trace-free Einstein equations and inflation, *Gen. Rel. Grav.* **46**, 1619 (2014), doi:10.1007/s10714-013-1619-5 [arXiv:1306.3021 [gr-qc]].

[28] R. Dawid, String theory and the scientific method, doi:10.1017/CBO9781139342513.

20

The Dangerous Irrelevance of String Theory

EVA SILVERSTEIN

20.1 Introduction

I was asked by R. Dawid to provide a perspective for the proceedings of the meeting "Why Trust a Theory?." I did not participate in this meeting, but am happy to comment, focusing on aspects not emphasized in other chapters of this book. Both sides of the debate [1, 2] start from the assumption that string theory is divorced from empirical observation. In this chapter, I describe a concrete role that string theory has been playing in the standard scientific method, in the context of early universe cosmology [3–7]. Along the way, I make some comments on the nature and role of the string landscape.

20.1.1 Dangerous Irrelevance and the Precision of Modern Data

Effective field theory provides a powerful method of characterizing observables, incorporating the limitations of accessible energy scales. In assessing the applicability of the physics of the ultraviolet (UV) completion, whatever it is, one must understand carefully the scope of the low energy theory. Obviously, the energy scale of new physics (e.g., masses of additional fields) may be substantially higher than those excited by terrestrial experiments or observable processes in the universe, leading to effects suppressed by this ratio.[1] One expresses this in terms of an effective action. For a scalar field ϕ, this is schematically

$$S = S_{kin} - \int d^4x \sqrt{-g} \{ \frac{1}{2}m^2(\phi - \phi_0)^2 + \lambda_1(\phi - \phi_0) + \lambda_3(\phi - \phi_0)^3 + \lambda_4(\phi - \phi_0)^4$$

$$+ \lambda_6 \frac{(\phi - \phi_0)^6}{M_*^2} + \lambda_{4,4} \frac{(\partial\phi)^4}{M_*^4} + \ldots \}, \tag{20.1}$$

---

[1] This was reviewed in Chapter 18 in this volume, while Chapter 22 describes various interesting phenomenological scenarios in string theory.

with effective dimensionless couplings for the irrelevant operators such as $\bar{\lambda}_6 = \frac{E^2}{M_*^2}$ decreasing at low energies. This makes it more difficult to access their effects at lower energies.

However, two standard caveats to this theory are relevant to the connection between string theory and observations.

One is known as *dangerous irrelevance*. An irrelevant operator can become important upon renormalization group (RG) flow, or over long timescales, large field ranges, or large parameter variations. Even if the input energy density is low, a system may develop sensitivity to the UV completion. In the previously mentioned action, this can be seen for example in the potential energy terms: If the field travels a distance $\geq M_*$, then we cannot neglect the higher-order terms. This applies in inflationary cosmology [11, 12], a point made forcefully in [13]. A more basic physical example is to consider charges in a weak electric field[2]: Over sufficiently long times, the charges will be accelerated, developing a much larger center of mass energy. A similar effect occurs in black hole geometries [14].

The second caveat has to do with the amount of available data. Roughly speaking, with N_m measurements, one can constrain a parameter ε down to

$$\Delta\varepsilon \sim \frac{1}{\sqrt{N_m}}. \tag{20.2}$$

With sufficiently many data points, this can lead to sensitivity to mass scales greater than those directly excited. This has sufficed to rule out proton decay in classic grand unified theory (GUT) models [10]. In current cosmology, this number is approximately $N_m \approx 10^6$ [4–7] (and growing), in a subject where the input energy densities can be rather large to begin with, and the timescales also imply UV sensitivity [11, 12] as just described.

Although the data are sufficient to constrain a variety of interesting possibilities for new physics, one could use up the data by searching for all possibilities allowed in an effective field theory (EFT). For example, in inflationary cosmology, this would occur without strong symmetry assumptions, even at the level of single-field inflation [15]). For this reason as well, UV-complete physics can play a useful part in suggesting, characterizing, and prioritizing analyses. I should stress that here I am alluding to many possibilities for effects that are subleading compared to the basic, empirically tested, features of inflationary cosmology [4–7].[3]

Here I do not mean to claim that the constraints from UV completion will turn out to be strong enough to whittle down the testable possibilities enough to avoid

---

[2] We thank S. Hartnoll and B. Swingle for discussion of such simple examples.

[3] These include the shape of the temperature and E mode power spectrum, indicating super-horizon perturbations [16] and a small tilt, as well as the roughly Gaussian shape of the histogram of temperature fluctuations, along with more precise constraints on several shapes of non-Gaussianity.

the problem of the look-elsewhere effect. Nevertheless, it can help make sense of the observational constraints in ways that would not follow from EFT alone. One of my favorite examples of this is the qualitative feature that even very massive (i.e., UV) degrees of freedom tend to adjust in an energetically favorable way to flatten the inflaton potential, driving down predictions for the tensor to scalar ratio compared to the corresponding models without such fields [17]. More generally, given the impossibility of testing all the parameters in the EFT, we can reduce the problem in several ways: (1) impose extra symmetry and/or minimal field content in the EFT [15], (2) test specific mechanisms that involve the interesting structures of the UV completion, and (3) do less specific (and hence less optimal) tests of qualitative features suggested by theoretical mechanisms. This combination leads to interesting constraints on early universe physics; I will give a current example illustrating this situation.

20.1.2 Spinoffs and Systematics of Effective Field Theory and Data Constraints

In a nutshell, a basic role that string theory plays is in its spinoffs for EFT and data analysis. The theory has stimulated numerous interesting phenomenological ideas worth testing. One classic example is low-energy supersymmetry. This symmetry is actually rare in the string landscape, but is a beautiful idea to test regardless. Further examples have emerged repeatedly in early universe cosmology. In addition to being tested in their own right, these models helped stimulate a more complete EFT treatment of inflation and its signatures.

20.1.3 Timescales, Null Results, and Information Theory

It is sometimes said that theory has strayed too far from experiment/observation. Historically, there are classic cases with long time delays between theory and experiment – Maxwell's and Einstein's waves being prime examples, at 25 and 100 years, respectively.[4] These are also good examples of how theory is constrained by serious mathematical and thought–experimental consistency conditions.

Of course, electromagnetism and general relativity are not representative of most theoretical ideas, but the point remains valid. When it comes to the vast theory space being explored now, most testable ideas will be constrained or falsified. Even there, I believe there is substantial scientific value to this exercise: We learn something significant by ruling out a valid theoretical possibility, as long as it is internally consistent and interesting. We also learn important lessons in excluding

---

[4] As well as plenty of delays going the other way (e.g., 31 years and counting from the discovery of high-T_C superconductors).

potential alternative theories based on theoretical consistency criteria. The pursuit of no-go theorems – and conversely the exceptions to their assumptions – is a standard and often useful part of theoretical physics research. Whether empirical or mathematical, constraints on interesting regions of theory space lead to valuable science. In this chapter, we focus on string theory's role in the former.

Since information theory is currently all the rage, it occurred to me that we can phrase this issue in that language. Information is maximized when the probabilities are equal for a set of outcomes, since one learns the most from a measurement in that case. The existence of multiple consistent theoretical possibilities implies greater information content in the measurements. Therefore, theoretical research establishing this (or constraining the possibilities) is directly relevant to the question of what and how much are learned from data. In certain areas, string theory plays a direct role in this process.

One thing that is certainly irrelevant to these questions is the human lifespan. Arguments of the sort "after X number of years, string theory failed to produce Y result" are vacuous. In any case, we are fortunate that the timescales for testing certain ideas are not so long.

20.1.4 The Many Facets of String Theory

Before going on, let us note that string theory has many motivations, developments, and potential applications. It is a strong candidate for a consistent UV completion of gravity. It is a rich framework for physical model-building (in high energy physics and condensed matter as well as in cosmology), a detailed source of mathematical ideas, an approach to black hole physics and other thought–experimental puzzles, and a source of insight into quantum field theory. In some ways its effectiveness is "unreasonable," although in other ways it has not developed the way some expected or hoped. Against this background, it would be difficult to support any simple ideological stand on the justification of string theory as a subfield, despite the tendencies of Internet discourse to try to do so. In any event, this chapter focuses on an important, but narrower question: How and why does string theory participate in current empirical science?

20.2 Case Study: Early Universe Cosmology

20.2.1 Empirical Observations

Recent decades have seen enormous progress in our understanding of cosmological evolution and the physical processes involved, raising deeper questions and stimulating further observations. In cosmology, as in many areas of science, we would like to optimally exploit the available data, especially given the enormous ingenuity and effort that goes into collecting and analyzing those data [4–7]. The first step in this enterprise, which has nothing directly to do with string theory, is to nail down the known cosmological parameters with greater accuracy and precision. This is a dry way of stating an enormously interesting process that has led, among many other things, to the discovery of accelerated expansion of the universe.

On top of that, we can use the data to test for new parameters and to interpret the resulting constraints. It is in this latter process that string theory already plays a significant role. There are by now many pertinent examples (see reviews such as [11, 12] for references), including DBI (Dirac–Born–Infeld) inflation, which broadened our understanding of inflationary dynamics while generating a now-standard shape of non-Gaussianity, and large-field inflation mechanisms relevant for B-mode searches (as well as more model-dependent structures in the scalar perturbations). We will not rehash these arguments, but rather make some brief comments and then spell out a relatively new example for illustration.

Additional fields, including very heavy ones, that are coupled to the inflaton have numerous effects, some of which are detectable or constrainable observationally. As mentioned previously, they adjust to the inflationary potential in an energetically favorable way, either destabilizing or flattening the inflaton potential depending on the details [17].[5] They renormalize the effective action for the inflaton, and they are subject to non-adiabatic production, slowing the homogenous rolling inflaton and leading to Bremsstrahlung emission of scalar and tensor perturbations.[6] Several of these basic effects were discovered in the context of string theoretic inflationary mechanisms, from which more model-independent lessons were abstracted.

A Current Example

Given an inflationary solution, the perturbations can be measurably affected by even heavy fields χ that couple to the inflaton. Such couplings produce a time-dependent mass. Consider the dominant Fourier component of this mass function, an interaction $\chi^2(\mu^2 + g^2 f^2 \cos\frac{\phi}{f})$. Expanding the inflaton into its background evolution and perturbations, $\phi = \dot\phi_0 t + \delta\phi$ generates a series of interaction terms, each of which has a sinusoidal time dependence at a frequency $\omega = \frac{\dot\phi_0}{f}$. For example, there is a three-point vertex $\int g^2 f \chi^2 \delta\phi \sin \omega t$. This causes non-adiabatic production that is exponentially suppressed, but only beyond the mass scale $\approx g\dot\phi$ (and the amplitude of these effects are enhanced by power law prefactors [21]). This, combined with the precision of the cosmic microwave background (CMB),

---

[5] The latter effect accounts for the continued viability of string-theoretic inflationary mechanisms as of this writing.

[6] They also feature in an interesting proposal for interpreting the low-multipole anomalies [20].

leads to sensitivity to mass scales up to two orders of magnitude above the Hubble scale of inflation, for a range of couplings consistent with perturbation theory.

The three-point correlation function of scalar perturbations is of a distinct shape from those previously analyzed in the data (as well as in theory). Moreover, there is a regime in which this effect has a strongly non-Gaussian shape, in contrast to all previously derived signatures. The $\sin \omega t$–dependent coupling enhances each leg of $\delta\phi$ emission from a given pair of produced χ particles. As a result, a regime emerges for $\frac{\omega}{H} \gg 1$ in which the signal/noise ratio grows with N for a range of N. Writing this in terms of the more directly observed perturbation $\zeta \approx \delta\phi H/\dot{\phi}$, this ratio is schematically

$$(S/N)^2 \sim \int_{\{k\}} \frac{|\langle \zeta_1 \ldots \zeta_N \rangle|^2}{N! \prod P(k_i)} \approx \frac{\left(g^2 \frac{f}{\mu} \sqrt{\frac{\omega}{H}}\right)^N}{N!} \tag{20.1}$$

in the regime where the Gaussian perturbations dominate the noise in the denominator. If one derives a histogram of temperature fluctuations generated by this effect, it is a strongly non-Gaussian shape in ζ space, convolved with a Gaussian distribution arising from the standard vacuum fluctuations of ζ. One can derive the effect this would have on the CMB map in position space: It gives a nearly scale-invariant pattern of defects of different sizes, with $\delta\phi$ perturbations radiating out from each χ production point, distorting the map within the light cone of each such event. This leads to several new types of non-Gaussianity searches under way using CMB data.

What does this have to do with string theory? There are several levels:

- First, this theoretical analysis grew out of a class of mechanisms for inflation in string theory (known as monodromy and trapped inflation [11]), where the ingredients leading to the effect arise naturally. This includes heavy fields coupling to the inflaton, in some cases sinusoidal couplings for string theory's analogue of axion fields. As such, the data analysis puts constraints on the parameter space of models for inflation in this class.

- Second, we can abstract from this the broader lesson that for a range of interesting masses and couplings, we must supplement the single-field EFT even if the extra particles are very heavy, and their effects can be constrained using existing CMB data [22]. (This lesson also applies to numerous other string-theoretic inflationary models, which also contain heavy fields.) In ongoing theoretical work, we are currently finding other roles for large-N point functions in non-Gaussianity, arising from combinatorial enhancements of such observables.

- Third, the theoretical demonstration that strongly non-Gaussian effects can arise in the primordial perturbations motivates a more extensive analysis of the theory and analysis of strongly non-Gaussian perturbations, which can have other origins.

- A similar effect arises from time dependent string tensions, with a related search strategy.

The middle two bullet points are logically independent of string theory. Nevertheless, as has happened several times in the past, we came to them via string theory (as in the first bullet point). The connection between string theory and early universe cosmology, which has been very active across the spectrum from theory to data analysis, was not covered in [1].

I should emphasize that these developments derive from the "landscape" of string theory solutions [18, 19], which I will discuss later in this chapter. Rather than representing an abdication of science or randomizing all physical observables, the landscape has led to new empirical information about the early universe and provided for a consistent interpretation of the dark energy.

20.2.2 *Thought Experiments*

The role of thought experiments is important as well. In the context of cosmology, one role that these experiments play is to constrain possible alternative scenarios for the initial perturbations, which is an interesting line of research. Some prominent examples can be excluded this way if one requires that black hole thermodynamics relations hold. Exotic forms of stress-energy required to introduce a bounce in the cosmological scale factor can violate these relations. In such cases, this leads to decreasing black hole mass as entropy increases, violating the second law.

20.3 Demystifying the Landscape

It is sometimes said that the landscape makes every parameter into an unexplainable selection effect. But even though there are many backgrounds of string theory, the landscape is still highly constrained. To begin with, there is not a hard cosmological constant term in the effective Lagrangian descending from string theory. The resulting metastability of de Sitter vacua fits with several conceptual (thought–experimental) and technical arguments It fits well with the observation of the dark energy. The middle ground between structure and variability complicates the task of modeling cosmology and particle physics in explicit detail, although as already noted it inspires mechanisms and dynamical effects testable in their own right. This program of research remains highly motivated, with new discoveries continuing to emerge in both directions: structures and mechanisms, and constraints.

20.3.1 The Role of Supersymmetry

In my view, the role of supersymmetry is chronically over-emphasized in the field, and hence understandably also in the article by Ellis and Silk [1]. The possibility of supersymmetry in nature is very interesting since it could stabilize the electroweak hierarchy, and extended supersymmetry enables controlled extrapolation to strong coupling in appropriate circumstances. Neither of these facts implies that low-energy supersymmetry is phenomenologically favored in string theory.

Almost every perturbative string limit has a positive dilaton potential, and almost every compactification geometry is negatively curved, leading to a positive potential from the internal curvature. The elegant mathematics of string theory applies to these cases, too. For example there is a generalization of T-duality that neatly relates the two; compact negatively curved spaces are supercritical as a result of their exponentially growing fundamental group [23]. To date, most work on string compactification presumes the extra dimensions to be six in number and built from a Ricci-flat (Calabi–Yau) manifold, chosen by hand to preseve supersymmetry below the compactification scale. Other, less supersymmetric mechanisms with positive tree level potential have also been studied, and ongoing work is uncovering interesting structure, and some simplifications, in the more generic setting. Perturbative control at large curvature radius and weak string coupling is available in both cases. Much further research, both conceptual and technical, is required to obtain an accurate assessment of the dominant contributions to the string landscape.

20.3.2 Is It Testable As a Whole?

The discussion presented here and more comprehensively in [12] emphasizes the testability of certain inflationary mechanisms descending from string theory, in modeling a process that ideally requires a quantum gravitational treatment (as well as the more nuanced role these mechanisms play in the interpretation of empirical observations). A somewhat separate question is the testability of the landscape itself, which leads to some of the more philosophical discussions in this book. But the two questions are not necessarily distinct. In principle, one could test string theory locally. In practice, this would require discovering a smoking gun signature (such as a low string scale at colliders, or perhaps a very distinctive pattern of primordial perturbations in cosmology), and nothing particularly favors such scenarios currently. But for the philosophical question of the empirical standing of the string landscape, this is an important point to include. Strong evidence for string theory locally would support its global predictions of a landscape.[7]

---

[7] As my colleague A. Linde frequently points out, the universe hypothesis is no more conservative than the multiverse hypothesis in the sense that both refer to physics outside of our empirical view.

This is an extreme example of a familiar chain of reasoning in science. Even in empirically established theories, we empirically test only a set of measure zero of their predictions. A plethora of such tests can provide compelling evidence for a theory, which makes further predictions beyond those explicitly tested. It is not ever the case that *all* of a theory's predictions are empirically verified. In any case, it is reasonable to test string theoretic physical models locally as far as possible, while continuing to assess the theory's implications more globally.

20.4 Summary

String theory participates in empirical science in several ways. In the context of early universe cosmology, on which we have focused in this chapter, it helped motivate the discovery and development of mechanisms for dark energy and inflation consistent with the mathematical structure of string theory and various thought–experimental constraints. Some of these basic mechanisms had not been considered at all outside of string theory, and some not quite in the form they take there, with implications for EFT and data analysis that go well beyond their specifics. Low-energy supersymmetry – a very special choice within the string landscape – is an earlier example of a major idea originating in string theory that is well worth testing, although it is not a general prediction of string theory as far as we know. A subset of models originating in string theory generate rich signatures, enabling direct constraints on their parameters, with some being falsifiable. A current example concerns an entirely new form of non-Gaussianity generated by very massive degrees of freedom in the early universe. As in previous examples, new constraints on these can be obtained from existing data, increasing our empirical knowledge of the universe via the standard scientific method, regardless of whether the analysis results in a discovery or a null result.

This is an active area in which big questions remain open, both theoretically and observationally, and it is too soon to draw conclusions about the ultimate level of empirical connection that string theory will attain. In any case, it already plays a useful role in interpreting empirical observations of significant interest.

Acknowledgments

I thank many collaborators and colleagues for sharing their insights over many years, and J. Polchinski for commenting on a draft. I would like to thank L. Senatore and KIPAC at Stanford University for an early discussion addressing the claims and proposals in [1]. The scientific research described here is supported in part by the National Science Foundation under grant PHY-0756174.

References

[1] G. Ellis and J. Silk, Scientific method: Defend the integrity of physics, *Nature*, **516**(7531), 321–3 (2014).

[2] R. Dawid, Non-empirical confirmation, American Astronomical Society, AAS Meeting no. 228, id.211.02.

[3] A. Albrecht and P. J. Steinhardt, Cosmology for grand unified theories with radiatively induced symmetry breaking," *Phys. Rev. Lett.* **48**, 1220 (1982).
A. H. Guth, The inflationary universe: A possible solution to the horizon and flatness problems," *Phys. Rev. D* **23**, 347 (1981).
A. D. Linde, A new inflationary universe scenario: A possible solution of the horizon, flatness, homogeneity, isotropy and primordial monopole problems, *Phys. Lett. B* **108**, 389 (1982).
A. A. Starobinsky, A new type of isotropic cosmological models without singularity, *Phys. Lett. B* **91**, 99 (1980).

[4] P. A. R. Ade et al. [Planck Collaboration], Planck 2015 results. XIII. Cosmological parameters, arXiv:1502.01589 [astro-ph.CO].
P. A. R. Ade et al. [Planck Collaboration], Planck 2015 results. XVI. Isotropy and statistics of the CMB, arXiv:1506.07135 [astro-ph.CO].
P. A. R. Ade et al. [Planck Collaboration], Planck 2015 results. XVII. Constraints on primordial non-Gaussianity, arXiv:1502.01592 [astro-ph.CO].
P. A. R. Ade et al. [BICEP2 and Planck Collaborations], Joint analysis of BICEP2/*Keck Array* and *Planck* data, *Phys. Rev. Lett.* **114**(10), 101301 (2015) [arXiv:1502.00612 [astro-ph.CO]].
P. A. R. Ade et al. [Planck Collaboration], Planck 2015 results. XX. Constraints on inflation, arXiv:1502.02114 [astro-ph.CO].
N. Aghanim et al. [Planck Collaboration], Planck 2015 results. XI. CMB power spectra, likelihoods, and robustness of parameters, submitted to *Astron. Astrophys.* [arXiv:1507.02704 [astro-ph.CO]].

[5] G. Hinshaw et al. [WMAP Collaboration], Nine year Wilkinson Microwave Anisotropy Probe (WMAP) observations: Cosmological parameter results, *Astrophys. J. Suppl.* **208**, 19 (2013) [arXiv:1212.5226 [astro-ph.CO]].

[6] K. T. Story et al., A measurement of the cosmic microwave background damping tail from the 2500-square-degree SPT-SZ survey, arXiv:1210.7231 [astro-ph.CO].

[7] J. L. Sievers et al., The Atacama Cosmology Telescope: Cosmological parameters from three seasons of data, arXiv:1301.0824 [astro-ph.CO].

[8] J. Polchinski, String theory to the rescue, arXiv:1512.02477 [hep-th].

[9] F. Quevedo, Is string phenomenology an oxymoron?, arXiv:1612.01569 [hep-th].

[10] K. Abe et al. [Super-Kamiokande Collaboration], Search for proton decay via $p \rightarrow e^{+}\pi^{0}$ and $p \rightarrow \mu^{+}\pi^{0}$ in 0.31 megaton years exposure of the Super-Kamiokande water Cherenkov detector, *Phys. Rev. D* **95**, no. 1, 012004 (2017) doi:10.1103/PhysRevD.95.012004 [arXiv:1610.03597 [hep-ex]].

[11] D. Baumann, TASI lectures on inflation, arXiv:0907.5424 [hep-th].
D. Baumann and L. McAllister, Inflation and string theory, arXiv:1404.2601 [hep-th].
D. Baumann and L. McAllister, Advances in inflation in string theory, *Ann. Rev. Nucl. Part. Sci.* **59**, 67 (2009) [arXiv:0901.0265 [hep-th]].
C. P. Burgess and L. McAllister, Challenges for string cosmology, *Class. Quant. Grav.* **28**, 204002 (2011) doi:10.1088/0264-9381/28/20/204002 [arXiv:1108.2660 [hep-th]].
A. Westphal, String cosmology: Large-field inflation in string theory, *Adv. Ser. Direct. High Energy Phys.* **22**, 351 (2015), doi:10.1142/97898146026860012.

[12] E. Silverstein, TASI lectures on cosmological observables and string theory, in *Proceedings, Theoretical Advanced Study Institute in Elementary Particle Physics: New Frontiers in Fields and Strings (TASI 2015)*, Boulder, CO, June 1–26, 2015 arXiv:1606.03640 [hep-th].

[13] S. Kachru, R. Kallosh, A. D. Linde, J. M. Maldacena, L. P. McAllister, and S. P. Trivedi, Towards inflation in string theory, *JCAP* **0310**, 013 (2003) doi:10.1088/1475-7516/2003/10/013 [hep-th/0308055].

[14] R. Ben-Israel, A. Giveon, N. Itzhakim and L. Liram, Stringy horizons and UV/IR mixing, *JHEP* **1511**, 164 (2015), doi:10.1007/JHEP11(2015)164 [arXiv:1506.07323 [hep-th]].
M. Dodelson and E. Silverstein, String-theoretic breakdown of effective field theory near black hole horizons, arXiv:1504.05536 [hep-th].
M. Dodelson and E. Silverstein, Long-range nonlocality in six-point string scattering: Simulation of black hole infallers, arXiv:1703.10147 [hep-th].
M. Dodelson, E. Silverstein, and G. Torroba, Varying dilaton as a tracer of classical string interactions, arXiv:1704.02625 [hep-th].
M. Dodelson et al., "Long-range longitudinal scattering in holographic gauge theory and black holes, in press.
D. A. Lowe, J. Polchinski, L Susskind, L. Thorlacius, and J. Uglum, Black hole complementarity versus locality, *Phys. Rev. D* **52**, 6997 (1995), doi:10.1103/PhysRevD.52.6997 [hep-th/9506138].
A. Puhm, F. Rojas, and T. Ugajin, (Non-adiabatic) string creation on nice slices in Schwarzschild black holes, *JHEP* **1704**, 156 (2017), doi:10.1007/JHEP04(2017)156 [arXiv:1609.09510 [hep-th]].
E. Silverstein, Backdraft: String creation in an old Schwarzschild black hole, arXiv:1402.1486 [hep-th].
L. Susskind, Strings, black holes and Lorentz contraction, *Phys. Rev. D* **49**, 6606 (1994), doi:10.1103/PhysRevD.49.6606 [hep-th/9308139].
D. Wenren, Hyperbolic black holes and open string Production, in press.

[15] C. Cheung, P. Creminelli, A. L. Fitzpatrick, J. Kaplan, and L. Senatore, The effective field theory of inflation, *JHEP* **0803**, 014 (2008), doi:10.1088/1126-6708/2008/03/014 [arXiv:0709.0293 [hep-th]].

[16] D. N. Spergel and M. Zaldarriaga, CMB polarization as a direct test of inflation, *Phys. Rev. Lett.* **79**, 2180 (1997), doi:10.1103/PhysRevLett.79.2180 [astro-ph/9705182].
N. Turok, A causal source which mimics inflation, *Phys. Rev. Lett.* **77**, 4138 (1996), doi:10.1103/PhysRevLett.77.4138 [astro-ph/9607109].

[17] X. Dong, B. Horn, E. Silverstein, and A. Westphal, Simple exercises to flatten your potential, *Phys. Rev. D* **84**, 026011 (2011), doi:10.1103/PhysRevD.84.026011 [arXiv:1011.4521 [hep-th]].

[18] R. Bousso and J. Polchinski, Quantization of four form fluxes and dynamical neutralization of the cosmological constant, *JHEP* **0006**, 006 (2000), doi:10.1088/1126-6708/2000/06/006 [hep-th/0004134].
S. B. Giddings, S. Kachru, and J. Polchinski, Hierarchies from fluxes in string compactifications, *Phys. Rev. D* **66**, 106006 (2002), doi:10.1103/PhysRevD.66.106006 [hep-th/0105097].
S. Kachru, R. Kallosh, A. D. Linde, and S. P. Trivedi, De Sitter vacua in string theory, *Phys. Rev. D* **68**, 046005 (2003), doi:10.1103/PhysRevD.68.046005 [hep-th/0301240].
A. Maloney, E. Silverstein, and A. Strominger, De Sitter space in noncritical string theory, hep-th/0205316.

[19] F. Denef, Les Houches lectures on constructing string vacua, arXiv:0803.1194 [hep-th].
M. R. Douglas and S. Kachru, Flux compactification, *Rev. Mod. Phys.* **79**, 733 (2007), hep-th/0610102.
A. R. Frey, Warped strings: Selfdual flux and contemporary compactifications, hep-th/0308156.
J. Polchinski, The cosmological constant and the string landscape, hep-th/0603249.
E. Silverstein, TASI/PiTP/ISS lectures on moduli and microphysics, hep-th/0405068.

[20] J. R. Bond, A. V. Frolov, Z. Huang, and L. Kofman, Non-Gaussian spikes from chaotic billiards in inflation preheating, *Phys. Rev. Lett.* **103**, 071301 (2009), doi:10.1103/PhysRevLett.103.071301 [arXiv:0903.3407 [astro-ph.CO]].

[21] R. Flauger, M. Mirbabayi, L. Senatore, and E. Silverstein, Productive interactions: Heavy particles and non-Gaussianity, arXiv:1606.00513 [hep-th].

[22] M. Mirbabayi et al., work in progress.

[23] D. R. Green, A. Lawrence, J. McGreevy, D. R. Morrison, and E. Silverstein, Dimensional duality, *Phys. Rev. D* **76**, 066004 (2007), doi:10.1103/PhysRevD.76.066004 [arXiv:0705.0550 [hep-th]].
J. McGreevy, E. Silverstein, and D. Starr, New dimensions for wound strings: The modular transformation of geometry to topology, *Phys. Rev. D* **75**, 044025 (2007), doi:10.1103/PhysRevD.75.044025 [hep-th/0612121].

21

String/M-Theories about Our World Are Testable in the Traditional Physics Way

GORDON L. KANE

21.1 Outline

Some physicists hope to use string/M-theory to construct a comprehensive under-lying theory of our physical world – a "final theory." Can such a theory be tested? A quantum theory of gravity must be formulated in 10 dimensions, so *obviously* testing it experimentally requires projecting it onto our 4D world (called "compact-ification"). Most string theorists study theories, including aspects such as AdS/CFT, not phenomena, and are not much interested in testing theories beyond the Standard Model about our world. Compactified theories generically have many realistic fea-tures whose necessary presence provides some tests, such as gravity, Yang–Mills forces like the Standard Model ones, chiral fermions that lead to parity violation, softly broken supersymmetry, Higgs physics, families, hierarchical fermion masses, and more. All tests of theories in physics have always depended on assumptions and approximate calculations, and tests of compactified string/M-theories do, too. String phenomenologists have also formulated some explicit tests for compactified theories. In particular, I give examples of tests from compactified M-theory (involv-ing Higgs physics, predictions for superpartners at LHC, electric dipole moments, and more). It is clear that compactified theories exist that can describe worlds like ours, and it is clear that even if a multiverse were real, this would not prevent us from finding comprehensive compactified theories like one that might describe our world. I also discuss what we might mean by a final theory and what we might want such a theory to explain, and comment briefly on multiverse issues from the point of view of finding a theory that describes our world.

Here is the outline for the chapter:

- Testing theories in physics – some generalities.
- Testing 10-dimensional string/M-theories as underlying theories of *our* world obviously *requires* compactification to four space-time dimensions.

- Testing of all physics theories requires assumptions and approximations, hopefully eventually removable ones.
- Detailed example: existing and coming tests of compactifying M-theory on manifolds of G_2 holonomy, *in the fluxless sector*, in order to describe/explain *our* vacuum.
- How would we recognize a string/M-theory that explains our world, a candidate "final theory"? What should it describe/explain?
- Comments on multiverse issues from this point of view: Having a large landscape clearly does not make it difficult to find candidate string/M-theories to describe our world, contrary to what is often said.
- Final remarks.

21.2 Long Introduction

The meeting "Why Trust a Theory" provided an opportunity for me to present some comments and observations and results that have been accumulating. The meeting was organized by Richard Dawid, author of *String Theory and the Scientific Method,* a significant book. Near the end of this chapter, I will comment on some of Dawid's points.

String/M-theory is a very promising framework for constructing an underlying theory that incorporates the Standard Models of particle physics and cosmology, and probably addresses all the questions we hope to understand about the physical universe. Some people hope for such a theory. A consistent quantum theory of gravity must be formulated in 10 or 11 dimensions. The differences between 10D and 11D are technical and we can ignore them here. (Sometimes the theory can be reformulated in more dimensions but we will ignore that, too.) Obviously, the theory must be projected onto four spacetime dimensions to test it experimentally; the jargon for such a projection is "compactification."

Remarkably, many of the compactified string/M-theories generically have the kinds of properties that characterize the Standard Models of particle physics and cosmology and their expected extensions! These include gravity, Yang–Mills gauge theories [such as the color $SU(3)$ and the electroweak $SU(2) \times U(1)$], softly broken supersymmetry, moduli, chiral fermions (that imply parity violation), families, inflation, and more. For some theorists, the presence of such features is sufficient to make them confident that the string/M-theories will indeed describe our world, and they don't feel the need to find the particular one that describes our vacuum. For others, such as myself, the presence of such features stimulates them to pursue a more complete description and explanation of our vacuum.

The cosmological constant (CC) problem(s) remain major issues, of course. We will assume that the CC issues are effectively orthogonal to the rest of the physics in finding a description/explanation of our world. That is, solving the CC problems will not help us find our underlying theory, and not solving the CC problems will not make it harder (or impossible) to find our underlying theory. The evidence we have is consistent with such an assumption. Ultimately, of course, it will have to be checked.

The value of the CC may be environmental, a point of view attractive to many. Some physicists have advocated that a number of the constants of the Standard Model and beyond are environmental. In particular, their values are not expected to be calculable by the normal methods of particle theory such as a compactified string/M-theory. One problem with that attitude is that apparently not all of the constants are environmental. Perhaps the strong charge conjugation parity (CP) angle, or the proton lifetime, or the top quark mass is not, and perhaps more fit this description. In particular, some theorists advocate that the Higgs boson mass is environmental, but (as described later in this chapter) we argue that in the compactified M-theory the Higgs mass (or at least the ratio of the Higgs mass to the Z mass and perhaps the Higgs vev) is calculable. It seems possible that the strong CP angle and the proton lifetime are also calculable. If so, it is still necessary to explain why some parameters seem to have to be in certain somewhat narrow ranges or the world would be very different. Note that the allowed ranges are quite a bit larger than the often stated ones – for example, if the forces are unified, one must change their strengths together, which in turn weakens typical arguments [1]. But the issue remains that some parameters need to lie in rather narrow ranges, and need to be understood. In a given vacuum, one can try to calculate the CC.

Much has been written and said about whether string/M-theories are testable. Much of that discourse is clearly not serious or interesting. For example, obviously you do not need to be somewhere to test a theory there. No knowledgeable person doubts our universe had a Big Bang, although no one was there to observe it. There are several very compelling pieces of evidence or relics. One is the expansion and cooling of the universe, a second the properties of the cosmic microwave background radiation, and a third the nucleosynthesis and helium abundance results. Amazingly, scientists probably have been able to figure out why dinosaurs became extinct even though almost everyone agrees that no people were alive 65 million years ago to observe the extinction. You don't need to travel at the speed of light to test that it is a limiting speed.

Is the absence of superpartners at the Large Hadron Collider (LHC) a test of string/M-theory, as some people have claimed? What if superpartners are found in the 2016 LHC run – does that confirm string/M-theory? Before a few years ago, there were no reliable calculations of superpartner masses in well-defined theories. An argument, called "naturalness," that if supersymmetry indeed solves the problems it is said to solve then superpartners should not be much heavier than the Standard Model partners, implies that superpartners should have been found already at the Large Elecron–Positron Collider (LEP) or the Fermi National Accelerator Laboratory (FNAL), and surely at Run I of LHC. All predictions up to a few years ago were based on naturalness, rather than on actual theories, and were wrong. It turns out that in some compactified string/M-theories, one can do fairly good generic calculations of some superpartner masses, and most of them (but not all) turn out to be quite a bit heavier than the heaviest Standard Model particles such as the top quark or the W and Z bosons. Generically, scalars (squarks, sleptons, and Higgs sector masses except for the Higgs boson) turn out to be a few tens of TeV (25–50 TeV), while gauginos (partners of gauge bosons such as gluinos, photinos, winos, and binos) tend to be of order one TeV. I will illustrate the mechanisms that produce these results later for a compactified M-theory.

Thus, actual compactified string/M-theories generically predict that superpartners *should not* have been found at LHC Run I. Typically some lighter gauginos do lie in the mass region accessible at Run II. It seems odd to call the results of good theories "unnatural," but the historical (mis-)use of the term "natural" has led to that situation.

Interestingly, many string theorists who work on gravity, black holes, anti-de Sitter/conformal field theory (AdS/CFT), amplitudes, and so on do not know the techniques employed to study compactified string/M-theories, and their comments may not be useful. Older theorists may remember those approaches. Much of what is written on the subject of testing string theory does not take into account the need for compactification, particularly the assertions made in blogs and some popular books, which are quite misleading. That's often also the case for what supposed experts say: In 1999, a well-known string theorist said at a conference that "string theorists have temporarily given up trying to make contact with the real world," and clearly that temporary period has not ended. Sadly, string theory conferences have few talks about compactified string/M-theories, and the few that might occur are mainly technical ones that do not make contact with experiments. String theorists seldom read papers about, or have seminars at their universities about, compactified string/M-theories that connect to physics beyond the Standard Model.

But I want to argue that string/M-theory's potential to provide a comprehensive underlying theory of our world is too great to ignore it. String/M-theory is too important to be left to string theorists.

Before we turn to actual compactified theories, let's look a little more at the meaning of testing theories. In what sense is "$F = ma$" testable? This expression is a claim about the relation between forces and particle properties and behavior. It might have not been correct. It can be tested for any particular force, but *not* in general. The situation is similar for the Schrödinger equation. For a given Hamiltonian, one can calculate the ground state of a system and energy levels, and make predictions. Without a particular Hamiltonian, no tests are possible. What do we mean when we ask to "test string theory"? The situation for string/M-theories is actually quite similar to $F = ma$. One can test predictions of particular compactified string theories, but the notion of "testing string theory" doesn't seem to have any meaning. If you see something about "testing string theory," beware.

Quantum theory has some general properties, basically superposition, that don't depend on the Schrödinger equation (or an equivalent). Similarly, quantum field theory has a few general tests that don't depend on choosing a force, such as that all electrons are identical (because they are all quanta of the electron field), or determining the connection between spin and statistics. Might string/M-theory have some general tests? Possibly black hole entropy might fit the bill, but I don't want to discuss that here, because it is not a test connected to data. Otherwise, there don't seem to be any general tests. Compactified string/M-theories generically give 4D quantum field theories, and they also imply particular Yang–Mills forces and a set of massless zero-modes or particles, so it seems unlikely they will have any general tests.

In all areas of physics, normally one specifies the "theory" by giving the Lagrangian. It's important to recognize *that physical systems are described not by the Lagrangian but by the solutions to the resulting equations.* Similarly, if string/M-theory is the right framework, our world will be described by a compactified theory, the projection onto 4D of the 10/11D theory, the metastable (or stable) ground state, called our "vacuum."

One also should recognize that studying the resulting predictions is how physics has always proceeded. All tests of theories have always depended on assumptions – from Galileo's use of inclined planes to slow falling balls so they could be timed, to the assumption that air resistance could be neglected in order to get a general theory of motion, to the determination of which corner of string theory and compactification manifold should be tried. Someday there may be a way to derive what is the correct corner of string theory or the compactification manifold, but most likely we will first find ones that work to describe/explain our world, and perhaps later find whether they are inevitable. Similarly, with a given corner and choice of manifold, we may first write a generic superpotential and Kähler potential and gauge kinetic function and use them to calculate testable predictions. Perhaps soon someone can calculate the Kähler potential or gauge kinetic function to a sufficient approximation that predictions are known to be insensitive to corrections.

It's very important to understand that the tests are tests of the compactified theory, but they depend on and are affected by the full 10/11D theory in many ways. *The curled-up dimensions contain information about our world* – the forces, the particles and their masses, the symmetries, the dark matter, the superpartners, electric dipole moments, and more. There are relations between observables. The way the small dimensions curl up tells us a great deal about the world.

Just as a Lagrangian has many solutions (e.g., elliptical planetary orbits for a solar system), so the string/M-theory framework will have several or many viable compactifications and many solutions. This is usually what is meant by "landscape." The important result to emphasize here is that the presence of large numbers of solutions is not necessarily an obstacle to finding the solution or set of solutions that describe our world, our vacuum. (See also Chapter 22 by Fernando Quevedo in this meeting.) That is clear because already people have found, using a few simple guiding ideas, a number of compactified theories that are like our world. Some generic features were described previously, and the detailed compactified M-theory described in the following sections will illustrate this better. *It is not premature to look for our vacuum.*

21.3 What Might We Mean by "Final Theory"? How Would We Recognize It?

In each vacuum, perhaps all important observables would be calculable (if enough graduate and postdoctoral students and were encouraged and supported to work in this area). What would we need to understand and calculate to think that an underlying theory is a strong candidate for a "final theory"? We don't really want to calculate hundreds of quantum chromodynamics (QCD) and electroweak predictions, or go beyond the Standard Model ones. We probably don't need an accurate calculation of the up quark mass, since at its MeV value there may be large corrections from gravitational and other corrections, but it is very important to derive $m_{up} < m_{down}$ and to known that m_{up} is not too large. It's interesting and fun to make such a list. Here is a good start.

✓ What is light?
- What are we made of? Why quarks and leptons?
- Why are there protons and nuclei and atoms? Why $SU(3) \times SU(2) \times U(1)$?
- Are the forces unified in form and strength?
- Why are quark and charged lepton masses hierarchical? Why is the down quark heavier than the up quark?
- Why are neutrino masses small and probably not hierarchical?
- Is nature supersymmetric near the electroweak scale?
- How is supersymmetry broken?
- How is the hierarchy problem solved? Hierarchy stabilized? Size of hierarchy?
- How is the μ hierarchy solved? What is the value of μ?
- Why is there a matter asymmetry?

- What is dark matter? Ratio of baryons to dark matter?
- Are protons stable on the scale of the lifetime of the universe?
- Quantum theory of gravity?
- What is an electron?
▷ Why families? Why 3?
▷ What is the inflaton? Why is the universe old and cold and dark?
◇ Which corner of string/M-theory? Are some equivalent?
◇ Why three large dimensions?
◇ Why is there a universe? Are there more populated universes?
◇ Are the rules of quantum theory inevitable?
◇ Are the underlying laws of nature (forces, particles, etc) inevitable?
◇ CC problems?

The first question, what is light, is answered: If there is an electrically charged particle in a world described by quantum theory, the phase invariance requires a field that is the electromagnetic field. Thus, this question is noted by a check, $\checkmark$. The next set of questions, marked the bullet •, are all addressed in the compactified M-theory; some are answered. They are all addressed simultaneously. The next two questions, indicated with the ▷, are probably also addressed in compactified string/M-theory. The last six questions are still not addressed in any systematic way, though some work on them is occurring.

The list of questions is presented somewhat technically, but the overall idea is probably clear to most readers. Other readers might construct a somewhat different list. I'd be glad to have suggestions. The most important point is that compactified string/M-theories do address most of the questions already, and will do better as understanding improves.

The compactified M-theory described in this chapter assumes that the compactification was to the gauge-matter content of the $SU(5)$ MSSM, the minimal supersymmetric Standard Model. Other choices could have been made, such as $SO(10), E_6, E_8$. So far, there is no principle to fix that content. There are probably only a small number of motivated choices.

21.4 Three New Physics Aspects

In compactified theories, three things emerge that are quite important and may not be familiar.

The *gravitino* is a spin 3/2 superpartner of the (spin 2) graviton. When supersymmetry is broken, the gravitino gets mass via spontaneous symmetry breaking. The resulting mass sets the scale for the superpartner masses and associated phenomena. For the compactified M-theory on a G_2 manifold, the gravitino mass is of order 50

TeV. That is the scale of the soft-breaking Lagangian terms, and thus of the scalars and trilinear couplings, as well as M_{Hu} and M_{Hd}. It also can contribute to the dark matter mass. Two different mechanisms (described later) lead to suppressions of the gaugino masses from 50 TeV to approximately 1 TeV, and to the Higgs boson.

The huge reduction from the Planck scale to the gravitino mass is essentially what solves the hierarchy problem and brings physics at the Planck scale (about 10^{19} GeV) down to the electroweak scale. One can understand this outcome qualitatively. The splitting between the graviton and the gravitino is due to supersymmetry breaking. At the Planck scale, the supersymmetry is not broken. To calculate predictions for our world, one has to use the quantum field theory equations to calculate what happens at lower scales. Some hidden sectors will have gauge groups with large charges, and those will run to become strong forces by about four orders of magnitude below the Planck scale. They form bound states and break the supersymmetry. The superpotential has dimensions of energy cubed, so it is proportional to four orders of magnitude cubed, giving 12 orders of magnitude. The gravitino mass is no longer zero but is proportional to the superpotential, smaller by a factor of about 100 for technical reasons, so the resulting gravitino mass is reduced from the Planck scale by about 10^{14}, resulting in about 10^5 GeV or about 100 TeV. Careful calculations give about 40 TeV, consistent with this qualitative estimate.

The second aspect is *moduli*, which have many physics effects, including leading to a "non-thermal" cosmological history. The curled-up dimensions of the small space are described by scalar fields that determine their sizes, shapes, metrics, and orientations. The moduli get vacuum expectation values, like the Higgs field does. Their vacuum values determine the coupling strengths and masses of the particles, and they must be "stabilized" so the laws of nature will not vary in space and time. The number of moduli is calculable in string/M-theories (the third Betti number); it is typically of order tens to even more than 200. In compactified M-theory, supersymmetry breaking generates a potential for all moduli and stabilizes them. The moduli fields (like all fields) have quanta, unfortunately also called moduli, with calculable masses fixed by fluctuations around the minimum of the moduli potential. In general, inflation ends when they are not at the minimum, so they will oscillate and dominate the energy density of the universe soon after inflation ends.

The moduli quanta couple to all particles via gravity, so they have a decay width proportional to the particle mass cubed divided by the Planck mass squared. Their lifetime is long, but one can show that generically the lightest eigenvalue of the moduli mass matrix is of order the gravitino mass, which guarantees the moduli decay before nucleosynthesis and do not disrupt nucleosynthesis. Their decay introduces lots of entropy, which in turn washes out all earlier dark matter, matter asymmetry, and so on. They then decay into dark matter and stabilize the matter asymmetry. That the dark matter and matter asymmetries both arise from the decay of the lightest modulus and can provide an explanation for the ratio of matter to dark matter, though so far only crude calculations have been done along these lines. When moduli are ignored, the resulting history of the universe after the Big Bang is a simple cooling as it expands, dominated by radiation from the end of inflation to nucleosynthesis, called a thermal history. Compactified string/M-theories predict instead a "non-thermal" history, with the universe matter dominated (the moduli are matter) from the end of inflation to somewhat before nucleosynthesis. All of these results were derived from the compactified M-theory before 2012.

Compactified string theories give us quantum field theories in 4D – but they also give us much more. They predict generically a set of forces, the particles the forces act on, softly broken supersymmetric theories, and the moduli that dominate cosmological history and whose decay generates dark matter and possibly the matter asymmetry.

The third aspect is that because there are often many solutions, we look for *generic* results. We have already used the term "generic" several times. Generic results are probably not a theorem, or at least not yet proved. They might be avoided in special cases. One has to work at constructing non-generic examples. Importantly, predictions from generic analyses are generically *not* subject to qualitative changes from small input changes. Most importantly, they generically have no adjustable parameters, and no fine-tuning of results is possible. Many predictions of compactified string/M-theories are generic, and thus powerful tests. When non-generic Kähler potentials are used, the tests become dependent on the model used and are much less powerful.

21.5 Compactified M-Theory on a G_2 Manifold ($11 - 7 = 4$)

In this section, we consider one compactified theory. Of course, I use the one I have worked on. The purpose here is pedagogical, not review or completeness, so references are not complete, and I apologize to many people who have done important work similar to what is mentioned here. References are given only so that those who want to can begin to trace the work. From 1995 to 2004, a set of results led to the establishment of the basic framework.

- In 1995, Witten discovered M-theory.
- Soon after, Papadopoulos and Townsend [2] showed explicitly that compactifying 11D M-theory on a 7D manifold with G_2 holonomy led to a 4D quantum field theory with $N = 1$ supersymmetry. Thus, the resulting world is automatically supersymmetric – that is not an assumption!
- Acharya [3] showed that non-Abelian gauge fields were localized on singular 3-cycles. The 3-cycles can be thought of as "smaller" manifolds within the 7D one.

Thus, the resulting theory automatically has gauge bosons, photons, and Z's and W's, and their gaugino superpartners.

- Atiyah and Witten [4] analyzed the dynamics of M-theory on G_2 manifolds with conical singularities and their relations to 4D gauge theories.
- Acharya and Witten [5] showed that chiral fermions were supported at points with conical singularities. The quarks and leptons of the Standard Model are chiral fermions, with left-handed and right-handed ones having different $SU(2)$ and $U(1)$ assignments, and giving the parity violation of the Standard Model. Thus, the compactified M-theory generically has the quarks and leptons and gauge bosons of the Standard Model, in a supersymmetric theory.
- Witten [26] showed that the M-theory compactification could be to an $SU(5)$ MSSM and solved the doublet-triplet splitting problem. He also argued that with a generic discrete symmetry, the μ problem would have a solution, with $\mu = 0$.
- Beasley and Witten [7] derived the generic Kähler form.
- Friedmann and Witten [8] worked out Newton's constant, the unification scale, proton decay, and other aspects of the compactified theory. However, in their work supersymmetry was still unbroken and moduli not stabilized.
- Lucas and Morris [9] worked out the generic gauge kinetic function. With this and the generic Kähler form of Beasley and Witten, one had two of the main ingredients needed to calculate predictions.
- Acharya and Gukov [10] brought together much of this work in a Physics Reports.

To extend previous work, we explicitly made five assumptions:

- Compactify M-theory on a manifold with G_2 holonomy, *in the fluxless sector.* This is well motivated. The qualitative motivation is that fluxes (the multidimensional analogues of electromagnetic fields) have dimensions, so they are naturally of string scale size. It is very hard to get TeV physics from such large scales. To date, relatively few examples of generic TeV mass particles have emerged from compactifications with fluxes. Using the M-theory fluxless sector is robust (see Acharya [3] and recent papers by Halverson and Morrison [11, 12]), with no leakage issues.
- Compactify to the gauge matter group $SU(5) - MSSM$. We followed Witten's path here. One could try other groups, such as $SU(3) \times SU(2) \times U(1), SO(10), E6, E8$. There has been some recent work on the $SO(10)$ case, and results do seem to be different (Acharya et al. [13]). Someday, hopefully there will be a derivation of what manifold and what gauge matter group to compactify to, or perhaps a demonstration that many results are common to all choices that have $SU(3) \times SU(2) \times U(1) - MSSM$.
- Use the generic Kähler potential and generic gauge kinetic function.

- Assume the needed singular mathematical manifolds exist. We have seen that *many* results do not depend on the details of the manifold (we provide a list later in this chapter). Others do. There has been considerable mathematical progress recently, such as a Simons Center semester workshop with a meeting, and proposals for G_2 focused activities. There is no known reason to be concerned about whether appropriate manifolds exist.

- We assume that cosmological constant issues are not relevant, in the sense stated earlier; that is, solving them does not help find the properties of our vacuum, and not solving them does not prevent finding our vacuum. Of course, we would like to actually calculate the CC in the candidate vacuum or understand that it is not calculable.

In 2005, we began to try to construct a full compactification. Since the LHC was coming, we focused first on moduli stabilization, how supersymmetry breaking arises, calculating the gravitino mass, and determining the soft-breaking Lagrangian for the 4D supergravity quantum field theory, which led to Higgs physics, LHC physics, dark matter, electric dipole moments, and so on, while leaving for later quark and lepton masses, inflation, and other topics. Altogether, this work has led to about 20 papers with about 500 arXiv pages in a decade.

Electric dipole moments (EDMs) are a nice example of how unanticipated results can emerge. When we examined the phases of the terms in the soft-breaking Lagrangian, all had the same phase at tree level, so it could be rotated away (as could the phase of μ); thus, there were no EDMs at the compactification scale. The low scale phase is then approximately calculable from known renormalization group equation (RGE) running, and indeed explains why EDMs are much smaller than naively expected [14, 15], a significant success.

One can write the moduli superpotential. It is a sum of exponential terms, with exponents having beta functions and gauge kinetic functions. Because of the axion shift symmetry, only non-perturbative terms are allowed in the superpotential – no constant or polynomial terms. One can look at early references (Acharya et al. [16, 18, 19, 21]) to see the resulting terms.

We were able to show that in the M-theory compactification, supersymmetry was spontaneously broken via gaugino and chiral fermion condensation, and simultaneously the moduli were all stabilized, in a de Sitter vacuum, unique for a given manifold. We calculated the soft-breaking Lagrangian, and showed that many solutions had electroweak symmetry breaking via the Higgs mechanism.

Thus, we have a 4D effective softly broken supersymmetric quantum field theory. It's important to emphasize that in the usual "effective field theory" the coefficients of all operators are independent and not calculable. *Here the coefficients are all related and are all calculable.* This theory has no adjustable parameters. In practice, some quantities cannot be calculated very accurately, so they can be allowed to vary a little when comparing with data.

21.6 Some Technical Details

In this section, for completeness and for workers in the field, we list a few of the most important formulae – in particular the moduli superpotential, the Kähler potential, and the gauge kinetic function. Readers who are not working in these areas can skip the formulae, but might find the words somewhat interesting.

The moduli superpotential is of the form

$$W = A_1 e^{ib_1 f_1} + A_2 e^{ib_2 f_2}. \tag{21.1}$$

The b's are basically beta functions. Precisely, $b_k = 2\pi/c_k$, where the c_k are the dual coxeter numbers of the hidden sector gauge groups. W is a sum of such terms. Each term will stabilize all the moduli – the gauge kinetic functions are sums of the moduli with integer coefficients, so expanding the exponentials generates a potential for all the moduli. With two (or more) terms, one can see in calculations that the moduli are stabilized in a region where supergravity approximations are good, while with one term that might not be so. With two terms, we can also get some semi-analytic results that clarify and help understanding, so we mostly work with two terms, though some features are checked numerically with more terms. This is not a "racetrack" potential; the relative sign of the terms is fixed by axion stabilization. The generic Kähler potential is

$$K = -3 \ln(4\pi^{1/3} V_7), \tag{21.2}$$

where the 7D volume is

$$V_7 = \sum_{i=1}^{N} s_i^{a_i} \tag{21.3}$$

with the condition

$$\sum_{i=1}^{N} a_i = 7/3. \tag{21.4}$$

The gauge kinetic function is

$$f_k = \sum_{i=1}^{N} N_i^k z_i \tag{21.5}$$

with integer coefficients. The $z_i = t_i + is_i$ are the moduli, with real parts being the axion fields and imaginary parts being the zero modes of the metric on the 7D manifold; they characterize the size and shape of the manifold.

Generically, two 3D sub-manifolds will not intersect in the 7D space, so no light fields will be charged under both the Standard Model visible sector gauge group and any hidden sector gauge group, and therefore supersymmetry breaking index-supersymmetry breaking will be gravity mediated in M-theory vacua. This is an example of a general result for the compactified M-theory, which is not dependent in any way on details of the Kähler potential. It is not automatic in other corners of string theory, and indeed often does not hold in others. The 11D Planck scale is

$$M_{11} = \sqrt{\pi} M_{pl} / V_7^{1/2} \tag{21.6}$$

and lies between the unification scale and the 4D Planck scale, which is related to the absence of fluxes. Acharya and Bobkov have calculated the cross term in the Kähler potential between the moduli and the matter sector. The only results currently sensitive to that term are the Higgs mass and the precise value of the gravitino mass, which have been included in their calculation [23].

21.7 Main Results, Predictions, and Tests of the Compactified M-Theory

The results listed here follow from the few discrete assumptions mentioned earlier. Note that the results hold simultaneously. The only dimensionful parameter is the Planck constant, which is related to Newton's constant G.

- All moduli are stabilized. Their vacuum expectation values (vevs) are calculated, and typically $\lesssim \frac{1}{10} M_{Pl}$. The moduli mass matrix is calculable and one can find its eigenvalues [21].
- The lightest moduli mass matrix eigenvalue has about the same mass as the gravitino, for general reasons [22].
- The gravitino mass is calculated approximately to be about 50 TeV, starting from the Planck scale; see Figure 21.1 [23].
- The supersymmetry soft-breaking Lagrangian is calculated at high and low scales. Scalar masses (i.e., squarks, sleptons, M_{Hu}, M_{Hd}) are heavy, about equal to the gravitino mass at the compactification scale. RGE running leads to the third family being significantly lighter than the first two at the electroweak scale, and M_{Hu} driven to approximately 1 TeV there [18, 19].
- Trilinear masses are calculated to be somewhat heavier than the gravitino mass.
- Gaugino masses are always suppressed since the visible sector gauginos get no contribution from the chiral fermion F-terms. This relationship is completely

general and robust and just follows from the the supergravity calculations. Since
the matter Kähler potential does not enter, the results are reliable. The suppression
ratio is approximately the ratio of 3-cycle volumes to the 7D volume (in dimensionless units), so the gluino mass is about 1.5 TeV ($\pm$10–15%). The wino mass
is about 640 GeV and the bino is the lightest supersymmetric particle (LSP), with
a mass of about 450 GeV [16].

- The hierarchy problem is solved as long as there are about 50 or more hidden
sectors, which is generically true. That is, all solutions have gravitino masses
in the tens of TeV. This is another result that does not depend on details or the
manifold. Technically, the number is given by the third Betti number, which Joyce
has shown ranges from a few tens to somewhat greater than 200 (for non-singular
manifolds) [21].

- When we set the potential to zero at its minimum by hand (since we do not
solve the CC problem), we find the gravitino mass is in the tens of TeV region
automatically. In other corners of string theory, this does not happen.

- The cosmological history should be non-thermal, with moduli giving a matter-
dominated universe from soon after inflation until the lightest modulus decays
somewhat before nucleosynthesis [24].

- The lightest modulus generates both the matter asymmetry and the dark matter
and, therefore, their ratio. Calculations are consistent with the observed ratio but
are very approximate at this stage [25].

- No approach could be complete without including μ in the theory. Witten took the
first step toward that goal by exhibiting a generic matter discrete symmetry that
led to $\mu = 0$ [26]. But the moduli are generically charged under that symmetry,
so it is broken when they are stabilized. We have not been able to calculate the
resulting value of μ after moduli stabilization. It is not clear whether the original
discrete symmetry is broken or perhaps a new discrete symmetry emerges. We
can estimate μ because we know that μ should vanish if either supersymmetry is
unbroken or the moduli are not stabilized, so μ will be proportional to a typical
moduli vev times the gravitino mass, implying $\mu \lesssim \frac{1}{10}M_{3/2}$ given the calculated
moduli vevs. Thus μ will be of order 3 TeV [27].

- Axions are stabilized [28], giving a solution to the strong CP problem, and a
spectrum of axion masses. The axion decay constant is allowed to be as high as
about 10^{15} GeV.

- We calculated the ratio of the Higgs boson mass to the Z mass, or equivalently
the coefficient λ of the h^4 term in the Higgs potential (before the LHC data
were reported). The calculation is done via the soft-breaking Lagrangian at the
compactification scale, and then the results are run down to the electroweak scale
(so there is no simple formula for M_h). One looks for all solutions that have electroweak symmetry breaking and calculates the resulting M_h. Because the scalars

are heavy, the theory lies in what is known as the supersymmetry decoupling Higgs sector; thus, the Higgs mass is the same for all solutions, 126.4 GeV. Because the top mass is not measured precisely and the RGE equations depend on the top Yukawa coupling (and α_3 somewhat less), the running introduces an error of about 1.2 GeV purely from Standard Model physics. In the decoupling sector, the Higgs boson properties are close to those of a Standard Model Higgs, so the branching ratios were predicted to be those of the Standard Model (to within a few percent, from loops), as is indeed observed. The Higgs potential is stable, with λ never falling below about 0.1. As a result, the vacuum instability is not an interesting question. The Higgs mechanism occurs because of radiative electroweak symmetry breaking and is generic.

The size of the Higgs vacuum expectation value is calculable as well, and it is one of the most fundamental quantities we want to understand. The mechanism for the Higgs getting a vev is fully understood. It is called "radiative electroweak symmetry breaking" (REWSB) because radiative corrections lead to a Higgs potential with a minimum away from the origin, giving a non-zero vacuum value. One might think that because the gravitino mass and the scalars are tens of TeV that the predicted vacuum value would also be that size, but in fact the corrections lead to a weak scale value of M_{Hu} of about a TeV near the weak scale. As just described, μ is at most a few TeV, so the naive REWSB would give a Higgs vev of a few TeV, a full order of magnitude smaller than the gravitino mass. This is an important success of the compactified theory. Further, the EWSB conditions imply a cancellation occurs. For a certain value of M_{Hu}, the theory would actually give the observed Higgs vev. The value of M_{Hu} is given by an expression of the form $f_{M_0} M_0^2 - f_{A_0} A_0^2$, where f_{M_0} and f_{A_0} are fully calculable Standard Model physics. M_0 and A_0 come from the soft-breaking Lagrangian calculated from the compactification; they are calculated at tree level. Unfortunately, M_0 and A_0 can have loop corrections and Kähler corrections that are large enough so that the degree of cancellation cannot currently be determined. It seems that the compactified theory might actually explain numerically as well as conceptually the Higgs vacuum value, but we don't yet know that for certain. This relationship can work only because the compactified theory predicted large trilinears, about 1.5 times the gravitino mass. That range had not been previously studied phenomenologically. In any case, the predicted value of the Higgs vacuum value is within about an order of magnitude of the observed value, so it is probably qualitatively understood, and not a mystery.

- Interestingly, electric dipole moments are calculable. At tree level, the non-perturbative superpotential leads to all soft-breaking terms having the same phase, which can therefore be rotated away. Similarly, μ and B have the same phase and a Peccei–Quinn rotation removes their phase. Thus, at the high scale,

EDMs are approximately zero. When running down to the weak scale, non-zero EDMs are generated, with the phases entering via the Yukawa couplings in the trilinears. If those were completely known, one could calculate the low scale EDMs precisely. Because some model-dependence persists in the high scale Yukawas, we can calculate only the upper limits on the EDM predictions. These are less than the current limits, by approximately a factor of 20. Thus, the compactified M-theory explains the surprisingly small EDMs and provides a target for future experiments [14, 15].

- There are no flavor or weak CP problems.
- Combining the electroweak symmetry breaking results and μ, $\tan\beta \approx 7$.
- LHC can observe gluinos, the lighter chargino, and the LSP. To see higgsinos and the scalars (via associated production with gluinos or winos), one needs a proton–proton (pp) collider with energy near 100 TeV [23].
- Many results, such as the gravitino mass of tens of TeV, the heavy high-scale scalars, the suppressed gaugino masses (gluinos, LSP), gravity mediation, and small EDMs, are generic and do not depend on details of the manifolds.
- Our main current work is on dark matter. As we discussed earlier, the physics of the hidden sectors plays several roles. We live on one that is our visible sector. Others with large gauge groups have couplings that run fast; they lead to gaugino condensation and associated supersymmetry breaking at about 10^{14} GeV (the scale at which F-terms become non-zero). Others have small gauge groups, so they run slowly and condense at MeV or GeV or TeV scales, perhaps giving stable particles at those scales. We can calculate the relic density of those light particles, which must be done in a non-thermal universe in compactified M-theory (and probably in string-theory worlds in general) [29]. We are doing this systematically, expecting to find some dark matter candidates. The bino LSP will decay into these lighter stable "wimps." We are investigating how generic kinetic mixing is in M-theory. While some people have looked at what they call "hidden sectors" mostly what they have addressed are not actual hidden sectors of a compactified string/M-theory (they should probably be called hidden valleys to distinguish them), and almost all of them have calculated the relic densities in a thermal history that is unlikely to be relevant for string/M-theory hidden sectors.

It's interesting to put together the various results on scales to see how the physics emerges and is connected from the Planck scale to the gaugino masses and the Higgs mass. This is shown in Figure 21.1.

It's worth emphasizing that the gluino (and wino and bino) mass prediction is not one "just above current limits" or a tuned calculation. It is actually very generic, robust, and simple to understand. *F*-terms are generated by gaugino (and chiral fermion) condensation, at about a scale $\Lambda \approx 10^{14}$ GeV, a generic result

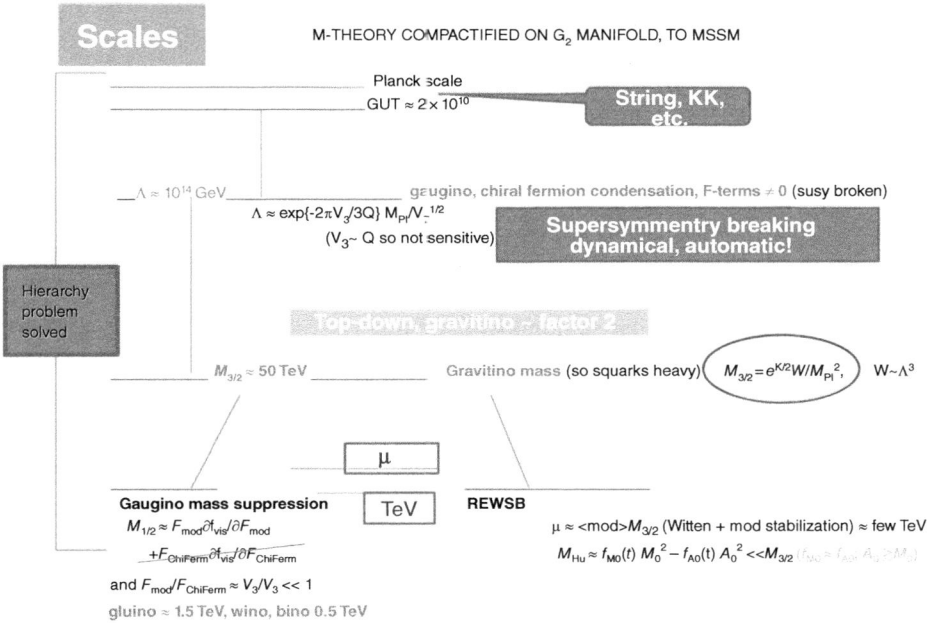

Figure 21.1 Figure showing the various scales in M-theory compactified on a G_2 manifold, with the MSSM as the low energy effective theory.

of the theory, from the running of the largest hidden sector gauge groups. The superpotential is then the ratio of this scale to the compactification scale cubed, and the gravitino mass is the superpotential times $e^{K/2}$; the latter factor is basically the inverse of the 3-cycle volume, as shown in Figure 21.1. The gaugino suppression is very general because of the absence of the chiral fermion contribution to the gaugino mass, since it is a derivative of the visible sector gauge kinetic function that has no dependence on the chiral fermion F-term [16]. All this is illustrated in Figure 21.1, on the top half and the left side. The Higgs mass suppression is illustrated on the right side.

The gluino production cross-section at LHC is 10–15 fb. Note that this is significantly smaller (because of the suppressed heavy quark contribution) than the cross-sections often quoted in the literature. Because the third family is lighter due to the RGE running, somewhat more than half of the gluino decays are to third family final states, and the rest to first + second families. Those signatures make detection more difficult than most LHC studies report for a given gluino mass, and they require larger luminosities than naively expected according to our background estimates, probably more than 40 fb^{-1}. The largest background is top pair production.

Of course, there is still a lot to do to complete the M-theory compactification, in terms of both the physics and the mathematics. It will be very interesting to try compactifications to other gauge matter groups in M-theory and to pursue other corners of string theory compactifications.

21.8 A Comment on Landscape and Multiverse Issues

There is clearly a landscape of string/M-theory solutions. The question is whether the resulting universes are viable ones or too short-lived to have galaxy formation (with the resulting solar systems). Some recent studies (e.g., Dine and Paban; Mersini and Perry; Greene et al.; Shiu et al.) suggest most of the landscape does not give viable universes. Even if there is a large landscape, compactification studies have demonstrated that it is not difficult to find vacua that are good candidates for describing our world and calculating its properties (except maybe for the CC). To date, such studies have been performed by Quevedo and collaborators; Nilles and collaborators; Acharya, Kane, and collaborators; and Vafa and Heckman. Sometimes it is argued that in a landscape, it will be hard to find our vacuum. It's now clear that finding such candidates is not an obstacle to finding a final theory for our world.

21.9 Final Remarks

Some of the remarks that follow address the topics of the meeting, "Why Trust a Theory?"; some summarize points made about compactified string/M-theories; and some focus on the M-theory G_2 example. I want to emphasize again that this is not a review, but rather has several pedagogical aspects. In particular, there are basically no references, but a handful are given (only to arXiv postings) to help people look further into topics they might want to pursue. There is some overlap among remarks since the point is communication rather than conciseness.

- String/M-theory is too important to be left to the string theorists.

- If you want an underlying theory that is a quantum theory and includes gravity and the other forces, as well as the quarks and leptons, a 10/11 dimensional theory with curled-up small dimensions is probably the simplest framework that could incorporate and explain all that you want to understand.

- The compactified M-theory on a manifold of G_2 holonomy is a promising candidate to describe our vacuum. It at least demonstrates that it is not premature to look for such theories, even in the presence of a landscape of solutions. Moduli are generically present in string theories and are inevitable in M-theory.

They imply a non-thermal cosmological history and may explain the ratio of matter to dark matter.

- The compactified M-theory anticipated the mass and decay branching ratios of the Higgs boson. It is the lightest eigenvalue of the Higgs mass matrix of a two-doublet decoupling supersymmetric Higgs sector that satisfies the electroweak symmetry breaking conditions. The Higgs potential does not vanish at any scale, and the universe is metastable. The vacuum value of the Higgs field is not a mystery. The compactified M-theory predicts that superpartners are too heavy to have been seen in LHC Run I, but gluinos and winos and the LSP bino can be seen in Run II at 13 TeV with sufficient luminosity. Backgrounds for gluinos, mainly from top quark production, imply more than 40 fb^{-1} is needed to see the signal.

- The discovery of the Higgs boson is evidence for supersymmetry. In a supersymmetric full theory, one computes the supersymmetry soft-breaking Lagrangian at the compactification scale. It contains a potential for scalars. The RGE running down to the TeV scale implies that potential has a minimum away from the origin. One can calculate accurately the ratio of Higgs mass to Z mass, with a value of 126.4 GeV being obtained for the Higgs mass that way. This result does not depend on parameters. The calculation was done before the data were available, but the calculation is determined so the answer would be the same whenever it is done.

- Compactified string/M-theories imply naturalness predictions are wrong, so superpartners should not have been seen in Run I. Some superpartners can be seen in LHC Run II with sufficient luminosity (i.e., gluinos, winos, binos). Full testing of the superpartner spectrum will require colliders with total energy in the 100 TeV region and sufficient luminosity. The squarks can only be directly seen at such a facility, in associated production.

- The LSP will decay into lighter stable particles from hidden sectors. Such candidates are generic and probably inevitable. We are doing systematic studies of such dark matter in non-thermal cosmological histories.

- The statement that estimates of the cosmological constant are off by 10^{120} is a red herring, or perhaps worse. First, in supersymmetric theories, the value of the potential at its minimum is generically $M_{3/2}^2 M_{pl}^2 \approx 10^{48}$ GeV^4. But this is the potential, and people think in terms of mass, so take the fourth root, giving 10^{12} GeV as the meaningful value. The Higgs potential and the QCD phase transition give similar values – still not so good. But the QCD strong CP problem requires setting a number naively of order unity to a value 10^{10} times smaller, almost the same! Why is there so much anguish over one and none over the other? Partly this debate arises because there are possible solutions to the strong CP problem (such as axions) and good models that solve it (such as compactified string theories), while we don't understand the CC problem. But partly it is just hype. It is reasonable to think that the CC problem(s) are orthogonal to the rest of physics, in the sense that solving the CC problem will probably not help us solve the rest of the issues about understanding our vacuum, and not solving the CC problem probably will not prevent us from making progress with understanding our vacuum.

- Are string/M-theories falsifiable? Yes, in the same sense as traditional theories: One makes predictions from the compactified theories and tests the predictions. Such predictions are tests of the full underlying theory plus the gauge-matter compactification group. You do not have to be there to test whether there was a Big Bang, and you do not have to be at Planck scale energies to test string/M-theories formulated there. There are always relics and implications. Are 10D theories science? Yes, the curled-up dimensions after compactification contain information about the forces and particles; such information is not lost, but characterizes the predictions. There are a number of predictions that test the theories, but not directly the 10D theories. What could it mean to test a 10D theory? Lots of people talk about that, but ask them what they mean.

- What does "empirically testable" mean for string/M-theories?

  - If it means project the theory onto a 4D world (i.e., compactify) and find generic predictions, then it is well defined, has been going on, and is the way traditional physics has worked. Compactified string/M-theories are not post-empirical science.
  - If it means including non-generic predictions with additional assumptions about (say) the Kähler potential, then it tests the cleverness of the people doing it as well as the theory, and is much less powerful.
  - If it is done for non-compactified theories, it predicts a landscape of solutions, but it cannot address the issue of how many of those solutions lead to viable worlds that live long enough to contain planets. Even if the non-viable ones are not universes with people, perhaps they could still have an effect on the wave function of the universe as part of a superposition.

- Are string/M-theories forever beyond the possibility of testing – for example, because they are formulated at energies too high to ever reach? No! Just as no one was at the Big Bang, every knowledgeable scientist knows it is well tested by the universe's expansion, by nucleosynthesis and the helium abundance, and by the embers of the cosmic microwave background and additional technical results. Are small extra dimensions a problem? No! The curled-up dimensions contain lots of information that implies testable predictions for the compactified theories, particularly determining the forces and particles. Do string/M-theories exist? Can anyone define what they are? This red herring should not be taken seriously: Recall that some of the main successes of quantum theory in the mid-1920s were achieved before rigorous definitions of quantum theory were given.

- Sometimes people say wrongly that the Standard Model offers no path forward. In fact, it tells us to focus on theories beyond the Standard Model that contain Yang–Mills gauge forces, quarks and leptons with hierarchal masses, one and only one fermion with a large Yukawa coupling of order gauge couplings, gravity, and so on.

- Is string/M-theory "the only game in town," Dawid's no-alternatives argument (NAA)? It's important to consider only games that are comprehensive and include all the issues – not only gravity, but all the Standard Model forces, existence of quarks and leptons, dark matter, a cosmic matter asymmetry, electroweak symmetry breaking, a stable hierarchy of the right size between atoms and the Planck scale, and more. Compactified string/M-theories are known to address all these issues, and indeed nothing else is known to do that, or even come close.

- Dawid's unexpected explanatory power argument is very strong. One of the best examples is supersymmetry, which was introduced for theoretical reasons, but turned out to provide possible explanations for the hierarchy problem, electroweak symmetry breaking, dark matter, matter asymmetry, and more. The unexpected explanatory power of string theory is impressive, too – look at the issues addressed by the compactified M-theory presented in this chapter.

- One can think of Dawid's book as describing how physicists use theory assessment at a given stage to decide what to work on. It does that very well. Alternatively, one could think of it as describing how non-physicists might evaluate the work of physicists (from the general public to philosophers). Then one could imagine controversy. It holds up well, with much of the confusion arising from books and articles and blogs that don't understand the string/M-theories, and that don't understand that only compactified theories can make contact with our world.

Acknowledgments

I'm grateful to my collaborators over the past decade on compactified M-theory, particularly Bobby Acharya and Piyush Kumar as well as Konstantin Bobkov, Sebastian Ellis, Eric Kuflik, Ran Lu, Jing Shao, Scott Watson, and Bob Zheng; to David Gross, Brent Nelson, Malcolm Perry, Joe Polchinski, and others for discussions; to Eric Gonzalez for help with the manuscript; and to Richard Dawid, Slava Mukhanov, and Gia Dvali for their hospitality at the Munich meeting.

References

[1] Kane, G. L., Perry, M. J., & Żytkow, A. N. (2005). An approach to the cosmological constant problem(s). *Physics Letters B, 609*(1–2), 7–12.

[2] Papadopoulos, G., & Townsend, P. K. (1995). Compactification of D= 11 supergravity on spaces of exceptional holonomy. *Physics Letters B, 357*, 300–306.

[3] Acharya, B. S. (1999). M theory, Joyce orbifolds and super Yang–Mills. *Advances in Theoretical and Mathematical Physics, 3*(2), 227–248.

[4] Atiyah, M., & Witten, E. (2002). M-Theory Dynamics On A Manifold Of G_2 Holonomy. *Advances in Theoretical and Mathematical Physics, 6*(1), 1–106.

[5] Acharya, B., & Witten, E. (2001). Chiral Fermions from Manifolds of G_2 Holonomy. *arXiv preprint hep-th/0109152.*

[6] Witten, E. (2001). Deconstruction, G (2) holonomy, and doublet triplet splitting. *Prepared for 10th International Conference on Supersymmetry*, pp. 472–491.

[7] Beasley, C., & Witten, E. (2002). A note on fluxes and superpotentials in M-theory compactifications on manifolds of G2 holonomy. *Journal of High Energy Physics, 2002*(07), 046.

[8] Friedmann, T., & Witten, E. (2003). Unification scale, proton decay, and manifolds of G2 holonomy. Advances in Theoretical and Mathematical Physics, 7(4), 577–617.

[9] Lukas, A., & Morris, S. (2004). Moduli Kähler potential for M theory on a G 2 manifold. *Physical Review D, 69*(6), 066003.

[10] Acharya, B. S., & Gukov, S. (2004). M theory and singularities of exceptional holonomy manifolds. *Physics reports, 392*(3), 121–189.

[11] Halverson, J., & Morrison, D. R. (2016). On gauge enhancement and singular limits in G 2 compactifications of M-theory. *Journal of High Energy Physics, 2016*(4), 100.

[12] Halverson, J., & Morrison, D. R. (2015). The landscape of M-theory compactifications on seven-manifolds with G 2 holonomy. *Journal of High Energy Physics, 2015*(4), 47.

[13] Acharya, B. S., Bożek, K., Romão, M. C., King, S. F., & Pongkitivanichkul, C. (2015). S O (10) grand unification in M theory on a G 2 manifold. *Physical Review D, 92*(5), 055011.

[14] Kane, G., Kumar, P., & Shao, J. (2010). C P-violating phases in M theory and implications for electric dipole moments. *Physical Review D, 82*(5), 055005.

[15] Ellis, S. A., & Kane, G. L. (2016). Theoretical prediction and impact of fundamental electric dipole moments. *Journal of High Energy Physics, 2016*(1), 77.

[16] Acharya, B., Bobkov, K., Kane, G., Kumar, P., & Vaman, D. (2006). M theory Solution to the Hierarchy Problem. *Physical review letters, 97*(19), 191601.

[17] Acharya, B. S., Bobkov, K., Kane, G. L., Kumar, P., & Shao, J. (2007). Explaining the electroweak scale and stabilizing moduli in M theory. *Physical Review D, 76*(12), 126010.

[18] Acharya, B. S., Bobkov, K., Kane, G. L., Shao, J., & Kumar, P. (2008). G 2-MSSM: an M theory motivated model of particle physics. *Physical Review D, 78*(6), 065038.

[19] Acharya, B. S., & Bobkov, K. (2010). Kahler independence of the G 2-MSSM. *Journal of High Energy Physics, 2010*(9), 1.

[20] Ellis, S. A., Kane, G. L., & Zheng, B. (2015). Superpartners at LHC and future colliders: predictions from constrained compactified M-theory. *Journal of High Energy Physics, 2015*(7), 81.

[21] Acharya, B. S., Bobkov, K., Kane, G. L., Kumar, P., & Shao, J. (2007). Explaining the electroweak scale and stabilizing moduli in M theory. *Physical Review D, 76*(12), 126010.

[22] Acharya, B. S., Kane, G., & Kuflik, E. (2014). Bounds on scalar masses in theories of moduli stabilization. *International Journal of Modern Physics A, 29*(11n12), 1450073.

[23] Ellis, S. A., Kane, G. L., & Zheng, B. (2015). Superpartners at LHC and future colliders: predictions from constrained compactified M-theory. *Journal of High Energy Physics, 2015*(7), 81.

[24] Acharya, B. S., Bobkov, K., Kane, G., Shao, J., Watson, S., & Kumar, P. (2008). Non-thermal dark matter and the moduli problem in string frameworks. *Journal of High Energy Physics, 2008*(06), 064.

[25] Kane, G., Shao, J., Watson, S., & Yu, H. B. (2011). The baryon-dark matter ratio via moduli decay after Afflec×-Dine baryogenesis. *Journal of Cosmology and Astroparticle Physics, 2011*(11), 012.

[26] Witten, E. (2001). Deconstruction, G (2) holonomy, and doublet triplet splitting. *Prepared for 10th International Conference on Supersymmetry*, pp. 472–491.

[27] B. S. Acharya, G. Kane, E. Kuflik, and R. Lu, Theory and phenomenology of μ in M theory. *Journal of High Energy Physics 1105* (2011) 033.

[28] B. S. Acharya, K. Bobkov, and P. Kumar, An M theory solution to the strong CP-problem, and constraints on the axiverse. *Journal of High Energy Physics 1011* (2010) 105.

[29] Kane, G. L., Kumar, P., Nelson, B. D., & Zheng, B, Dark matter production mechanisms with a nonthermal cosmological history: A classification. *Physical Review D, 93*(6) (2016) 063527.

22

Is String Phenomenology an Oxymoron?

FERNANDO QUEVEDO

22.1 Introduction

Although the scope of the conference 'Why Trust a Theory?' was much broader and touched general aspects of the philosophy of science, the main motivation that triggered this meeting was the debate about string theory and the difficulty of testing this theory experimentally. Over a period of more than 30 years, a sub-field of string theory has developed that aims to actually address potential observational issues of string theory. This sub-field is known as string phenomenology. The aim of string phenomenology is well defined and very ambitious: to uncover string theory scenarios that satisfy all particle physics and cosmological observations and, one hopes, that lead to measurable predictions in the short, medium, or long term. For a comprehensive treatment of the field with a very complete set of references, see [1].

Most string theorists do not work directly on string phenomenology since string theory, being a theory under development, has many other open questions that range from conceptual to technical and computational challenges. We may distinguish string phenomenology from what could be called 'string noumenology' (as introduced in [2] and appropriate for the Munich conference) using the difference that philosophers like Kant make between *the noumenon* (the thing in itself) and *the phenomenon* (the thing as it manifests). Given the fact that string theory has been questioned precisely for its lack of concrete predictions that can be tested experimentally, it is natural to wonder about the term 'string phenomenology', which is what motivates the title of this chapter. We will argue that the answer to the question in the title is *No*. But before giving the arguments, we will recall some basic facts.

During the past 25 years, we have witnessed at least four major discoveries that have had a big influence in understanding our universe:

1. The discovery of the density perturbations of the cosmic microwave background in 1992 by COBE, which has been followed by the impressive precision of

the WMAP and PLANCK satellites manifested by the by-now famous plot of the power spectrum of the primordial density perturbations giving rise to precise values of physical observables such as the spectral index $n_s = 0.968 \pm 0.006$.

2. The surprising discovery in 1997 that our universe is not only expanding but also accelerating, indicating the existence of an unknown subject, dark energy, that overcomes the well-known gravitational attraction and forces the universe to expand faster. The simplest manifestation of dark energy is a positive vacuum energy or cosmological constant (CC) that is extremely small $\Lambda \approx 10^{-120} M_{Planck}^4$ [1]. The dark energy equation of state is currently given by $w = p/\rho = -1.006 \pm 0.045$. A positive cosmological constant giving rise to de Sitter space would imply $w = -1$.

3. The discovery of the Higgs boson in 2012, closing the particle content of the Standard Model (SM) of particle physics. The Higgs is responsible for the symmetry-breaking mechanism that drives the electroweak unification and provides the mass scale for the standard model $m_{Higgs} = 125$ GeV (approximately $10^{-16} M_{Planck}$).

4. The LIGO interferometers' spectacular discovery of gravitational waves produced by merging black holes 1.2 billion years ago.

These few major discoveries may be complemented by a few others, albeit probably less spectacular ones, such as the discovery of the top quark, the accumulated evidence for non-vanishing neutrino masses, and the precision tests of the SM by LEP giving rise to the limit of three light neutrinos and providing indirect evidence for an intriguing apparent unification of the three gauge couplings of the SM at a scale of order $M_{GUT} \approx 10^{-2} M_{Planck}$ if the SM is extended by a supersymmetric version. Also, the indirect accumulation of experimental evidence for the existence of dark matter (although its nature is not yet uncovered has important implications). It should be emphasised that these success stories reflect a community effort involving theory, computational physics, and experiment. There have also been so far unsuccessful attempts such as the search for proton decay, cosmic strings (both by direct search through gravitational lensing and by their potential contribution to the cosmic microwave background [CMB] power spectrum), dark matter particles, axions, and more.

The discoveries mentioned here illustrate several important lessons regarding scientific methodology. The acceleration of the universe was hardly foreseen. Physicists had been trying for decades to address what now seems to be the wrong

---

[1] Here, $M_{Planck} = \sqrt{\hbar c / G} \approx 10^{19}$ GeV is the natural scale for relativistic (c) quantum ($\hbar$) gravity (G) known as the Planck scale.

question – why the energy of the vacuum vanishes – with the remarkable exception of Weinberg's 'prediction' in 1987 of the approximately observed experimental value through an anthropic argument [3]. Explicit calculations of the density perturbations were done only less than 15 years before their discovery, whereas the potential existence of the Higgs particle was predicted 50 years before and gravitational waves 100 years before their discovery. Indeed, Einstein predicted gravitational waves without having any idea about how to detect them. Only one year later, he theoretically discovered 'stimulated emission', which is the physical basis behind the lasers. Nevertheless, it would take 40 years for the first laser to be built. Einstein could not have dreamed that these devices, built following his idea and which he did not live to see, would be used to discover the gravitational waves he had predicted. This is a good example to keep in mind when discussing potential experimental tests of string theory. We have to be patient and we may not even imagine the technology that will eventually be used to test it.

On the theoretical side, however, we cannot identify a similar list of greatest discoveries in the past 25 years. The nature and level of maturity of the field itself make it difficult to state with certainty which theoretical developments will have the biggest impact in the relatively short term. Perhaps only the gauge/gravity and possibly other string dualities will survive a future assessment of important recent developments in high energy physics. Indeed, it could be that discoveries in other fields, such as condensed matter or quantum information, or even other attempts to address the quantum gravity problem, may turn out to be more relevant.

Even if we go one decade further, to the 1980s, we may identify only cosmological inflation (including the calculation of density perturbations) and the emergence of string theory as a fundamental theory of nature as the main candidates to survive as great discoveries. Just before that, developments in gravity such as the Bekenstein–Hawking work in black holes are good contenders. But we have to go to the early 1970s to identify theoretical discoveries (such as the renormalisation of gauge theories and asymptotic freedom) that we are sure will remain in the long term.

Clearly, big discoveries in theory are very rare and hard to identify in the short term. But these discoveries come from an accumulated effort by the whole community with small steps (and some failed attempts) building up before a major breakthrough. In the end, experimental evidence is the final judge and the timing depends on many factors, most of them beyond the control of the theorists. Therefore other criteria, such as mathematical consistency and well-posed questions play an important role in guiding the research. These have also been the working premises in string theory so far.

22.2 Basic Theories

Let us start with an overview of the basic principles that are considered as the pillars of our current fundamental understanding of nature. The basic theories in physics so far are special relativity and quantum mechanics. In 1939, Wigner took precisely these two theories and studied the representations of the Poincaré group (basis of special relativity) to identify the fundamental quantum states. They are classified by the quantum numbers specified first by the eigenvalues of the Casimir operators $C_1 = P^\mu P_\mu$ and $C_2 = W^\mu W_\mu$, where P^μ, $\mu = 0, 1, 2, 3$ is the momentum operator generator of spacetime translations, and $W^\mu = \varepsilon^{\mu\nu\rho\sigma} P_\nu M_{\rho\sigma}$, where $M_{\rho\sigma}$ are the generators of rotations and Lorentz boosts and W^μ is the Pauli–Ljubanski vector. Besides the eigenvalues of C_1 and C_2, other labels identifying quantum states are given by the eigenvalues of generators that commute with C_1, C_2 and among themselves – namely, the momentum operators and other operators that commute with them. This naturally leaves three classes of states, depending on the sign of C_1.

If $C_1 = m^2 > 0$, we can choose a frame for which the eigenvalues of the momenta are in the rest frame $p^\mu = (m, 0, 0, 0)$ and the rest of the labels are determined by representations of the Little or Stability group H defined as the subgroup of the Poincaré group that leaves invariant the corresponding momenta. In this case, it is easy to identify $H = O(3)$ as the group of rotations in three spatial dimensions. The finite dimensional unitary representations of $O(3)$ are well known to be determined by the eigenvalues of spin $J^2 \propto C_2$ with integer or half-integer values and its third component $s = -J, -J+1, \ldots, J-1, J$. Therefore the quantum states are the one-particle states $|m, J; p^\mu, s\rangle$ essentially defining a massive particle.

Massive particle states $\qquad |m, J; p^\mu, s\rangle, \qquad s = -J, -J+1, \ldots, J-1, J$
$$\tag{22.1}$$

Further labels can be added corresponding to conserved quantities associated to internal symmetries, such as electric charge, colour, and so on.

A similar analysis could be done for $C_1 < 0$. In this case, the states are tachyonic, corresponding to particles moving faster than the speed of light or $m^2 < 0$. These particles are not realised in nature but when appearing in a physical theory can usually be understood as perturbations of a field around a maximum (unstable) rather than a minimum vacuum state.

Finally, there are the massless states $C_1 = 0$ for which the momenta can be chosen as $p^\mu = (E, 0, 0, E)$. In this case, the Little group is the full Euclidean group in two dimensions, which has unitary *infinite* dimensional representations. However, we do not observe such a big degeneracy of massless particles. Imposing finite dimensional representations, the Little group reduces to $O(2)$, which then adds one label to the representations corresponding to helicity λ. Topological conditions restrict helicity to be $\lambda = 0, \pm\frac{1}{2}, \pm1, \pm\frac{3}{2}, \pm2, \ldots$. In this case, the eigenvalues of both C_1 and C_2 vanish and the states are simply labelled by $|p^\mu, \lambda\rangle$.

$$\text{Massless particle states} \qquad |p^\mu, \lambda\rangle \qquad \lambda = 0, \pm\tfrac{1}{2}, \pm1, \pm\tfrac{3}{2}, \pm2, \ldots \quad (22.2)$$

The theory that describes interactions for massive and massless particles up to helicity $|\lambda| \leq 1$ is quantum field theory (QFT). Gauge field theories in their different phases provide the consistent framework to describe interactions of particles of helicities $\lambda = \pm1$. For higher helicities, we need effective field theories (EFT). EFTs are well-posed QFTs that are valid up to a cut-off scale M. General relativity describing $\lambda = \pm2$ is an example. Supergravity, which is also needed to describe helicities $\lambda = \pm3/2$, is another example. In the 1960s, Weinberg proved that particles of helicities $|\lambda| \geq 1$ need to couple to conserved currents. Particles of helicity $\lambda = \pm1$ couple to gauge currents, those with $\lambda = \pm2$ couple to the stress energy tensor, and those with $\lambda = \pm3/2$ couple to the supersymmetry current. There are no further conserved quantities (from the Coleman–Mandula theorem and extensions), so massless particles of higher helicities are not expected to exist, or at least do not mediate long-range interactions.

These "soft theorems" of Weinberg are very powerful and only rely on the basic understanding of special relativity and quantum mechanics. In particular, they set the general relativity description of gravity at the same level of other interactions as just the relativistic theory describing massless particles of helicity $\lambda = \pm2$. This is the formulation of gravity that fits with string theory. For a detailed description of this perspective, see Weinberg's QFT textbook, in particular chapter 13 of volume 1 [4] (see also the last 2012 Salam Lecture of N. Arkani-Hamed, www.ictp.it/about-ictp/salam-lecture-series.aspx).

22.3 General Predictions of Quantum Field Theories

By now, there is no doubt that QFT is the right framework to address interactions of particles of spin 1 or less. Most of the work on QFT has been based on concrete models such as quantum electrodynamics (QED) and quantum chromodynamics (QCD). Nevertheless, a handful of general predictions can be extracted from all relativistic QFTs.

- *Identical particles.* All particles of the same type are indistinguishable from each other. Every electron is identical to any other electron and if one replaces one by the other in a material object, no change should be noticed. This is simply because the electrons are all excitations of the same field (electron field); the same is true for photons (electromagnetic field) and other particles. This is essentially a tautology since a particle is defined by its labels (e.g., mass, spin).

- *Antiparticles.* Every field that creates a particle, annihilates another particle with the same mass but opposite electric charge or other quantum numbers, in the sense that they can annihilate each other in radiation. Some particles such as the photon are their own antiparticles. This general implication of field theories has been verified experimentally since the discovery of the positron in 1939.

- *Spin statistics.* Particles of integer spin (bosons) behave very differently from those of half-integer spin (fermions) that are constrained by the Pauli exclusion principle. Bosons satisfy the Bose–Einstein statistics, allowing them to be in the same state as in a laser beam, whereas fermions are subject to the Fermi–Dirac statistics. The Pauli exclusion principle does not allow two of them in the same physical state, a restriction that explains the rich atomic structure of matter.

- *The CPT theorem.* Physics is the same under a combination of time translations (T), space inversions (P), and charge conjugation (C).

- *Decoupling and RG.* Physical processes are understood by different scales, following the Wilsonian approach to field theory in which integrating out high scales describes an effective field theory for the lightest states. Couplings (such as gauge couplings and masses of particles) change with energy according to the renormalisation group (RG) equations.

These are very robust general predictions of relativistic quantum field theories. However, since they are very few in number, most of our knowledge of field theories is actually based on concrete examples such as QED, QCD, and low-dimensional field theories.

22.4 The Standard Model and Beyond

The SM is perhaps the greatest theoretical development of the past 75 years. It is a very particular QFT based on the gauge theory of the symmetry group $SU(3) \times SU(2) \times U(1)$ with spin $\frac{1}{2}$ matter fields (quarks and leptons) in three families fitting in bi-fundamental representations of the factor groups. The theory is described by a simple Lagrangian that can be written on a T-shirt or a mug such as Figure 22.1. The experimental evidence is simply spectacular – not only the existence of the relevant particles but the values of the physical observables have been confirmed with incredible precision over the past 40 years from many experiments. There is no other theory in history that has been tested with such high precision. Quantities such as the anomalous magnetic moment of the electron fit the experiments with 11 decimal figures. The equivalence principle has also been tested with similar precision, with no parallels in any other field of science. The robustness of the

Figure 22.1 The standard model on a mug. The first row has the Einstein–Hilbert term for gravity ($\lambda = 2$) and the kinetic and topological terms for the gauge fields ($\lambda = 1$) describing the electromagnetic, weak, and strong interactions. The second line has the kinetic energy for the matter fields: quarks and leptons $\lambda = \frac{1}{2}$ as well as their (Yukawa) couplings to the Higgs field H ($\lambda = 0$). The third line is the kinetic and potential energy for the Higgs field.

SM is simply impressive. Other features to emphasise about this model include the following:

- *The SM is an EFT.* The non-gravitational part of the Lagrangian is renormalisable and therefore quantum mechanically complete (up to Landau poles). The inclusion of gravity makes it into an EFT, which is well defined up to scales close to the Planck scale $M_{Planck} = \sqrt{\hbar c / G} \approx 10^{19}$ GeV. The fact that the non-gravitational part of the SM is renormalisable used to be regarded as a positive property. However, it is because of this property that we do not know at which scale the SM ceases to be valid and, therefore, we have less guidance as to what lies beyond the SM. In this sense, it could be possible that new physics may only manifest at or close to the Planck scale.

- *The SM is simple but not the simplest.* The structure of gauge fields and matter content of the SM is relatively simple (e.g., small-rank simple gauge groups, matter in bi-fundamental representations). However, there are simpler gauge symmetries [such as just abelian $U(1)$ symmetries or an $SO(3)$ group] and matter content (only singlets, a single family, and so on), but they do not fit the experiments.

- *The SM has a rich phase structure.* The SM is actually rich enough to illustrate most of the theoretically known phases of gauge theories: Coulomb phase (electromagnetism), Higgs phase (electroweak), and confinement phase (strong interactions). In particular, the three gauge couplings g_i ($i = 1, 2, 3$) measuring the strength of the interactions are such that well-controlled perturbative expansions can be defined as long as the quantity $\alpha_i = g_i^2/(4\pi) \ll 1$ (for electromagnetism, $\alpha_{em} \approx 1/137$ at low energies is the celebrated fine structure constant). This is satisfied for the electroweak interactions at low energies and, thanks to asymptotic freedom, for the strong interactions at higher energies. Extrapolating with the RG equations to high energies, the three gauge couplings tend to approximately unify at energies smaller but close to the Planck scale and at weak coupling ($\alpha_i < 1$). See Figure 22.2. Even though simple extensions of the SM at intermediate scales may modify this behaviour, it is a good hint that an ultraviolet completion of the SM at weak coupling could exist, such that a perturbative expansion can hold. Notice that the couplings g_i themselves are not too different from one.
- *The SM includes a range of mass hierarchies.* The SM has a wide range of mass scales: from $\Lambda^{1/4} \approx 10^{-3}$eV $\approx 10^{-30} M_{planck}$, which is relevant for dark energy and the approximate range of neutrino masses; to $m_e \approx 10^{-22} M_{planck}$ for the electron mass; to $m_{t,H} \approx 10^{-15} M_{planck}$ for the top and Higgs masses to the Planck scale M_{planck}. There is no explanation for this huge hierarchy of mass scales extending for at least 30 orders of magnitude. The only 'understood' small mass scale relative to the Planck scale is the QCD scale $\Lambda_{QCD} \approx 10^{-20} M_{planck}$, in which the small factor is determined by dimensional transmutation coming from the logarithmic RG running of the gauge coupling, which determines the infrared cut-off $\Lambda_{QCD} \propto e^{-b/g_3^2}$ with b being the QCD beta function coefficient. Here the existence of the large hierarchy is determined by the logarithmic running of the coupling. However, there is an assumption that the coupling is small but of order 1 (small changes in g_3 generate big changes in Λ_{QCD}). So far, there is no other independent mechanism known to explain the emergence of a small mass scale and all other hierarchies are left unexplained.
- *The SM is ugly.* It is important to understand that even though the symmetry principles behind the SM are very elegant, the concrete experimental realisation that is the SM itself is ugly in the sense that there are many (approximately 20) arbitrary parameters, mostly related to the Higgs couplings, which are not fixed. In addition, there exist apparently unnecessary particles expanding a large range of different mass scales, and so on.
- *The SM has a landscape of universes.* The SM including gravity implies the existence of a landscape of vacua. The Lagrangian of the SM has a unique solution in four dimensions describing the physics that we know. However, this same Lagrangian allows for an essentially infinite number of solutions in which one of

the spatial dimensions is curled into a circle so that the space, instead of being the Euclidean $\mathbb{R}^3$, is $\mathbb{R}^2 \times S^1$ with S^1 as a circle. In [5], explicit solutions were found fixing the value of the radius of the circle from the parameters of the SM and using well-understood quantum corrections. This provides a concrete realisation of a 'landscape' of a huge number of (2 + 1 dimensional) universes or a multiverse. Usually, the existence of a landscape is associated with string theory or higher-dimensional gravitational theories that are not yet confirmed by experiment and for which the existence of a multiverse is too speculative. Especially after the discovery of the Higgs, essentially no one has questioned the validity of the SM, yet this experimentally confirmed theory also implies the existence of a landscape of vacua, with each vacuum describing a different universe. This makes the idea of the multiverse far less speculative than it is usually presented.

- *The SM is incomplete.* The SM is almost certainly not complete. It cannot by itself explain dark matter, the density perturbations of the CMB, and baryogenesis, for instance. Moreover, the value of the many parameters of the SM is not understood. In particular, the mass of the Higgs is not protected under quantum corrections, which tend to bring it to be as high as the limit of validity of the effective field theory, namely M_{planck}. The nature of dark energy responsible for the current accelerated expansion of the universe is not understood, especially the fact that it seems to indicate a vacuum energy as small as $\Lambda \approx 10^{-120} M_{planck}^4$. Furthermore, gravity is described only at the classical or EFT level. Thus, the SM is not ultraviolet complete. This is the best evidence we have for the need to go beyond the SM.

To search for the new physics that will supersede the SM, we have to explore experimentally all possibilities, increasing the energy, intensity, and reach to the highest possible limits. The history of science tells us we are bound to find something. For theorists, we can follow several directions:

1. *Simplicity.* Add the simplest possible component to the SM (e.g., one extra neutral fermion or boson to be dark matter and/or drive inflation) and contrast the result with observations. This is a way to start, at least to eliminate the simplest cases and start building up a more meaningful theory.
2. *Follow your nose.* Follow aesthetic arguments (usually subjective) as a guideline (e.g., add extra symmetries or dimensions to address dark energy, dark matter, or the flavour structure of the SM; consider mechanisms such as the see-saw mechanism to explain smallness of neutrino masses).
3. *Bottom-up.* Use any experimental hint to introduce new particles or modifications of the SM that fit data and then use the result as a guide toward model building (e.g., attempts to explain some astrophysical events from fundamental

physics such as a concrete dark matter candidate, attempts to explain some deviations form the SM at colliders data).

4. *Top-down.* Start with a basic theory at high energies and then deduce its implications at lower energies to describe the SM and any physics beyond (e.g., grand unified theories [GUT] of the past, extra dimensional theories and supersymmetric theories that address some problems of the SM and/or add the extra possibility of unification of interactions). Since the mid-1980s, string theory has become the prime example of this approach: In addition to addressing those problems, it includes unification with gravity at the quantum level, providing the explicit ultraviolet complete framework to address physics beyond the SM.

Given the fact that we do not have a clue of what lies beyond the SM, *all* of these avenues and potential combinations have to be explored. This is what makes our current state of affairs so interesting. We are facing the problems created by the fact that the SM is so successful experimentally. Once and again, the experiments tend to verify its validity with unprecedented precision and all potential hints of physics beyond the SM (BSM) have turned out to be statistical fluctuations, misinterpretation of data, or possible to explain within the SM (e.g., faster-than-light neutrinos, several astrophysical hints for dark matter particles, potential evidence of cosmic strings, primordial gravitational waves, the disappeared 750 GeV resonance).

22.5 General Predictions of String Theory

Following the case of QFT, we may try to extract the general predictions of string theory as known so far (see, for instance, [6]).

- *One theory, no free parameters.* Even though over the years five consistent supersymmetric string theories were identified, in the mid-1990s it was understood that all of them are different manifestations of one single underlying theory, M-theory, that also includes as another weak coupling limit 11-dimensional supergravity. Uniqueness is an important property that a candidate for a fundamental theory should have. Furthermore, each of the different string theories has one single parameter corresponding to the string length that defines the units; otherwise, there are no free parameters, another desired property of a fundamental theory. This immediately raises the question of how to determine all the 20 or so free parameters of the SM taking values in such a huge range. The answer could be dynamical, which creates a big challenge to identify the dynamics that selects all the observables we see in nature.

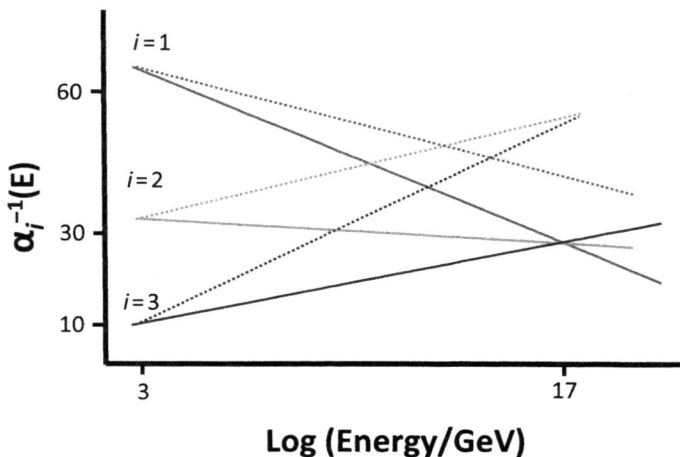

Figure 22.2 A sketch of the running of the SM gauge couplings with energy through the renormalisation group equations. At the TeV scale the three gauge couplings are very different, but with increasing energy they tend toward unification. We plot the quantities $\alpha_i^{-1} = \frac{4\pi}{g_i^2}$ against (log of) energy. Dashed lines correspond to the SM and solid lines to the supersymmetric extension of the SM. Two points are usually emphasised: Supersymmetry remarkably unifies the three interactions at an energy close to the Planck scale and even without supersymmetry the running is toward weaker couplings, raising expectations for a weakly coupled ultraviolet completion. Gravity is not included in the plot.

• *Extra dimensions.* Another well-known property of consistent string theories is their dimensionality.[2] A positive point is that contrary to most theories of physics for which the spacetime dimension is assumed from the start, in string theory this dimension is determined by the condition of criticality of the underlying conformal field theory. The fact that the critical dimension is 10 and not 4 creates a challenge for how we obtain our 4-dimensional world. It immediately indicates that even though the theory may be unique, the number of solutions (giving rise to different universes) most likely cannot be unique. In particular, flat supersymmetric 10-dimensional spacetime should be a solution. Furthermore, the shape and size of the extra dimensions (known as *moduli*) should be determined dynamically. Since they will naturally affect the physics of the 4-dimensional solution, the same dynamics that selects them should also select the physical

---

[2] Strictly speaking, the prediction is on the central charge c of the corresponding conformal field theory (CFT) and the extra contributions to c may not have spacetime interpretations. Also, supercritical strings can be considered but proper attention should be paid to the dilaton tadpole. In fact, spacetime may not even be the proper arena of the theory beyond the string scale, though its manifestation in terms of 10- or 11-dimensional theories may still be the relevant starting point for exploring low-energy physics with all its hierarchies of scales and weak couplings.

parameters. Therefore, having more dimensions in a fundamental theory without free parameters could be seen as a positive property that allows the possibility of fixing the many observables of the SM from the dynamics of the theory. Similar comments apply to the 11-dimensional limit of M-theory.

- *Supersymmetry.* The third property that is generic in string theories is supersymmetry. Notice that these theories have either 16 or 32 supercharges (supersymmetry generators), which happens to be the maximum number of supersymmetries that allow both just one single graviton and no massless states of helicity larger than 2 (consistent with Weinberg's results mentioned earlier). Supersymmetry is good for consistency and the absence of tachyons (although non-tachyonic non-supersymmetric solutions are known), but it also raises the challenge of how to find a mechanism to break supersymmetry. Furthermore, supersymmetric solutions of the theory tend to keep the moduli unfixed. Since moduli represent massless scalar fields in four dimensions, these solutions are very predictive and have been already tested experimentally. Notably, they predict long-range interactions that should and have not been found and, therefore, are ruled out by experiment. Furthermore, supersymmetry would imply the existence of supersymmetric partners of the particles we know; such partners have also not been observed. Thus, only non-supersymmetric solutions can be realistic. The scale at which supersymmetry should be broken could be as large as the string scale – but could also be smaller, so it should be determined dynamically. We can then conclude that determining the value of the moduli and breaking supersymmetry dynamically is a key question for string theory to make contact with the real world.

- *Gravity, gauge interactions.* As can be seen in Table 22.1, all string theories and 11-dimensional supergravity have in the massless spectrum a symmetric tensor g_{MN} with $M, N = 1, 2, \ldots 10(11)$, which, as is well known from field theory, corresponds to a massless particle of helicity 2 describing gravitational interactions. This is the basis of the general statement that string theory predicts gravity in the sense that the existence of a helicity 2 particle is an output, rather than an input, of the theory. Upon compactification to 4 dimensions, the components $g_{\mu\nu}$, $\mu, \nu = 1, \ldots 4$ correspond to the 4-dimensional graviton, whereas the g_{mn}, $m, n = 1, \ldots 6(7)$ are scalar fields in 4 dimensions that correspond to some of the moduli fields mentioned previously.

   In Table 22.1, it is seen that gauge fields A_M^{ij} are also present in the heterotic and type I strings. Upon compactification, these gauge fields can give rise to gauge and matter fields similar to the supersymmetric partners of quarks and leptons. This justifies the fact that before the mid-1990s only these string theories were considered for the search of realistic models.

- *Antisymmetric tensors.* The new ingredients in the bosonic spectrum of string/M-theory models are the antisymmetric tensors of different indices. They correspond to the only extra fields that give rise to helicities less than 2. In 4 dimensions, they are usually not mentioned since they are either equivalent (dual) to scalar fields or do not correspond to propagating degrees of freedom. But starting from 6-dimensional field theories, these fields correspond to independent propagating fields. They play a role in string theory in at least three ways.

First, upon compactification, they give rise to pseudoscalar fields or axion-like particles that are the imaginary components of complex moduli fields. This opens up a big window of opportunity for low-energy physics since axion-like particles are associated with shifts symmetries of the Peccei–Quinn type that tend to survive at low energies.

Second, they indicate the necessary presence of p-branes (either D-branes, NS-branes, or M-branes). In the same way that the worldline of a particle couples to gauge fields A_M by $\int dx^M A_M$, so a string couples to a 2-index tensor $\int dx^M \wedge dx^N B_{MN}$ and a p-brane to a $p+1$ antisymmetric tensor:

$$\int dx^{M_1} \wedge dx^{M_2} \wedge \ldots dx^{M_{p+1}} C_{M_1 \ldots M_{p+1}} = \int C_{(p+1)}. \tag{22.1}$$

In the case of Dirichlet or D-branes (coupling to R-R tensors in Table 22.1), they host in their worldvolume open strings with both ends attached to the brane that correspond to $U(1)$ gauge fields. Depending on the boundary conditions, a set of K overlapping D-branes can give rise to $U(K), O(K)$, or $Sp(K)$ gauge fields and chiral matter, thereby making the type IIA and IIB string theories viable candidates to have realistic models. Notice in IIA strings there are only even-dimensional D-branes, whereas in IIB strings there are only odd-dimensional D-branes (in this case, D7 and D3 – or equivalent D9 and D5 – branes can host the SM). Just like the case of charged particles, D-branes, being charged under antisymmetric tensor fields, have anti-branes, which carry the same tension and opposite charge. D-branes and anti-D-branes preserve different supersymmetries. Therefore the presence of both type of branes in a compactification would break supersymmetry.

The third important role of antisymmetric tensors is that since they are generalisations of gauge fields, they can give rise to non-trivial fluxes. Similar to the Dirac magnetic monopole, which implies a quantisation of magnetic fluxes over a 2-sphere $\int_{S^2} F_{\mu\nu} dx^\mu dx^\nu = \int_{S^2} B \cdot dS = n$, whenever there is a homological non-trivial surface γ_i of dimension $p+2$ there can be fluxes of the field strength $F_{(p+2)} = dC_{(p+1)}$:

$$\int_{\gamma_i} F_{(p+2)} = n_i. \tag{22.2}$$

Table 22.1 *Massless Spectrum of String/M Theories*

Theory	Dimensions	Supercharges	Bosonic Spectrum	
Heterotic $E_8 \times E_8$	10	16		g_{MN}, B_{MN}, ϕ A_M^{ij}
Heterotic $SO(32)$	10	16		g_{MN}, B_{MN}, ϕ A_M^{ij}
Type I $SO(32)$	10	16	NS-NS R-R	g_{MN}, ϕ, A_M^{ij} C_{MN}
Type IIA	10	32	NS-NS R-R	g_{MN}, B_{MN}, ϕ C_M, C_{MNP}
Type IIB	10	32	NS-NS R-R	g_{MN}, B_{MN}, ϕ C, C_{MN}, C_{MNPQ}
11D Supergravity	11	32		g_{MN}, C_{MNP}

The five consistent string theories and 11-dimensional supergravity. The number of dimensions and supersymmetry generators is given as well as the bosonic massless content. These theories all include ($\lambda = \pm 2$) gravity ($g_{\mu\nu}$), antisymmetric tensors of different ranks (the B and C fields with $\lambda = \pm 1, 0$), and gauge fields ($\lambda = \pm 1$).

The integers n_i provide an important example of a potentially large set of discrete parameters that can play a role in determining physical observables.

- *Infinite tower of massive states.* The spectrum of string theories also includes an infinite tower of massive states, with masses that are multiples of the fundamental string scale, and with higher spins. All states are related by string creation and annihilation operators. If the string scale is substantially larger than the TeV scale, these states would not correspond to any observable particle in the forseeable future. If we are extremely lucky and the string scale is close to the TeV scale, then these states can be the smoking gun for string scenarios to be tested in the near future.

- *No continuous spin representations.* As mentioned in Section 2, special relativity and quantum mechanics also predict the existence of continuous spin representations (CSR), which have not been observed. Wigner immediately realised the problem and came up with an (anthropic) argument to explain why they are not seen in nature: If they existed, it would require infinite heat capacity to excite them. Also, Weinberg in his first field theory textbook simply states (phenomenologically) that since these states have not been observed, he concentrates only on the finite dimensional unitary representations that reduce the Little group to $O(2)$, which has no CSRs. String theory, at least in its current perturbative formulation, offers a simple explanation for the absence of these states. All string states are related by creation and annihilation operators. We can get massless states applying annihilation operators to massive states. Since massive states fit in finite dimensional representations, then massless states should also be in finite dimensional representations and, therefore, there are no CSRs in perturbative string theory [7]. Notice that standard field theory does not provide an explanation for the absence of CSRs. In fact, over the past 75 years, different proposals have been made to find the physical meaning of CSRs in field theory. For a very recent interesting analysis, see [8].

If we judge progress in theory by the development of better explanations, this is another success of the theory. If a single piece of evidence for a CSR is found, all classes of string models studied so far would be ruled out. This reminds one of the old statement attributed to J. B. S. Haldane, who said that the theory of evolution could be ruled out if a single fossil of, say, a rabbit is found in the wrong archeological stratum. CSRs may still be present in a full-fledged non-perturbative string theory and it may be interesting to investigate their potential implications.

22.6 Four-Dimensional Strings

As in the case of field theories, to make closer contact to experimental observables, we have to concentrate on more concrete string models with realistic perspectives.

Given the fact that string theories are formulated in 10 (11) dimensions, to obtain something like our universe, solutions (or vacua) have to be found that include the SM at low energies. On this basis, a spacetime of the form

$$\mathcal{M}_4 \otimes X_6 \tag{22.1}$$

should be selected, where $\mathcal{M}_4$ is essentially Minkowski space and X_6 is an internal space. Also, the number of supersymmetry generators in 4 dimensions could be from 32, 16, 8, 4, 0 (corresponding to $\mathcal{N} = 8, \ldots, 1, 0$ supersymmetries, respectively). It is well known that only $\mathcal{N} = 1, 0$ supersymmetries allow for chirality, one of the most important properties of the SM. Thus, this eliminates many possibilities for the extra-dimensional space X_6 (such as the simplest cases like toroidal manifolds $X_6 = T_6$). Solutions with $\mathcal{N} = 0$ are physically possible but difficult to handle from our current understanding of string theory. This leaves $\mathcal{N} = 1$ as the preferred option. Since we do not observe supersymmetry in nature, this supersymmetry has to be broken, but the breaking scale is not fixed. It may well be that the scale is as high as the compactification scale and essentially captures the physics of $\mathcal{N} = 0$ compactifications. But it might also be as low as our experimental limits of TeV. Most of the work on string compactifications has concentrated on $\mathcal{N} = 1$ vacua.

This process of eliminating solutions that are automatically ruled out by experiment is usually known as 'vacuum cleaning'. Following the original Kaluza–Klein ideas, the size of X_6 is expected to be so small as to not be detectable experimentally.

Progress in the past 10 years has allowed us to actually determine dynamically the shape and size of the extra dimensions and obtain the small sizes expected in the Kaluza–Klein scenario.

It is natural to concentrate on 4-dimensional models (guided by observations and, some could argue, for anthropic reasons). Over the years, a great deal of work has been devoted to string model building. It is fair to say that at the moment there is not a single fully realistic string model. Even so, progress has been made in different directions starting from the different known string theories and 11D supergravity.

Clearly, the most developed class of models includes the heterotic string compactifications on Calabi–Yau (CY) manifolds. At present, tens of thousands of models are known with the massless spectrum of the minimal supersymmetric standard model (MSSM) (see, for instance, [9]). This is non-trivial progress. However, in this theory, given the complicated topology of Calabi–Yau manifolds and the absence of knowledge of explicit metrics, it is very difficult to compute couplings such as physical Yukawa couplings for which not only holomorphic quantities such as superpotentials are needed but also kinetic terms that are non-holomorphic and more difficult to compute. The main problem, however, is a concrete scenario for moduli stabilisation.

On this front, type II models are better suited since turning on fluxes goes a long way toward stabilising the moduli. Particularly promising are type IIB compactifications for two main reasons: (1) Turning on fluxes keeps the compact manifold as a (conformal) Calabi–Yau and all the knowledge acquired in the past 30 years can be used and (2) the fact that there are two different three-index tensors $F_3 \approx dC_{(2)}$ and $H_3 \approx dB_{(2)}$ gives rise to a superpotential that fixes most of the moduli. Quasi realistic type II models have been constructed from either intersecting D-branes or D-branes at singularities, including three family models containing the SM spectrum and a few extra particles. Only in the past few years have fully consistent compact quasi-realistic models been constructed that included moduli stabilisation. Much room remains for finding more realistic models.

In a more general class of type IIB models known as F-theory, the dilaton ϕ is considered as a modulus for an auxiliary torus and, therefore, can extend to strong coupling. These models have the interesting property that allow grand unified groups such as $E_6, SO(10), SU(5)$ and have potential realistic properties. The mathematics required to fully investigate F-theory models is being uncovered. Athough some progress is expected in a few years, at present no concrete models can be considered realistic. Furthermore, the computational advantage of weak coupling type IIB models is not maintained in these case, so moduli stabilisation is not under control.

An even less developed class of models corresponds to G_2 holonomy compactifications in which the SM has to be at a singularity but not much is known about constructing explicit models. However, educated guesses have led to interesting developments and concrete scenarios as discussed by Gordy Kane at the Munich meeting.

In summary, even though progress has been made and continues to be made, there are clearly open challenges. Rather than concentrating our efforts on the search for realistic models, it is better to try to address model-independent issues and extract general scenarios that can be contrasted with experiments.

22.6.1 Model-Independent Results

Regarding model-independent results, we can mention the following points:

- *No global symmetries.* In field theoretical models, we can have local and global symmetries. Local symmetries determine the interactions, and global symmetries may be used to address some questions such as flavour structure, baryon and lepton number conservation, and so on. In string theory, there are no global symmetries except for non-compact symmetries such as Poincaré invariance or shifts associated with axion fields. The proof is very elegant and based purely on conformal field theory. It essentially states that any potential global symmetry will imply a state in the spectrum with the properties of the gauge field for that symmetry, such that the symmetry is actually local [10]. This does not apply to D-branes in flat space. Thus, if two sets of D-branes intersect locally, the local symmetries of one set would be seen as the global symmetries of the other. However, once they are embedded in a proper compactification, all symmetries are local and may remain only approximate but not exact global symmetries, depending on the structure of the extra dimensions (large volume or large warping). Notice that this absence of global symmetries is consistent with standard claims about the absence of global symmetries in theories of gravity coming from the no-hair theorems.
- *Small matter representations.* In string models, essentially only fundamental, bifundamental, adjoint, symmetric and antisymmetric elementary representations are realised and (as in the SM) higher-dimensional representations are not possible.[3] So again, if any experiment discovers a representation of dimension, say, 6 of $SU(2)_L$, that finding would rule out most string constructions known so far.
- *Anomalous $U(1)$s.* A very general low-energy implication of string theory comes from the celebrated Green–Schwarz anomaly-cancelling mechanism that triggered the theory as the main candidate for a consistent theory of gravity. Upon

---

[3] Higher-level Kac–Moody representations and/or highly Higgsed quiver and particular F-theory models may lead to higher-dimensional representations, but they are clearly very rare.

compactification to 4 dimensions, this mechanism manifests as an extra term in the action that cancels the anomaly of a $U(1)$ gauge group. The net result is that the corresponding $U(1)$ gauge field obtains a mass by eating an axion field in a realisation of the Stuckelberg (but not Higgs) mechanism. As a consequence, the symmetry can be broken without a charged field getting a vev. This process provides a low-energy effective action mechanism to have approximate global symmetries in string compactifications. Alternatively, anomalous and non-anomalous $U(1)$ symmetries may be broken by fields with higher than minimal charge giving rise to local discrete symmetries.

- *Moduli stabilisation and supersymmetry breaking.* Independent of the particular constructions, 4-dimensional string models have moduli fields (measuring, for instance, the size and shape of the extra dimensions) that need to be stabilised. Without a concrete mechanism for supersymmetry breaking, these moduli fields are massless and would mediate new long-range interactions that are ruled out by experiments. Thus, most string models constructed before 2003 were actually very predictive but ruled out by experiment. Together with moduli, axion-like particles are very generic (coming from antisymmetric tensors) and abound because their number depends on the topology of the extra dimensions, in particular the number of non-trivial homological cycles. Without a mechanism to break supersymmetry, they are also massless.

Independent of the particular string model of particle physics, since 2003 general mechanisms have been found, using fluxes of the antisymmetric tensors mentioned in the previous section together with quantum corrections to the EFT, that allow for fixing the moduli and breaking supersymmetry. These mechanisms allow physicists to consider general classes of string models without contradicting experiments (no particles mediating long-range interactions and no supersymmetric particles with the same masses as quarks and leptons).

The breaking of supersymmetry may be direct from compactification. In turn, the scale of supersymmetry breaking can be as large as the string or Kaluza–Klein scale, which can be very close to the Planck mass. In this case, it is difficult to have the low-energy EFT under control (this is not wrong in itself; it is our own limitation). The more standard construction is to start with $N = 1$ supersymmetry compactifications (4 of the 16 or 32 supercharges preserved), using the fact that this is the largest amount of supersymmetry allowed by the chirality (difference between left and right) observed in nature. In this case, the EFT is under better control and supersymmetry may play a role at low energies. Yet the definition of low energy is not specified; it may be 1 TeV and accessible to the Large Hadron Collider (LHC), but also any other higher scale below the Planck mass.

After supersymmetry breaks, some of the axions also become massive with mass of order of the gravitino mass. Others may remain as essentially massless.

- *Cosmological moduli problem.* In the general class of models in which super-symmetry breaking and moduli stabilisation are addressed (by any unspecified method), there is a very clean prediction: Moduli obtain a mass but survive at low energies and may cause serious cosmological problems [11]. Typically, their mass is of the order of the gravitino mass (and even smaller), but they would overclose the universe and/or ruin nucleosynthesis if they are lighter than 30 TeV. That is, their mass should be $m > 30$ TeV. This indicates that the supersymmetry breaking scale cannot be that low (which is not bad news, given the fact that supersymmetric particles have not yet been observed). At first sight, it seems unfortunate that the most generic implication of low-energy supersymmetry breaking and moduli stabilisation is actually a problem. Conversely, a positive point is that moduli naturally survive at low energies. Even though they couple only gravitationally to the SM, the fact that they survive at low energies means that there are 'stringy' remnants later in the history of the universe that may have some effects in post-inflationary cosmology. Moduli change the standard cosmological scenario in which it is assumed that a short period of inflation gives rise to reheating and then the radiation-dominated era before Big-Bang nucleosynthesis. That is, after inflation the energy density is dominated by the moduli oscillations around their minima, which behave like matter rather than radiation domination, and the true reheating is the late decay of the modulus field. Moduli can contribute and modify standard scenarios of dark matter, baryogenesis, dark radiation, and more.

- *Gauge couplings unification.* Four-dimensional strings usually give rise to unifi-cation of the gauge couplings, even though the gauge group may not necessarily be a simple GUT group. That is, even if the gauge group is a product of simple gauge groups, the gauge couplings of non-Abelian factors tend to be unified. This is the case in heterotic and models of branes at singularities and most F-theory models (but not necessarily for intersecting brane models) [12]. Notice that string theories add an extra component to the unification issue since both the unification scale and the gauge coupling itself are quantities that depend on the moduli and have to be determined dynamically. Obtaining precisely the values that fit with the RG running that agrees with the observed gauge couplings at low energies is a very difficult challenge, since since these values must agree with the independent calculation based on the low-energy spectrum and the scale of supersymmetry breaking (also determined after moduli stabilisation).

22.6.2 *The Landscape*

The landscape and its anthropic implications are the standard source of debate in this field. The existence of energy landscapes is very generic in other areas of science, such as in studies of structure and dynamics of atomic and molecular clusters, in protein folding, and in glasses and supercooled liquids [13]. We argued earlier that even the SM has a big landscape. For the string theory landscape, I do not have much to add to Joe Polchinski's chapter. What I find important to emphasise here is not the philosophical and anthropic discussions on this issue that have dominated the debate for the past decade, but rather the concrete achievements coming out of explicit calculations that address physical questions.

First, the origin of the landscape was addressing a very important question for string theory models: how to stabilise the moduli [14, 15]. This question had been open for 20 years. Many of us thought that since the starting point was a supersymmetric compactification, the celebrated non-renormalisation theorems of supersymmetry would imply that only non-perturbative effects could break supersymmetry and most probably we would have to wait for a proper non-perturbative formulation of string theory to make progress. The fact that turning on fluxes of antisymmetric tensor fields achieves part of the job is remarkable. These fluxes (in IIB, IIA string compactifications) being quantised give rise to the huge number of solutions.

Briefly speaking, CY manifolds have usually hundreds of moduli. They can be classified based on the size of 4-cycles (or their dual 2-cycles), which are called Kähler moduli and are labelled by the Hodge numbers h_{11}, and based on the complex structure moduli corresponding to the size of the non-trivial 3-cycles (and their dual 3-cycles) labelled by the Hodge numbers h_{21}. Both h_{11} and h_{12} rank at least in the hundreds for the known CY manifolds. In IIB strings, since there are two 3-index antisymmetric tensors, their fluxes can thread the corresponding 3-cycles and fix the size of complex structure moduli. For Kähler moduli, fluxes do not seem to help. However, precisely for IIB compactifications, there are D7-branes that can wrap these 4-cycles. The size of these cycles is inversely proportional to the size of the gauge coupling of that theory, and non-perturbative effects induce a scalar potential that fixes the moduli.

Two main scenarios have been developed over the past decade or so that differ in terms of the value of the main quantity in the EFT known as the flux superpotential, W_0[4]. The original KKLT scenario [16] works in the regime $W_0 \ll 1$, whereas the large volume scenario (LVS) [17] works for the regime $W_0 \approx 1 - 100$. Both are based on the same ideas but give rise to vey different physics. Even though the moduli are fixed with a negative value of the vacuum energy (cosmological constant) in both cases, the vacuum preserves supersymmetry in KKLT and breaks it in LVS. Both scenarios need new ingredients to obtain positive vacuum energy

---

[4] Further, less-developed scenarios have been proposed and there may be room for additional scenarios to emerge in the future, especially in other string theories. In particular, non-geometrical fluxes may deserve more studies and fully stringy scenarios of moduli stabilisation may be welcome.

(de Sitter apace). In KKLT, the presence of anti-D3-branes breaks supersymmetry and adds a positive term to the vacuum energy that can do the uplift. In LVS, besides anti-D3-branes there are other options such as T-branes. In both cases, the Bousso–Polchinski (BP) [18] approach to the cosmological constant problem is at work.

Independent of the concrete scenario of moduli stabilisation, we may inquire about general properties of the landscape. The most concrete prediction of the landscape is that our universe is the outcome of a tunneling event similar to the Coleman–de Luccia bubble nucleation. A direct consequence of this process is that our universe should be open instead of flat or closed unless a long period of inflation occurs. This has been an open question for many years. Currently, as we know, the data tell us that the universe is almost flat and a future precise measure of the curvature of the universe could, in principle, rule out this scenario. This is a very concrete prediction [19]. Achieving the right sensitivity to determine if the universe is open is an important experimental challenge. It may also have implications for the low-scale imprints of the CMB.

One general question is if the landscape is finite or infinite. Clearly, for classes of vacua with moduli, there is a continuum of solutions. But for physically relevant vacua, it is still not clear if the number of CY manifolds is finite, although general statements are known (the number of elliptically fibered CY are finite and an overwhelming majority of the known CY manifolds are fibrations). Furthermore, general arguments based on studies of sequence of vacua and general mathematical theorems have been given to support the fact that the number of vacua with 'realistic' properties is finite [20].

Another general point is to determine if some field theories cannot be included in the landscape (the 'swampland') [21]. Even though the landscape of string theories seems to be huge, it is interesting to study cases that will not belong to the landscape in an effort to extract general results (absence of global symmetries and CSRs could be seen as examples of the existence of the swampland). One of the most popular of these cases is the weak gravity conjecture [22]. It suggests that gauge-like interactions, including brane–brane interactions, are all limited to be stronger than gravity. This is motivated by examples in string theory, the known absence of global symmetries, and black holes physics. Although this idea is still only a conjecture, if valid it would not allow many effective field theories – in particular, potential candidates to large field inflation – to belong to the landscape. Also, some popular proposals for BSM physics to address concrete open questions may be questioned as to whether they could actually have a realisation within string theories. For instance, models for which the so-called null energy condition is violated seem difficult to embed in string theoretical models. This group includes proposals for alternatives of inflation and for dark energy. Moreover, some of the $f(R)$ scenarios for modifications of gravity can be contrasted with string theory effective actions. Even the original Starobinski proposal for inflation might not be incorporated in the landscape. If any of these proposals is found to be in the string swampland, it does not rule out the proposal, but it does create a big challenge to find an ultraviolet completion of the model.

Supersymmetry Breaking

Scenarios for moduli stabilisation can be used to concretely study the breaking of supersymmetry and, therefore, the spectrum of the supersymmetric particles. See, for instance [23]. The picture that emerges for KKLT and LVS is as follows:

1. *Split spectrum:* Mass $M_{1/2} \geq 1$ TeV for the fermionic partners of gauge bosons and Higgs (gauginos and Higgsinos) and $m_0 \approx (10 - 1,000)M_{1/2}$ for the scalar partners of quarks and leptons (squarks and sleptons).
2. *High-scale supersymmetry:* All superparticles and gravitino as heavy as $m \approx 10^{11}$ GeV.

The split supersymmetry scenario is realised in KKLT for the SM living on D3- or D7-branes (with a small split $m_0 \approx 30M_{1/2}$) and in the LVS for the SM on D3-branes with a bigger split ($m_0 \approx 10^3 M_{1/2}$). High-scale supersymmetry is realised in LVS for the SM living on D7-branes. This implies that if, say, light stop particles are detected at LHC, it would rule out all these scenarios. The only particles reachable at LHC are the gauginos and Higgsinos. There are also limiting cases in which the spectrum is not necessarily split but contracted; these cases can also be tested in the next colliders.

The landscape has positive and not-so-positive properties for low-energy supersymmetry:

- *The good.* The good news is that the strongest arguments against supersymmetry were always related to the cosmological constant problem (supersymmetry had a chance to solve the CC problem and did not do so. When supersymmetry is used for the hierarchy problem, it neglects the CC problem, and any other mechanism that would solve the CC problem may also modify the effects of supersymmetry for the hierarchy problem). The landscape just tells us that the Bousso–Polchinski mechanism takes care of the CC problem for anthropic reasons but does not affect any supersymmetry calculations that seek to address the hierarchy problem. In particular, soft terms can be trusted.

- *The bad.* The bad news is that we always hoped that any solution of the CC problem would have further implications at low energies that could be tested (e.g., modifications of gravity). Unfortunately, the landscape solution does not provide any experimental test at low energies.

- *The ugly.* The ugly part is that if anthropic arguments are used for the CC problem, then they may also be used for other problems, such as the hierarchy problem, and the need for low-energy supersymmetry may be essentially irrelevant. At this stage, however, nature may not care what we consider ugly. Actually, the two scenarios mentioned previously happen to be precisely split and high-scale supersymmetry would also rely, at least partially, on the landscape to address the hierarchy problem.

Models of Inflation

Similarly, a few scenarios have been developed for cosmology, in particular to realise cosmological inflation (see, for instance, [24]). The inflaton field distinguishes between brane separation to a Kähler or complex structure modulus, axions in different guises, and so on. They all fit well with the most recent constraints from Planck (see Table 22.2). However, they disagree on the predictions for the tensor-to-scalar ratio r. If the claim from BICEP 2 ($r \approx 0.2$) would have been correct, then all models would have been ruled out except for 'axion monodromy inflation'. This implies that these scenarios will be put to the test shortly. Again, this would not be a test of string theory, but rather would test string theory scenarios in a similar way to the relation of the standard model and its extensions to field theory. This is not as exciting and ambitious as testing string theory but it follows the same track that physics has worked in the past.

The concrete realisation of large field inflation leading to tensor-to-scalar ratio r is very difficult to achieve. String models of inflation have been the main source for candidates. In particular, axion monodromy is considered the benchmark model that experimentalists analysed and compared with data regarding values of $r > 10^{-3}$. Furthermore, a Bayesian analysis (very popular at the Munich conference) of the almost 200 different models of inflation proposed over the past 35 years has selected as the 'winner' a class of string theory models known as Kähler inflation. Moreover, a cosmologist working in string models of inflation first performed the detailed study of the foreground for the BICEP2 experiment that led to the conclusion that BICEP's results were not actually due to tensor modes [25]. This study was later confirmed by the detailed analysis of Planck.

Another effect that could be relevant for string models is dark radiation (relativistic particles such as neutrinos that decouple from thermal equilibrium with standard matter before nucleosynthesis). String models tend to have several hidden sectors, and the last modulus to decay after inflation can give rise to observable and hidden sector matter. Relativistic particles in hidden sectors could dominate this decay and then contribute to dark radiation, measured in terms of the effective number of neutrinos N_{eff}. We know that N_{eff} is very much constrained by CMB and

Table 22.2 *Representatitive Models of String Inflation (from [26])*

String Scenario	n_s	r
D3/$\overline{D3}$ inflation	$0.966 \le n_s \le 0.972$	$r \le 10^{-5}$
Inflection point inflation	$0.92 \le n_s \le 0.93$	$r \le 10^{-6}$
DBI inflation	$0.93 \le n_s \le 0.93$	$r \le 10^{-7}$
Wilson line inflation	$0.96 \le n_s \le 0.97$	$r \le 10^{-10}$
D3/D7 inflation	$0.95 \le n_s \le 0.97$	$10^{-12} \le r \le 10^{-5}$
Racetrack inflation	$0.95 \le n_s \le 0.96$	$r \le 10^{-8}$
N – flation	$0.93 \le n_s \le 0.95$	$r \le 10^{-3}$
Axion monodromy	$0.97 \le n_s \le 0.98$	$0.04 \le r \le 0.07$
Kahler moduli inflation	$0.96 \le n_s \le 0.967$	$r \le 10^{-10}$
Fibre inflation	$0.965 \le n_s \le 0.97$	$0.0057 \le r \le 0.007$
Poly – instanton inflation	$0.95 \le n_s \le 0.97$	$r \le 10^{-5}$

nucleosynthesis observations with $\Delta N_{eff} < 1$. Conversely, if any deviation ΔN_{eff} from the SM value $N_{eff} = 3.04$ is observed in the near future, this could be a hint for stringy proposals for dark radiation (such as the axionic partner of the volume modulus).

22.7 Criticisms of String Phenomenology

• *Not real string theory*. From a more formal perspective, string theory has many other challenges. It is natural to claim that, before we understand the theory properly, it is better not to address the potential low-energy physical implications, many of them based on effective field theory techniques. This is a valid argument for some string theorists not to concentrate their research on phenomenological aspects of string theory. However, it is not an argument against other members of the string theory community to do it. It is well known that the history of scientific discoveries does not follow a logical structure as it happened with the SM – the model was proposed even before we knew that gauge theories were renormalisable. Nevertheless, we may understand the theory better than we think (at low energies and weak couplings) using all foreseeable ingredients: geometry, branes, fluxes, perturbative and nonperturbative effects, and topological and consistency arguments.

String phenomenologists often follow major developments on the formal understanding of the theory and incorporate them into concrete physical models (e.g., CFT techniques for model building and string amplitudes, D-brane/orientifold constructions, F-theory phenomenology, brane/antibrane inflation, physical implications of the open string tachyon, phenomenological and cosmological implications of anti-de Sitter/conformal field theory [AdS/CFT]). What is less appreciated is that by addressing phenomenological questions of string theory, string phenomenologists have actually achieved some of the most important developments in string theory. First and foremost is mirror symmetry, which was essentially identified as a geometrical manifestation of stringy conformal field theories and later found 'experimentally' by just enumeration of Calabi–Yau manifolds relevant for phenomenology (Figure 22.3) [27]. Moreover, the attempt to use mirror symmetry to compute some couplings has led to fundamental discoveries within pure mathematics. Mirror symmetry, after more than 25 years, remains a very active field of pure mathematics.

Second, T-duality, identifying large to small distances in string theory, was discovered by attempts to find interesting cosmological solutions of string theory [28]. Also the strong–weak coupling S-duality was proposed with the motivation to address the moduli stabilisation problem of string phenomenology [29] (the widely used terms 'T-duality' and 'S-duality' were introduced for the first time in reference [29]). Later studies of the implications of these two dualities were the source of the so-called second string revolution of the mid-1990s. This illustrates that the study of string phenomenology has been crucial to better understanding of the theory itself.

Moreover, some of the most important purely stringy calculations (not just EFT) have been performed in the context of string phenomenology. For instance, the string loop corrections to gauge couplings led to the much studied holomorphic anomaly that has had much impact in formal string and field theory [30].

- *Not real phenomenology.* If there is a sudden experimental discovery or hint of a discovery, it will be the theorists working closer to the experimentalists who will make a difference immediately after. String phenomenologists will usually not be the first ones to react (based on what has happened on several occasions). In that case, if a young scientist wants to do phenomenology very close to experiments, it may be wiser not to enter the complications of string theory but instead to concentrate on the simplest ways to address the relevant physical questions. This is also a valid argument for part of the community to concentrate purely on phenomenology. Nevertheless, with the current stage of affairs, it is better to have all different specialisations and tools at hand since we do not know in which direction things will develop. In particular, this has been illustrated with the recent advances in the study of the CMB and cosmic inflation. Some of the most popular models to compare with experiment are string models (e.g., axion monodromy and Dirac–Born–Infeld [DBI] inflation). Furthermore some, of the most popular ideas of BSM physics were inspired by string theory (e.g., large and warped extra dimensions, split supersymmetry, axiverse).

Mirror Symmetry

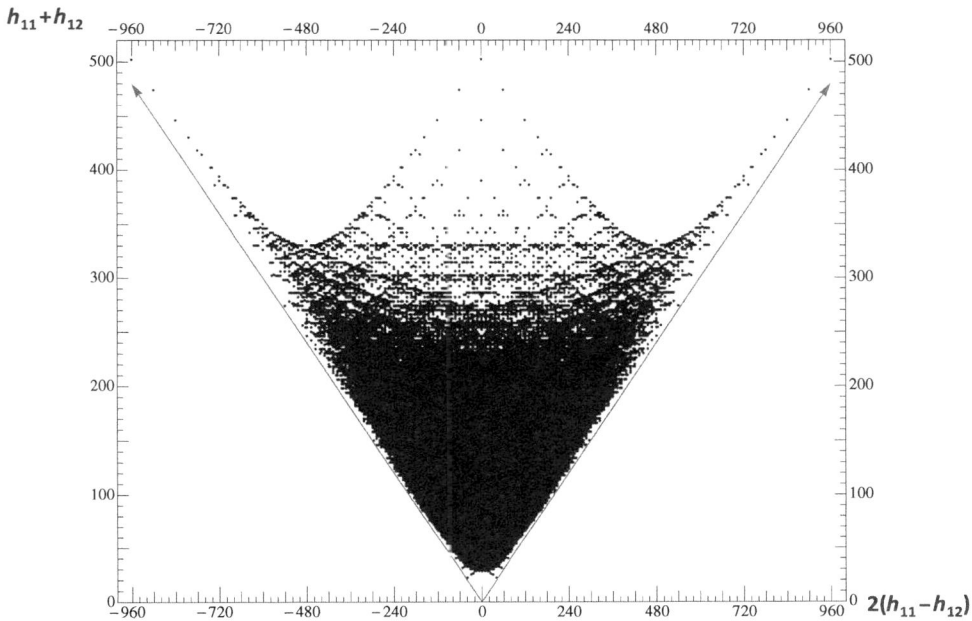

Figure 22.3 Evidence for mirror symmetry. The plot is essentially symmetric among interchange of Hodge numbers h_{11} and h_{12}. See, for instance, [31].

Finally, on a more sociological note, for training young generations who like the challenge of formal and mathematical questions but at the same time like to keep track of physics developments, this field suits very well. Practitioners enjoy very much this line of research. which is a must in the choice of research subjects. Also, even if, in the worst-case scenario, string theory is proved wrong or inconsistent in the future, string phenomenologists of all generations will have enough knowledge to optimise their versatility and move to other fields. This career path has actually been tested over the years by some string phenomenologists who either stayed in the field or moved to pure phenomenology, pure cosmology (mostly mathematics), astrophysics. Monte Carlo modelling of LHC data, and even experimental high-energy physics (besides those who moved to other fields such as finance or biology). The training they receive is ample preparation to enable them to adapt to other research areas.

- *Not even wrong.* For non-string theorists, the main criticism of the theory itself regards to the questions addressed by string phenomenology(not a testable prediction, failure to detect supersymmetry or other BSM physics hinted by string theory so far, and so on). Following the famous Pauli expression, it may be believed that the theory is not even wrong. In principle, this criticism seems to be valid. However, as we have discussed, there are actually testable predictions that could rule out the theory, albeit in a far future. Also, string phenomenologists see these problems as challenges and interesting questions to address in their research, which actually have made the field active and physically interesting over the years. It is standard practice in science to fully explore the physical implications of a given theory, and string theory has proved that it more than deserves to be explored in this regard, despite the fact that it is so difficult to disentangle its physical implications at low energies. In fact, one of the key recommendations of Steven Weinberg to young generations [32] is precisely to do research in the field that looks not closed and elegant, but rather open and messy. Given the nature of the problem (to explain all fundamental physical phenomena), we have to be patient and think about the long term. Concrete progress may take several decades or even centuries, given the experimental difficulties, unless we are lucky and several indications materialise that would support the theory or some particular string scenario.

Regarding time, the recent case of gravitational waves discovery offers an excellent example. Its time frame spanned 100 years from prediction to experimental confirmation, with the techniques for the waves' discovery being developed some 40 years after the prediction of their existence. The 50 years between prediction and discovery of the Higgs particle is another example. Other theories, notably the theory of evolution and the atomic theory, took many decades or centuries before they were accepted by the scientific community and even longer to be fully accepted.

At the other extreme, consider the case of superconductivity. Historically, this was an opposite example, in the sense that it was first discovered experimentally (Omnes, 1911) and it took almost 50 years for a proper microscopic theory to be developed (Bardeen, Cooper, and Schriefer, 1957). But it is possible to imagine that the events could have happened in the opposite order. With the knowledge of quantum mechanics and condensed matter systems, theorists could have, in principle, predicted a superconducting phase of some materials at low temperatures. It did not happen in this way, however, because at that time experiments were easier to perform and it was a difficult challenge to extract macroscopic implications of quantum mechanics from a non-perturbative effect such as the physics of Cooper pairs. Exploring potential physical implications of string theories may be comparable to a potential world in which the study of materials at extremely low temperatures is not available and very talented theorists eventually come up with a prediction of quantum mechanics that implies superconductivity. We can always get lucky.

22.8 Final Remarks

The title of the Munich conference was 'Why trust a Theory?' The direct answer to this question is clearly that as scientists, we do not trust any theory until it is supported by experimental evidence. But the point of the question, as I understood, actually relates to why theorists (working on string or any other theory) keep working on a theory that has no foreseeable experimental verification. To answer it, we must separate two issues: if the theory does not have predictions in principle or if its predictions are not achievable in the short term. In the first case, even though the main goal of science is to have better explanations rather than to make predictions, a theory can be questioned by calling it scientific. String theory is clearly not of this type. It actually makes many predictions, especially at or close to the string scale (Kaluza–Klein tower of states, massive string states with specific behaviour under high energy scattering, and so on), which should be the natural case for a theory of quantum gravity. But as we have seen, being less ambitious, some string scenarios have many other potential implications at much lower energies, which is encouraging.

To eventually be able to rule out the theory, knowing if the theory is weakly or strongly coupled in the UV range is important. If the theory is weakly coupled and volumes are large compared to the string scale, the perturbative string theory and effective field theory calculations that have been done so far are valid. But if the theory is strongly coupled in the UV range, we do not understand it well enough to make predictions, except in the regimes left open by the strong–weak duality symmetries. Fortunately, from what we observe, it is not unreasonable to expect the UV theory to be weakly coupled. The naive extrapolation of the gauge couplings from the TeV scale up, even using only the SM particles, tends to point in that direction. Furthermore, the quartic coupling of the Higgs potential is also running toward weak coupling. Even though this understanding may be modified by new physics at scales higher than the TeV scale, it is reasonable to study string theory in the weak coupling regime. Luckily enough. this is the regime in which calculations are under control. Moreover, having these small parameters (weak couplings and inverse volumes) fits with the general structure of hierarchies we observe in nature.

String theory has all the desired ingredients for a fundamental theory of nature (unique; no independent parameters; includes gravity, gauge interactions, and matter fields; and so on). It provides a very rich mathematical structure capable of describing our world and deserves to be explored to its limits. There has been continuous progress in 30 years, albeit not as fast as originally expected. Many open questions remain that touch upon issues of gravity, cosmology, phenomenology, astrophysics, and mathematics. Thus, it is a fruitful field in which to do research. We have to be patient and humble and not expect spectacular results that test the theory definitively in a short term. The set of questions to be answered is very large, but exploring different general scenarios and extracting potential observable implications can give us some guidance.

If in the relatively short term new physics is discovered experimentally, it may give indications and guidance for string scenarios, but most probably will not give a definitive positive test for any particular string scenario. Since string models have to address all BSM questions, searching for correlations among observations may prove useful. It may take several discoveries in different directions to get closer to test string scenarios. For instance, discovery of low energy supersymmetry and large tensor modes may eliminate most, if not all, string scenarios proposed so far. Conversely, indications of the existence of split supersymmetry and non-thermal dark matter or dark radiation and power loss at large angular scales of the CMB and/or ultra-light dark matter (e.g., $m \approx 10^{-22}$ eV) or keV astrophysical photon excess, axion–photon conversion evidence, or evidence for very weak gauge interactions would support some concrete scenarios. Also, some general scenarios have very concrete predictions that could be tested. For example, in LVS, the axion partner of the volume modulus is always present and is almost massless, providing a model-independent candidate for dark radiation and ultra-light dark matter or dark energy.

Even though predictability is an important aspect to assess for a theory, it is not the theory's only *raison d'être*. The main aim of a theory is to provide better explanations of natural phenomena; this is how progress has been made in science over the centuries. Also, internal consistency is a crucial component for a theory to be seriously considered. As we have seen, string theory ranks high on the better explanation aspect. Also, since its beginnings string theory has been subjected to many consistency checks, from anomaly cancellations to many tests of dualities, formal properties of D-branes, and connections with mathematics such as K-theory, derived categories, modular forms, mock modular forms, and wall crossing.

Probably some of the most important recent results have been related to the conjectured AdS/CFT duality, its many checks, and potential applications to different areas of physics, from addressing the information loss paradox, to quark-gluon plasma, cold atoms, and quantum criticality in condensed matter systems. To date, the most concrete achievement is the explicit calculation of the black hole entropy for a class of black hole solutions and their comparison with the famous Bekenstein–Hawking formula $S_{bh} = A/4$ with exact agreement. These developments and many others, including the progress described here in obtaining quasi-realistic string compactifications, give circumstantial evidence and reassures the researchers working in this field that this theory has deep implications; they also support the expectations that the theory has the potential to eventually provide the ultraviolet completion of the SM. They partially explain why a group of theorists are committed to do research in this theory despite the fact that experimental verification will be needed, which may take a long time.

Let us finish on an optimistic note. Early ideas on black holes go back to the eighteenth century (Michell, Laplace). Black holes were not seriously considered as physical objects even after Einstein's theory of general relativity was established and predicted their existence. Following an almost adiabatic change of perspective during the past 40 years, supported by several pieces of observational evidence, now their existence is no longer questioned. Similarly, quarks started life as mathematical entities, later proposed to correspond to physical particles but with the understanding that from their own structure, free quarks cannot be observed in nature, Yet physicists were creative enough to identify ways to detect their existence. Quarks are now seen, together with leptons, as the fundamental degrees of freedom for visible matter. It is worth recalling Weinberg's statement about the early development of the Big Ban model in his celebrated book *The First Three Minutes* [33], which may apply as well to string theory scenarios today:

This is often the way it is in physics – our mistake is not that we take our theories too seriously, but that we do not take them seriously enough. It is always hard to realise that these numbers and equations we play with at our desks have something to do with the real world.

Acknowledgements

I am grateful to the organisers, in particular Slava Mukhanov and Richard Dawid, for the kind invitation to the Munich meeting. I thank the string phenomenology/cosmology community, especially my collaborators over many years, for shaping my views on the subject. Thanks to Cliff Burgess, Shanta de Alwis, Keith Dienes, Anne Gatti, Luis Ibáñez, Damián Mayorga Peña, and Pramod Shukla for comments on the manuscript and to Giovanni Villadoro for discussions.

Appendix A Timeline of Developments in String Phenomenology

To illustrate how the field has evolved over the past three decades, it may be useful to collect some of the main achievements for each decade. This illustrates that while the field may be evolving slowly, it has produced concrete results and calculations (see also the String Phenomenology 2014 summary talk by L. E. Ibáñez at http://stringpheno2014.ictp.it/program.html).

First Decade: 1985–1994

- Heterotic string compactifications: (Calabi–Yau) × (flat 4D spacetime) with $N = 1$ supersymmetry.
- $E6$ string phenomenology.
- Simplest dimensional reduction to 4D.

- Gaugino condensation and supersymmetry breaking.
- Explicit 4D strings (orbifolds, free fermions, bosonic lattice, Gepner, Kazama–Suzuki models) from exact conformal field theories.
- First quasi-realistic string models (CY, orbifolds, free fermions) with three families and including the SM group.
- Evidence for and applications of mirror symmetry.
- Anomalous $U(1)$'s and field-dependent Fayet–Iliopolus terms.
- Computation of threshold corrections to gauge couplings and gauge coupling unification.
- Introduction of T- and S-dualities and first attempts at moduli stabilisation.
- Identification of the cosmological moduli problem as a generic low-energy feature of supersymmetric string models.
- First study of soft terms induced from moduli and dilaton supersymmetry breaking.

Second Decade: 1995–2004

- Dualities indicating one single string theory.
- D-branes model building: local models (branes at singularities plus intersecting branes).
- Stringy instanton techniques to explicitly compute non-perturbative superpotentials W.
- Large extra dimensions.
- Expanding techniques for model building (heterotic orbifolds, general CYs, G_2 holonomy manifolds).
- Fluxes and moduli stabilisation: the landscape (BP, GKP, KKLT).
- Metastable de Sitter (KKLT).
- First computable models of string inflation.
- String-inspired alternatives to inflation.

Third Decade: 2005–2014

- Large volume scenario (LVS) of moduli stabilisation.
- Instanton calculation tools generalised and expanded (and used for neutrino masses, among other applications).
- Local F-theory phenomenology developed reviving string model building with GUT groups (SU(5), SO(10), and so on).
- Big data heterotic models constructions (including many models with the spectrum of the minimal supersymmetric SM).
- Discrete symmetries studied in detail and classified as remnants of local gauge symmetries.
- Classes of models of string inflation developed, some including moduli stabilisation and some including the possibility of large tensor modes.
- Global model building developed including first compact CY compactifications with moduli stabilisation (including de Sitter vacua)
- Concrete studies of global F-theory models (including SM-like models, extra $U(1)$'s, and more).
- Concrete calculations regarding quantum (loop and α') corrections to the Kähler potential and their phenomenological and cosmological implications.
- Realisations of de Sitter vacua from purely supersymmetric effective field theories (including constrained superfields).
- Exploring general constraints to large field inflation, including weak gravity conjecture.

Appendix B Ten Open Questions for BSM

Any attempt to move physics beyond the Standard Model (BSM) has to address at least one of the questions presented here. What makes string models so challenging is that if a model fails in *one* of these issues, then the model is ruled out. Even though some string models address several of these questions, a fully realistic string model does not exist yet. Non-stringy models usually concentrate on a small subset of the questions; in particular, question 1 is often ignored.

1. Ultraviolet completion.
2. Gauge and matter structure of the SM (origin of basic degrees of freedom and corresponding quantum numbers).
3. Origin of values of SM parameters, including hierarchy of scales, masses (including neutrino masses), quark mixing (CKM matrix), lepton mixing (PMNS), and the right amount of CP violation, avoiding unobserved flavour-changing neutral currents.
4. Strong CP problem.
5. Hierarchy of gauge couplings at low energies potentially unified at high energies.
6. Almost stable proton with a quantitative account of baryogenesis (including appropriate CP violation).
7. Source of density perturbations in the CMB (realisations of inflation or alternative early universe cosmologies).
8. Explanation of dark matter (avoiding over-closing).
9. Absence or small amount of dark radiation consistent with observations ($4 \geq N_{eff} \geq 3.04$).
10. Explanation of dark energy (with equation of state $w = p/\rho \approx -1$).

Appendix C Some Open Challenges for String Phenomenology

1. Computation of (holomorphic) Yukawa couplings and quantum (string and α') corrections to Kähler potentials for moduli and matter fields for different string theories and F-theory.
2. Moduli stabilisation for heterotic and F-theory models.
3. Stringy control of moduli stabilisation (including both open and closed string moduli) and de Sitter uplift.
4. Explicit model of inflation with calculational control, especially for the case with large tensor modes.
5. Better control on non-supersymmetric compactifications as well as supersymmetry breaking terms.
6. Explicit constructions of G_2 holonomy chiral models from 11 dimensions.
7. Model-independent information related to flavour issues, including neutrino masses.
8. Stringy mechanisms of baryogenesis.
9. Model-independent studies of stringy dark matter candidates (from visible and hidden sectors).
10. Explicit construction of a fully realistic string model or scenario (including early universe cosmology, dark matter candidates, mechanism for baryogenesis, and so on) that reduces to the SM at low energies.

References

[1] L. E. Ibáñez and A. M. Uranga, *String Theory and Particle Physics: An Introduction to String Phenomenology,* Cambridge: Cambridge University Press (2012); B. Acharya, G. Kane, and P. Kumar (eds.), *Perspectives on String Phenomenology,* World Scientific (2015).

[2] F. Quevedo, Phenomenological aspects of D-branes, in *Superstrings and Related Matters: Proceedings*, Spring School, Trieste, ed. C. Bachas, E. Gava, J. Maldacena, K.S. Narain, and S. Randjbar-Daemi. Trieste: ICTP, 2003. (ICTP lecture notes series; 13).

[3] S. Weinberg, Anthropic bound on the cosmological constant, *Phys. Rev. Lett.* **59** (1987) 2607, doi:10.1103/PhysRevLett.59.2607.

[4] S. Weinberg, *The Quantum Theory of Fields. Vol. 1: Foundations,* Cambridge: Cambridge University Press (1995).

[5] N. Arkani-Hamed, S. Dubovsky, A. Nicolis, and G. Villadoro, Quantum horizons of the Standard Model landscape, *JHEP* **0706** (2007) 078, doi:10.1088/1126-6708/2007/06/078 [hep-th/0703067 [HEP-TH]].

[6] E. Witten, Reflections on the fate of space-time, *Phys. Today* **49N4** (1996) 24.

[7] A. Font, F. Quevedo, and S. Theisen, A comment on continuous spin representations of the Poincare group and perturbative string theory, arXiv:1302.4771 [hep-th].

[8] P. Schuster and N. Toro, A gauge field theory of continuous-spin particles, *JHEP* **1310** (2013) 061, doi:10.1007/JHEP10(2013)061 [arXiv:1302.3225 [hep-th]].

[9] L. B. Anderson, A. Constantin, J. Gray, A. Lukas, and E. Palti, A comprehensive scan for heterotic SU(5) GUT models, *JHEP* **1401** (2014) 047, doi:10.1007/JHEP01(2014)047 [arXiv:1307.4787 [hep-th]].

[10] T. Banks and L. J. Dixon, Constraints on string vacua with space-time supersymmetry, *Nucl. Phys. B* **307** (1988) 93; C. P. Burgess, J. P. Conlon, L-Y. Hung, C. H. Kom, A. Maharana, and F. Quevedo, Continuous global symmetries and hyperweak interactions in string compactifications, *JHEP* **0807** (2008) 073, [arXiv:0805.4037 [hep-th]].

[11] G. D. Coughlan, W. Fischler, E. W. Kolb, S. Raby, and G. G. Ross, Cosmological problems for the Polonyi potential, *Phys. Lett. B* **131** (1983) 59; T. Banks, D. B. Kaplan, and A. E. Nelson, Cosmological implications of dynamical supersymmetry breaking, *Phys. Rev. D* **49** (1994) 779 [hep-ph/9308292]; B. de Carlos, J. A. Casas, F. Quevedo, and E. Roulet, Model independent properties and cosmological implications of the dilaton and moduli sectors of 4-D strings, *Phys. Lett. B* **318** (1993) 447, [hep-ph/9308325].

[12] K. R. Dienes, String theory and the path to unification: A review of recent developments, *Phys. Rept.* **287** (1997) 447 doi:10.1016/S0370-1573(97)00009-4 [hep-th/9602045]; A. Maharana and E. Palti, Models of particle physics from type IIB string theory and F-theory: A review, *Int. J. Mod. Phys. A* **28** (2013) 1330005, doi:10.1142/S0217751X13300056 [arXiv:1212.0555 [hep-th]].

[13] D. J. Wales, *Energy Landscapes, with Applications to Clusters, Biomolecules and Glasses*, Cambridge: Cambridge University Press (2003).

[14] S. B. Giddings, S. Kachru, and J. Polchinski, Hierarchies from fluxes in string compactifications, *Phys. Rev. D* **66** (2002) 106006, [hep-th/0105097].

[15] M. R. Douglas and S. Kachru, Flux compactification, *Rev. Mod. Phys.* **79** (2007) 733, [hep-th/0610102].

[16] S. Kachru, R. Kallosh, A. D. Linde, and S. P. Trivedi, De Sitter vacua in string theory, *Phys. Rev. D* **68** (2003) 046005 [hep-th/0301240].

[17] V. Balasubramanian, P. Berglund, J. P. Conlon, and F. Quevedo, Systematics of moduli stabilisation in Calabi–Yau flux compactifications, *JHEP* **0503** (2005) 007, [hep-th/0502058]; J. P. Conlon, F. Quevedo, and K. Suruliz, Large-volume flux compactifications: Moduli spectrum and D3/D7 soft supersymmetry breaking, *JHEP* **0508** (2005) 007, [hep-th/0505076].

[18] R. Bousso and J. Polchinski, Quantization of four form fluxes and dynamical neutralization of the cosmological constant, *JHEP* **0006** (2000) 006, [hep-th/0004134].

[19] B. Freivogel, M. Kleban, M. Rodriguez Martinez, and L. Susskind, Observational consequences of a landscape, *JHEP* **0603** (2006) 039, doi:10.1088/1126-6708/2006/03/039 [hep-th/0505232].

[20] B. S. Acharya and M. R. Douglas, A finite landscape?, hep-th/0606212.

[21] C. Vafa, The string landscape and the swampland, hep-th/0509212.

[22] N. Arkani-Hamed, L. Motl, A. Nicolis, and C. Vafa, The string landscape, black holes and gravity as the weakest force, *JHEP* **0706** (2007) 060, doi:10.1088/1126-6708/2007/06/060 [hep-th/0601001].

[23] L. Aparicio, F. Quevedo, and R. Valandro, Moduli stabilisation with nilpotent goldstino: Vacuum structure and SUSY breaking, *JHEP* **1603** (2016) 036, doi:10.1007/JHEP03(2016)036 [arXiv:1511.08105 [hep-th]].

[24] D. Baumann and L. McAllister, Inflation and string theory, arXiv:1404.2601 [hep-th].

[25] R. Flauger, J. C. Hill, and D. N. Spergel, Toward an understanding of foreground emission in the BICEP2 region, *JCAP* **1408** (2014) 039, doi:10.1088/1475-7516/2014/08/039 [arXiv:1405.7351 [astro-ph.CO]].

[26] C. P. Burgess, M. Cicoli, and F. Quevedo, String inflation after Planck 2013, *JCAP* **1311** (2013) 003, doi:10.1088/1475-7516/2013/11/003 [arXiv:1306.3512 [hep-th]].

[27] L. J. Dixon, Some world sheet properties of superstring compactifications: On orbifolds and otherwise, in *Superstrings, Unified Theories and Cosmology 1987*, ed. G. Furlan, J. C. Pati, D. W. Sciama, E. Sezgin, and Q. Shafi, Singapore: World Scientific, 1988; P. Candelas, M. Lynker, and R. Schimmrigk, Calabi–Yau manifolds in weighted P(4), *Nucl. Phys. B* **341** (1990) 383, doi:10.1016/0550-3213(90)90185-G; B. R. Greene and M. R. Plesser, Duality in Calabi–Yau moduli space, *Nucl. Phys. B* **338** (1990) 15, doi:10.1016/0550-3213(90)90622-K; P. Candelas, X. C. De La Ossa, P. S. Green, and L. Parkes, A pair of Calabi–Yau manifolds as an exactly soluble superconformal theory, *Nucl. Phys. B* **359** (1991) 21, doi:10.1016/0550-3213(91)90292-6.

[28] K. Kikkawa and M. Yamasaki, Casmir effects in superstring theories, *Phys. Lett. B* **149** (1984) 357, doi:10.1016/0370-2693(84)90423-4; N. Sakai and I. Senda, Vacuum energies of string compactified on torus, *Prog. Theor. Phys.* **75** (1986) 692; erratum: [*Prog. Theor. Phys.* **77** (1987) 773], doi:10.1143/PTP.75.692.

[29] A. Font, L. E. Ibáñez, D. Lüst, and F. Quevedo, Strong–weak coupling duality and nonperturbative effects in string theory, *Phys. Lett. B* **249** (1990) 35, doi:10.1016/0370-2693(90)90523-9.

[30] L. J. Dixon, V. Kaplunovsky, and J. Louis, Moduli dependence of string loop corrections to gauge coupling constants, *Nucl. Phys. B* **355** (1991) 649, doi:10.1016/0550-3213(91)90490-O.

[31] P. Candelas, Max Kreuzer's contributions to the study of Calabi–Yau manifolds, doi:10.1142/9789814412551, arXiv:1208.3886 [hep-th].

[32] S. Weinberg, Scientist: Four golden lessons, *Nature* **426** (2003) 389, doi:10.1038/426389a.

[33] S. Weinberg, *The First Three Minutes: A Modern View of the Origin of the Universe*, Muenchen: Basic Books, 1977.

Index

ΛCDM model, 8, 234, 253, 254, 260–262, 265, *see also* standard model of cosmology

abduction, 9, 306, 307, *see also* inference to the best explanation
accelerated expansion of the universe, 369
accuracy, 306, *see also* theoretical virtues
ad-hoc hypothesis, 90
AdS/CFT correspondence, 3, 344, 359, 377, 380, 424, 428
aesthetic criteria, 24, 125, 408
alternative theories, 9, 36, 128, 129, 131, 138, 146, 148, 150, 157, 158, 266, 276, 317, 325, 326, 368
    active search for, 157, 158
    unconceived, 324
analogue black hole experiments, 8, 198, *see also* Bose-Einstein condensate analogue black hole
analogue experiments, 7, 8, 68, 76, 79, 82, 185, 186, 189, 198
analogue gravity, 3, 186
analogue model, 76, 78, 79, 203, 219
anarchism, 55
anomaly, 133
anthropic argument, 3, 287, 290, 402, 413, 418, 421
anthropic principle, 301, 308
area theorem, 206–208
argument of unexpected explanatory interconnections, 114, 116, 117, *see also* unexpected explanatory coherence argument
arguments by analogy, 185, 187, 189
asymmetry of time, 206
atomism, 5
auxiliary hypotheses, 89, 90, 275

B-mode polarization, 227, 234, 241
Bacon, Francis, 5
baryogenesis, 408
baryonic acoustic oscillations, 233, 234, 242
basic principles of scientific reasoning, 9

Bayes' Theorem, 71
Bayesian confirmation theory, 7, 37, 68, 70–72, 74, 79–82, 104–108, 120, 121, 123, 155, 186, 189, 276, 317
Bayesian inference, 307, 309
Bayesian Network, 72, 73, 75, 78–80, 82
Bayesianism, 71
beauty, 31, 52, 109, 176, 184, 303, 310, 316, *see also* theoretical virtues
Bekenstein's argument for the entropy of black holes, 203, 205, 206, 208, 210, 211, 218–220
Bekenstein-Hawking formula, 3, 8, 202, 203, 209, 219, 346, 428
Bianchi-identity, 255, 256, 258
black hole entropy, 169, 203, 208–210, 217, 218, 220, 381, 428
black hole firewall, 361
black hole thermodynamics, 3, 203, 204, 371, 402
Bohr, Niels, 5
Boltzmann, Ludwig, 5
Bose-Einstein condensate, 76–78, 186–188, 190, 193, 195, 196, 203
Bose-Einstein condensate analogue black hole, 186, 194–196, 198, *see also* analogue black hole experiments
Bose-Einstein condensate Hawking effect, 192–194, 198
Bose-Einstein condensate Hawking radiation, 195, 196
BSM physics, 156, 380, 406, 408, 409, 420, 424, 425, 428, 431

coherence, 36, 46, 54, 174, 176
compactification, 380
compatibility, 316
competing theories, 15, 129, 136, 145, 151, 158, 159, 325, 326, 329, *see also* rival theories
computer simulation, 184, 185
conditionalization, 71

confirmation, 104, 121, 309
  conclusive, 103, 105, 108, 118, 121
  conclusive empirical, 103
  degree of, 70, 186, 198, 315
  empirical, 1, 4, 9, 67, 73, 99, 100, 107, 120, 122, 173, 174
  indirect, 68, 70, 73, 82, 315
  non-empirical, 6, 101, 122
  significant, 108, 113, 116, 118
consilience, 36, 46, 54, *see also* theoretical virtues
consistency, 2, 176, 289, 306, 307, *see also* theoretical virtues
context of discovery, 41, 56, 132, 166
context of justification, 41, 132, 133
continuous spin representations, 413, 414
convergence argument, 174, 182
Copernican principle, 258, 259, 261
cosmic censorship hypothesis, 207
cosmic inflation, 1, 3, 4, 100, 154, 244, 261, 268, 310, 402
  discovery of, 230
cosmic microwave background, 230, 245, 253, 260, 261, 263, 267, 400, 408, 420, 422, 424
cosmic microwave background fluctuations, 232
cosmic microwave background radiation, 229, 281, 330, 379
cosmological constant, 22, 231, 235, 242, 259–262, 287–289, 291, 297, 302, 308, 310, 311, 343, 347–349, 356–360, 362, 378, 379, 387, 401, 420–422
cosmological multiverse, 300, *see also* multiverse
cosmological origin of lithium, 229
cosmophysics, 21, 22
credibility, 318, *see also* theoretical virtues
crisis of empirical testing, 5

dangerous irrelevance, 366
dark energy, 9, 227, 231, 233, 242, 401, 408
  dynamical, 304, 305
dark energy density, 259
dark matter, 230, 232–234, 236, 238, 243, 245, 261, 267, 268, 270, 300, 408
  cold, 231, 236, 239
  hot, 231, 236
  Neutrino, 231
  nonbaryonic, 230
  warm, 236, 239
definiteness, 303, *see also* theoretical virtues
demarcation problem, 7, 14, 17, 85, 89–91, 302, 303, *see also* falsificationism
Descartes, Rene, 5
distance-minimization methods, 71
distinct predictions, 68
dual resonance model, 177–181
dual theory of strong interactions, 176
duality in quantum phyiscs, 345
duality principle, 177
duality relations, 2, 3

dumb hole, 192
Dutch Book arguments, 71

early string theory, 174, 176, 177, 182
early universe cosmology, 367, 371, 373
effective field theory, 133, 365, 367, 373, 404, 406, 408
Einstein, Albert, 5, 342
elegance, 109, 174, 176, 316, *see also* theoretical virtues
eliminative program, 321, 325, 326, 328, 329
empirical adequacy, 70, 75, 77, 113, *see also* theoretical virtues
energy-momentum conservation, 257
entanglement entropy, 220
environmental theory, 356, 358, 360
epistemic shift, 14, 15
epistemic status of scientific results, 5
epistemic utility theory, 71
equivalence principle, 254
eternal inflation, 3, 9, 315, 331, 332
  false-vacuum, 305
evidence, 7, 67, 184
  conclusive empirical, 4, 99, 103, 106
  empirical, 4, 7, 81, 122, 166
  from authority, 48–50, 53, 54
  indirect, 68, 72, 81, 312, 401
  meta-inductive, 47–50, 53, 54
  non-empirical, 99, 111, 121
expanding universe, 227
explanatory connection, 49–54
explanatory power, 5, 17, 137, 138, 176, 291, 414, 428, *see also* theoretical virtues
extra dimensions, 2, 3, 164, 165, 182, 372, 410, 415
  discovery of, 174, 176, 178

fallibilism, 135
falsifiability, 14, 35, 84, 85, 89, 92, 291, 301–307, 396
falsification, 87, 89–91, 246, 301
falsificationism, 36, 68–70, 72, 81, 85, 90, 123, 127, 149, 155, 301, 302, 307, *see also* demarcation problem
family resemblance, 91
fertility, 109, 112, 176
Feyerabend, Paul, 36, 54, 56, 130
fine-tuning, 3, 287, 347
fluid mechanical experiments, 184
fruitfulness, 306, 307, 318, *see also* theoretical virtues
functional RG approaches, 3
fundamental physics, 1, 2, 4–6, 15, 20, 27, 92, 93, 99–101, 103, 106, 125, 133, 154, 156, 217, 254, 268, 354
fundamentality, 21, 59, 60, 142, 344, 348, *see also* theoretical virtues

galaxy formation, 227, 232, 239, 240, 244, 267, 269, 270, 394
Galilei, Galileo, 5
general relativity, 8, 31, 68, 84, 86, 89, 99, 203, 253, 256, 257, 260–262, 269, 302, 343–345, 348
generality, 176
ghost elimination programme, 178, 180
grand unified theory, 3, 123, 159, 409
gravitational wave background, 234

Hawking effect, 195
    hydrodynamic, 192
Hawking radiation, 76–79, 82, 169, 186, 190, 194, 196, 202, 203, 206, 220, 221
    gravitational, 185
    in analogue black holes, 8, 196, 203
Hawking, Stephen, 76
Heisenberg, Werner, 5
helium and deuterium synthesis, 229
Hempel, Carl, 69
hierarchy problem, 165, 384, 390, 421
Higgs mechanism, 118, 387, 391
Higgs-Boson
    discovery of, 4, 119, 121, 349, 357, 395, 401, 408, 426
high energy physics, 4
history of science, 23, 27, 93, 135, 138, 176, 245, 360, 408, 423
holographic principle, 344
Hume's problem of induction, 69
hypothesis testing, 6
    empirical, 9
hypothetico-deductive model, 68–70, 72, 81
hypothetico-deductivism, 46, 47, 52, 56, 155

inaccessible target system, 8, 76, 79
incommensurability, 15
independent warrant, 59
inference to the best explanation, 47, 52, 294, *see also* abduction
inflationary cosmology, 1, 228, 229, 310, 330, 366
inflationary models, 243, 246
information theory, 8, 368
information-theoretic entropy, 210–212, 218, 219
internal and external validity, 188
invisible axion, 230

Kuhn, Thomas, 306

lack of empirical data, 9, 156, 157, 168, 219, 291, 315
last scattering surface, 281
Lemaître, Georges, 227, 228
LHC, 4, 67, 238, 377, 379, 380, 387, 392, 393, 395, 421
light
    particle theory of, 30
    wave theory of, 30
loop quantum gravity, 3
Lorentz-covariance, 257

Lorentz-invariance, 167, 343
Lotka-Volterra equations, 210
luminiferous ether, 30, 31

Mach's principle, 257
Mach, Ernst, 5
marketplace of ideas, 127, 129
mathematical universe hypothesis, 22
Maxwell, James Clerk, 5, 57, 58, 60, 61
meta-inductive argument, 114, 116–118, 136–139, 147, 148, 174, 293, 294, 317, 329
meta-level hypothesis, 111–113, 116, 118
Mill, John Stuart, 129, 156
minimal models, 238
model and theory, 276, 278
modified gravity, 242
multiverse, 1, 3, 8, 9, 14, 15, 17, 18, 23, 84, 100, 227, 246, 275–277, 279, 281–285, 290–292, 294, 301–303, 305–312, 331, 333, 347–349, 354, 355, 357, 361, 362, 408, *see also* cosmological multiverse
multiverse cosmology, 16, 27

natural philosophy, 21, 22
naturalness, 288, 379
neutrinos, 236
    sterile, 236, 237
Newton, Isaac, 5, 34, 41, 44, 155
Newtonian gravity, 256
Newtonian inductivism, 46, 52
No Alternatives Argument, 68, 74, 75, 80–82, 114–117, 122, 136–139, 141, 143–148, 150, 157, 289, 291, 292, 294, 317, 325, 328, 397, *see also* only-game-in-town evidence
No-Ghost Theorem, 180, 181
no-go theorems, 162, 368
non-empirical confirmation, 101, 104–108, *see also* confirmation
non-empirical falsificationism, 135
non-empirical in theory assessment, 175
non-empirical theory assessment, 4, 6–9, 17, 20, 68, 74, 75, 92, 100, 102–104, 123, 125–127, 130, 131, 136, 138, 145, 166, 169, 276, *see also* theory assessment
non-empirical theory confirmation, 7, 100, 101, 105–108, 110, 112, 113, 116, 118, 277, 290, 292, 309, 362
non-gaussianity, 227, 242, 243, 245, 246, 263, 268–270, 366, 369, 370, 373
nonrenormalizability, 343, 345
nucleosynthesis, 229, 230, 236, 278, 282, 379, 384, 385, 390, 418, 423

observation about the research process, 74, 112
Occam's razor, 266, 280, 349
only-game-in-town evidence, 48–50, 53, 54, *see also* No Alternatives Argument
overdetermination of constants, 327–329

paradigm of scientific reasoning, 5
paradigm shift, 14
parametrized post-Newtonian formalism, 326
plausibility, 187, *see also* theoretical virtues
Poincaré, Henri, 5
Popper, Karl, 7, 35, 69, 85, 86, 89, 90, 92, 156, 301–303
post-empirical physics, 330, 331, *see also* supra-empirical physics
posterior probability, 71, 81
potential of a theory, 320
pre-established harmony, 23
precision cosmology, 8, 227, 229, 234
predictable black hole, 207
predictive power, 5, 276, 428, *see also* theoretical virtues
primordial black holes, 237, 246, 267
primordial gravitational waves, 270
principle of plentitude, 23
principle of proliferation, 128–131, 134, 136–138, 140, 141, 146–150
principle of tenacity, 129
principles of scientific reasoning, 4, 5, 9
provisional acceptance of a theory, 318, 319

quantum electrodynamics, 327, 328
quantum field theory, 105, 203, 344, 368, 384, 404, 409
    relativistic, 404
quantum gravity, 3, 4, 125, 126, 133, 156, 202, 206, 341, 343, 345, 348
    theories of, 7, 8, 169, 350
quantum mechanics, 5, 31, 67, 86, 89, 345, 348

reliability, 7, 67, 123, 316, 322, *see also* theoretical virtues
research programme, 8, 17, 131, 133, 292
research tradition, 139, 144
Ricci-curvature, 255–257, 260
Riemann-curvature, 256
rival research traditions, 324, 325, 329, 330
rival theories, 9, 324, 325, 329, 330, 332, *see also* competing theories

scarce empirical data, 2, 7
Schwarzschild black holes, 186
scientific community, 74, 75
scientific development, 18, 131
scientific discovery, 302
scientific methodology, 1, 5, 6, 125
scientific practice, 8, 87, 120, 135, 154, 156, 157, 166, 167, 169, 174, 301, 305, 322
scientific progress, 23, 35, 94, 127, 144, 148, 176, 306, 341, 358
scientific revolution, 2, 14, 15
second law of thermodynamics, 203, 204
    generalized, 205, 206, 209–211, 219

reformulated, 208
second superstring revolution, 359, 360
Shapiro-Virasoro Model, 177
simplicity, 17, 47, 59, 109, 174, 176, 266, 271, 306, *see also* theoretical virtues
simplified models, 134
speculation, 7, 21, 29–32, 35, 39–41, 43, 45, 56, 63, 65
    evaluation of, 29
    legitimacy of, 29
    scientific, 54
    truth-irrelevant, 33, 57, 61, 62
    truth-relevant, 32, 33, 35, 39, 48, 54, 57–59, 62
stability, 322, *see also* theoretical virtues
standard model of cosmology, 8, 234–236, 253, 254, 260–262, 278, 315, 378, *see also* ΛCDM model
standard model of particle physics, 2–4, 68, 99, 105, 123, 160, 161, 167, 173, 238, 246, 264, 315, 378, 405, 407, 408
string phenomenology, 9, 400, 423–427
string theory, 1–4, 7–9, 13, 16–18, 22–24, 26, 27, 31, 68, 73, 84, 86, 92, 99–101, 120–123, 137, 154, 173, 277, 279, 280, 285, 290, 310, 315, 342–344, 348, 350, 354–357, 361, 362, 365–369, 371–373, 377, 394, 400, 409–411, 413, 418, 419, 423, 426
    compactified, 9, 377, 379–386, 389, 392, 394–397
    discovery of, 341
    history of, 173, *see also* early string theory
    uniqueness of, 344
supermassive black holes, 233
superstrings, 15, 18, 22
supersymmetry, 3, 23, 154, 230, 238, 239, 372, 379, 383, 384, 386, 397, 411, 421
    low-energy, 367
supersymmetry breaking, 384, 387, 392, 411, 412, 417, 418, 421
supra-empirical physics, 318, 325, 330
symmetry principles, 321

tentative theories, 121
testability, 8, 9, 14–16, 18, 20, 23, 24, 32, 36, 65, 68, 85, 90, 128, 134, 146, 185, 271, 276, 278, 283, 286, 289–291, 296, 304, 367, 372, 380, 381, 396, 400, *see also* theoretical virtues
    indirect, 16, 17
    principle lack of, 6
theoretical discovery, 128
theoretical pluralism, 151
theoretical virtues, 174, 176, 291, *see* acuracy, aesthetic criteria, beauty, co-herence, consilience, consistency, credibility, deniteness, elegance, empirical adequacy, explanatory power, falsiability, fruitfulness, fundamentality, plausibility, predictive power, reliability, simplicity, stability, testability, truth, uniformity, uniqueness, viability
theory assessment, 14, 79, 135, *see also* non-empirical theory assessment
    alternative methods of, 81, 82

theory assessment *(cont.)*
    empirical, 8, 132
    indirect, 68, 69, 82
theory building, 133, 173, 175, 176, 178
theory comparison, 142, 151, 325
theory dependence, 134, 320, 329
theory development, 17, 142, 144, 146, 149, 165, 166, 169
theory justification, 128, 132
theory proliferation, 54, 127
theory selection, 127, 132
theory space, 75, 82, 130, 154, 157, 159, 162, 166, 167, 325, 328, 329, 367
    reliability of the assessment of, 162
    assessment of, 162, 164, 166, 168, 169
        inductive justification of, 166
    exploration of, 157–159, 170
    necessary constraints of, 160
theory-mediated measurements, 319, 320, 322, 324, 325, 327, 328, 331
thought experiments, 16, 371
truth, 105, 113, 127, 306, *see also* theoretical virtues

underdetermination, 9, 75, 114, 130, 131, 154, 293, 316, 318, 321–323, 329

transient, 316
unlimited, 115
unexpected explanatory coherence argument, 136–138, 174, 294, 317, 328, 397, *see also* argument of unexpected explanatory interconnections
unification, 23, 32, 47, 61, 62, 160, 182, 350, 409, 418
uniformity, 300
uniqueness, 17, 160, 257, 262, 269, 276, 409
unitarity, 177
unitarization programme, 179
UV completion, 366, 367, 421, 428
    of gravity, 368

verification, 24, 32, 89, 429
verificationism, 135
viability of a theory, 99, 102–104, 106, 108, 109, 113, 134, 140, 142, 145, 168, *see also* theoretical virtues
view of science, 126
violation of the null energy condition, 207
von Neumann entropy, 221
vortex theory of atoms, 13, 23–26

WIMP miracle, 238